Artificial Intelligence and Machine Learning Algorithms for Engineering Applications

This book comprehensively covers core algorithms and techniques used in artificial intelligence (AI) and machine learning (ML) for engineering applications. It further explores the use of AI in civil and structural engineering, quality control, and product design.

- Presents autonomous robots using onboard computing and AI algorithms to process the data from their sensors and make real-time decisions.
- Discusses nature-based optimization-based computing techniques to enhance the computational speed for solving engineering problems.
- Provides conceptual and practical knowledge about the design of modern computation techniques with advanced tools and methodologies.
- Highlights the importance of using smart techniques including AI and ML in product design and development.
- Covers time series analysis and forecasting in engineering, robotic process automation, and autonomous robots in manufacturing.

The text is primarily written for senior undergraduates, graduate students, and academic researchers in the fields of electrical engineering, electronics and communications engineering, computer science and engineering, manufacturing engineering, and environmental engineering.

Artificial Intelligence and Machine Learning Algorithms for Engineering Applications

Edited by
Krishan Arora, Himanshu Sharma, and
Aeidapu Mahesh

CRC Press
Taylor & Francis Group
Boca Raton London New York

CRC Press is an imprint of the
Taylor & Francis Group, an **informa** business

Front cover image: orbcat/Shutterstock

First edition published 2026
by CRC Press
2385 NW Executive Center Drive, Suite 320, Boca Raton FL 33431

and by CRC Press
4 Park Square, Milton Park, Abingdon, Oxon, OX14 4RN

CRC Press is an imprint of Taylor & Francis Group, LLC

ISBN: 978-1-032-90052-0 (hbk)
ISBN: 978-1-032-91458-9 (pbk)
ISBN: 978-1-003-56344-0 (ebk)

DOI: 10.1201/9781003563440

Typeset in Times
by Newgen Publishing UK

Contents

Preface

The rapid advancements in artificial intelligence (AI) and machine learning (ML) have transformed the way engineers approach problem-solving, automation, and optimization across various domains. From predictive maintenance in manufacturing to intelligent control systems in robotics and data-driven decision-making in civil and electrical engineering, AI and ML have become indispensable tools for modern engineering applications.

This book, *Artificial Intelligence and Machine Learning Algorithms for Engineering Applications*, aims to bridge the gap between theory and practical implementation by providing a comprehensive exploration of AI and ML algorithms tailored for engineers. It is designed to serve as a guide for professionals, researchers, and students who seek to apply these cutting-edge techniques in real-world engineering challenges.

The book begins with a foundational introduction to AI and ML, covering essential concepts such as supervised and unsupervised learning, neural networks, optimization techniques, and deep learning architectures. As the chapters progress, we delve into domain-specific applications, including signal processing, structural analysis, control systems, and energy management, highlighting how AI-driven solutions enhance efficiency and accuracy.

A key focus of this book is the practical implementation of AI and ML algorithms using widely adopted programming languages and tools such as Python, TensorFlow, and Scikit-Learn. Through step-by-step examples and case studies, readers will gain hands-on experience in developing intelligent models and integrating them into engineering workflows.

By the end of this book, readers will have a solid understanding of how AI and ML can be leveraged to tackle complex engineering problems, optimize processes, and drive innovation. Whether you are an aspiring engineer, a seasoned practitioner, or an academic researcher, this book will equip you with the knowledge and skills to harness the power of AI and ML in your respective fields.

We hope this book serves as a valuable resource and inspires further exploration into the fascinating intersection of AI and engineering.

About the Editors

Krishan Arora is associated with Lovely Professional University as a Professor with more than 17 years of experience in academics and research. He currently serves as the Department Head of Power Systems in the School of Electronics and Electrical Engineering. He has completed his PhD in Electrical Engineering from MM (Deemed to be) University, Ambala. He did his M.Tech in Electrical Engineering from IKG Punjab Technical University, Punjab and BTech in Electrical and Electronics Engineering from IKG Punjab Technical University, Punjab. He has published over 85 research papers in refereed science journals and conferences. He has organized several workshops, summer internships, International Conferences, special sessions, and expert lectures. He has also published six Indian patents and one copyright. He has supervised more than 10 Postgraduate Theses and more than 30 undergraduate student projects. He has taken and completed 15 Non-Government and Consultancy Projects. He has attended/Participated in 24 National/International Online webinars. His areas of expertise include Electrical Machines, Non-Conventional Energy Sources, Load Frequency Control, Automatic Generation Control, and Modernization of Smart Grids. He has edited five books in different areas of Electronics and Electrical engineering for various publishers, and has contributed to editing works with top academic publishers. He has taught various courses to UG and PG levels like Power Electronics, Non-Conventional Energy Sources, Electric Drives, Induction and Synchronous Machines, and Digital Electronics.

Himanshu Sharma has completed his PhD in Power Electronics from Punjab Engineering College Chandigarh. He did his MTech in Electrical Engineering from Punjab Engineering College Chandigarh. He is currently associated with GH Raisoni institute of Engineering and Management, Nagpur, India, as an Associate Professor with more than seven years of experience in academics. He has published over ten research papers in refereed science journals and conferences. He has organized several workshops, summer internships, International Conferences and expert lectures for students. He has supervised more than five postgraduate theses and more than ten undergraduate student projects. He has attended/Participated in twenty National/International Online webinars. His area of expertise includes power electronics, machine learning, and optimization techniques. He has taught various courses to UG and PG levels like Power Electronics, Non-Conventional Energy Sources, and Electric Drives, etc. He has edited many books for reputed publishers.

Aeidapu Mahesh has completed his ME in Power Electronics and Electric Drives from PEC University of Technology Chandigarh and PhD from NIT Kurukshetra in 2012 and 2018, respectively. Presently, he is working as Assistant Professor at SVNIT, Surat. Before joining SVNIT, he worked as Assistant Professor at NIT Kurukshetra from 2013 to 2019. He has published many journal papers and presented papers at various national and international conferences. His current research interests include hybrid renewable energy systems, evolutionary optimization techniques, power electronics, and drives. He is a reviewer for many reputed journals and a senior member of IEEE.

Contributors

Anju
Department of Computer Science
Dr B R Ambedkar NIT Jalandhar
Punjab, India

Ankit Aggarwal
Neelkanth Institute of Technology
Meerut, Uttar Pradesh, India

Krishan Arora
School of Electronics and Electrical Engineering
Lovely Professional University
Punjab

Ranjit Kumar Bindal
Electrical Engineering Department
Chandigarh University
Mohali, Punjab, India

Harpreet Kaur Channi
Department of Electrical Engineering
Chandigarh University
Gharuan, Mohali, Punjab, India

Nisha Chaurasia
Department of Information Technology
Dr B R Ambedkar NIT Jalandhar
Punjab, India

Shakti Raj Chopra
School of Electronics and Electrical Engineering
Lovely Professional University
Punjab, India

Niraj Kumar Choudhary
Department of Electrical Engineering
MNNIT
Allahabad, Prayagraj

Sandeep Chouhan
School of Computer Science and Engineering
Lovely Professional University
Punjab, India

Pritam Das
Mechanical Engineering Department
National Institute of Technology Agartala
Barjala, Jirania, Tripura (W), India

Deepika Ghai
School of Electronics and Electrical Engineering
Lovely Professional University
Punjab, India

Mansi Gupta
Department of Computer Science and Engineering
Vidya College of Engineering
Meerut

Amit Kohli
University Canada West
Vancouver, BC, Canada

Vishal Kohli
Neelkanth Institute of Technology
Meerut, Uttar Pradesh, India

Ashwani Kr. Vashistha
Vidya College of Engineering
Meerut, Uttar Pradesh, India

Manjit Kaur
School of Computer Science and Engineering
Lovely Professional University
Punjab, India

Anit Kumar
Mechanical Engineering Department
National Institute of Technology Agartala
Barjala, Jirania, Tripura (W), India

Ankur Kumar
Department of Electrical Engineering
MNNIT
Allahabad (Prayagraj), Uttar Pradesh, India

Gaurav Kumar
Department of Mechanical Engineering
Vidya College of Engineering
Meerut, Uttar Pradesh, India

Sassmit Laal
School of Computer Science and Engineering
Bennett University
Greater Noida, India

Hansinie Lorenzuhewa
University Canada West
Vancouver, BC, Canada

Amee Madhani
School of Computer Science and Engineering
Bennett University
Greater Noida, India

Manpreet Singh Manna
Electrical Engineering Department
Chandigarh University
Mohali, Punjab, India

Arindam Majumder
Mechanical Engineering Department
National Institute of Technology Agartala
Barjala, Jirania, Tripura (W), India

Parijata Majumdar
Department of Computer Science and Engineering
Indian Institute of Information Technology Agartala
Jirania, India

Praveen Kumar Malik
School of Electronics and Electrical Engineering
Lovely Professional University
Punjab, India

Sanjoy Mitra
Department of Computer Science and Engineering
Tripura Institute of Technology
Agartala, Narsingarh, India

Saraswathy Nagendran
SVKM`s Mithibai College of Arts
Chauhan Institutes of Science and Amrutben
Jivanlal College of Commerce and Economics
Mumbai, India

Madhu Naik Nenavath
School of Electronics and Electrical Engineering
Lovely Professional University
Punjab, India

Diego Oliva
Department of Computer Science and Engineering
Universidad de Guadalajara
CUCEI, Mexico

Leonardo Ramos Rodrigues
Department of Electronics
Institute of Aeronautics and Space
Sao Paulo, Brazil

Jeet Shah
SVKM's Mithibai College of Arts
Chauhan Institute of Science and Amrutben
Jivanlal College of Commerce and Economics
Mumbai, India

Chandan Kumar Sharma
Jigyasa University
Dehradun, Uttarakhand

Himanshu Sharma
GH Raisoni Institute of Engineering and Management
Nagpur, India

Ranjana Sharma
Chandigarh University
Gharuan, Mohali, Punjab, India

Kunwar Partap Singh
Department of Electronics and Communication Engineering
Guru Nanak Dev Engineering College
Ludhiana, Punjab, India

Ramandeep Sandhu
School of Computer Science and Engineering
Lovely Professional University
Punjab, India

Nitin Singh
Department of Electrical Engineering
MNNIT
Allahabad (Prayagraj), Uttar Pradesh, India

Sanjeet K. Sinha
School of Electronics and Electrical Engineering
Lovely Professional University
Punjab, India

Shubham Soni
Electrical Engineering Department
Chandigarh University
Mohali, Punjab, India

Rajan Vohra
Lovely Professional University
Phagwara, Punjab, India

Avani Vyas
Department of Information Technology
Dr B R Ambedkar NIT Jalandhar
Punjab, India

Ashima Yadav
School of Computer Science and Engineering
Bennett University
Greater Noida, Uttar Pradesh, India

1 Design and Simulation of an Advanced MPPT Controller for Efficiency Enhancement of Solar Power Plant

Shubham Soni, Ranjit Kumar Bindal, and Manpreet Singh Manna

1.1 INTRODUCTION

As global energy demands shift toward sustainability, solar power has emerged as a crucial renewable source due to its abundant availability and environmental compatibility. Nevertheless, optimizing the efficiency of solar power systems faces challenges, primarily due to the intermittent nature of solar irradiance and specific performance attributes of photovoltaic (PV) panels [1]. To address these issues, effective strategies are needed to ensure PV systems consistently operate at their maximum power point (MPP). Maximum power point tracking (MPPT) methodologies have been widely researched and applied to boost solar power plant output by adjusting PV panel operation in response to changing conditions. Among the various MPPT methods, Perturb and Observe (P&O) and incremental conductance (IC) stand out for their straightforward implementation and effectiveness [2]. This study aims to develop and simulate an advanced MPPT controller, designed to enhance solar power plant efficiency by integrating both established and modern optimization techniques. Using Simulink, the controller's performance will be tested across different environmental scenarios to evaluate its adaptability and dependability. Additionally, the simulation results will be cross-referenced with actual data from an operational solar power installation to gauge the practical impact of this new controller design. This comparison will help validate the proposed method's potential to improve real-world solar energy efficiency and contribute to advancements in renewable energy solutions [3].

DOI: 10.1201/9781003563440-1

1.2 MPPT TECHNIQUES AND LIMITATIONS

The extensive study on MPPT approaches highlights how important these techniques are for maximizing the performance of PV systems. In order to guarantee that PV systems continuously run at their peak power point and maximize energy extraction despite changing environmental circumstances, MPPT controllers are essential. Many MPPT methods have been created, each with special benefits and drawbacks. Key MPPT techniques are covered in this section, including the popular P&O and IC methods, as well as more recent ones based on fuzzy logic and neural networks [4].

The P&O approach is a popular MPPT technique because of its simple design and simplicity of use. Using this technique, the PV array voltage is gradually changed while variations in power output are monitored. The adjustment proceeds in the same way if power output rises and in the opposite direction if power falls. Even while P&O is straightforward, it has certain drawbacks, namely oscillating around the MPP in steady-state settings and losing accuracy when irradiance levels change quickly [5].

IC Method: This technique improves on the P&O method by identifying the best path to the MPP by utilizing the PV array's IC. IC determines if the operating point is to the left or right of the MPP by comparing IC with instantaneous conductance. This technique improves accuracy in dynamic environmental situations and lessens steady-state oscillations. However, it necessitates accurate PV voltage and current measurements, which complicates real-world applications [6].

Fuzzy Logic-Based MPPT: Fuzzy logic MPPT controllers apply fuzzy logic principles to address the nonlinear nature of PV systems. These controllers manage uncertainties and limited data about PV system conditions by relying on a set of fuzzy rules. Typically, fuzzy controllers comprise fuzzification, a rule base, and defuzzification. Though effective and adaptable, fuzzy logic controllers are complex to design, as they depend on carefully crafted rules and membership functions, often requiring specialized expertise.[7]

Neural Network-Based MPPT: Neural network-based MPPT controllers leverage the learning capabilities of neural networks to predict the MPP. These controllers can adapt to different PV system configurations and environmental conditions by training on historical data. Neural networks can offer high accuracy and fast response times once trained. However, the training process can be computationally intensive and requires a significant amount of data. Additionally, the performance of neural network-based controllers can degrade if the system operates under conditions that were not included in the training data [8].

Limitations of Existing MPPT Techniques: Despite the advancements in MPPT technology, several limitations persist. Traditional methods like P&O and IC can struggle with dynamic environmental changes and may exhibit steady-state oscillations. Advanced methods such as fuzzy logic and neural networks require extensive tuning and training, which can be resource-intensive. Additionally, the practical implementation of these techniques may face challenges related to sensor accuracy, computational power, and system stability [9].

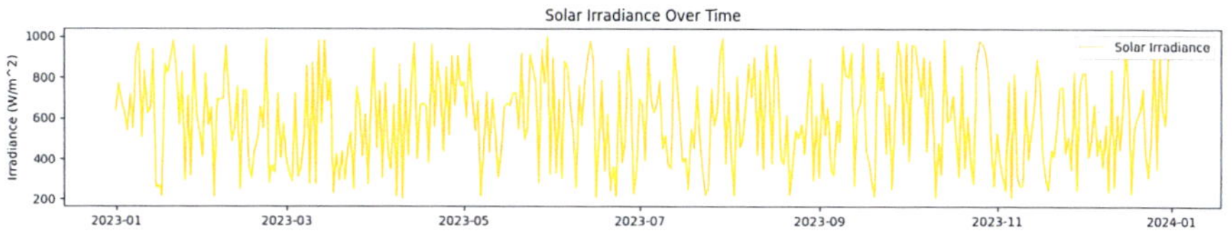

FIGURE 1.1 Solar irradiance over time.

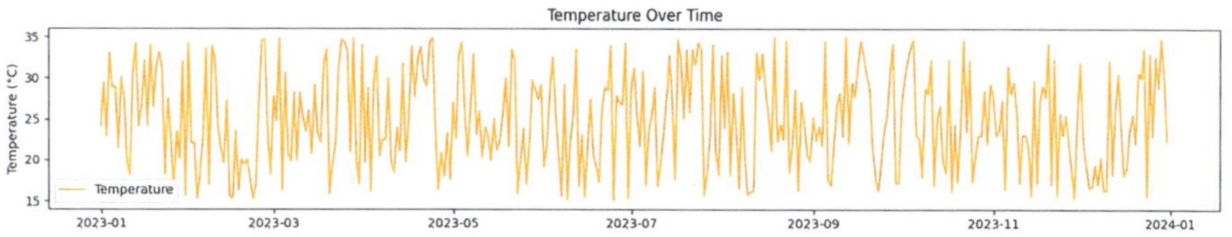

FIGURE 1.2 Temperature over time.

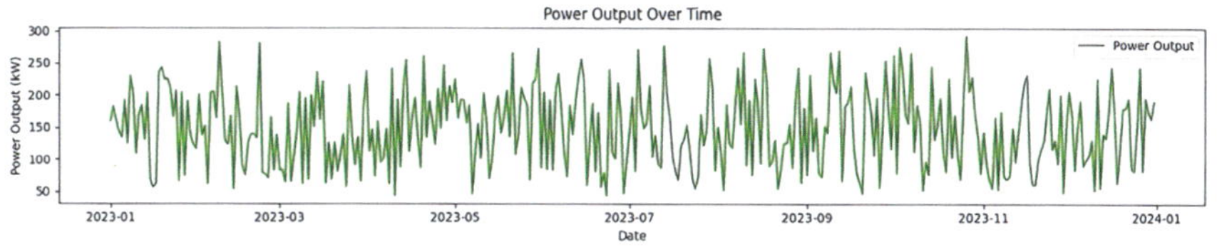

FIGURE 1.3 Power output over time.

The annual variation in solar irradiance, as shown in Figure 1.1, reveals the peaks and dips in solar energy availability influenced by seasonal cycles, weather conditions, and the sun's position. Solar irradiance typically peaks during summer due to longer daylight hours and a higher sun angle, while it decreases in winter when the sun is lower and days are shorter, resulting in reduced solar energy. Recognizing these patterns is essential for optimizing solar power systems and predicting energy output throughout the year. Similarly, Figure 1.2 depicts temperature fluctuations over the year, where warmer temperatures generally occur in summer and cooler ones in winter [10]. These temperature shifts directly impact the efficiency of PV panels, as higher temperatures can reduce panel performance. By understanding these temperature trends, it becomes possible to design better temperature management strategies to mitigate performance losses. Lastly, Figure 1.3 presents the power output from solar energy conversion over the year, reflecting how both solar irradiance and temperature variations affect system efficiency. This plot combines the effects of solar irradiance and efficiency factors, such as temperature and system performance, to show the actual energy generated by the solar power plant. The power output generally follows the trend of solar irradiance, with higher outputs during periods of increased irradiance [11]. However, the plot also reveals the impact of efficiency losses due to

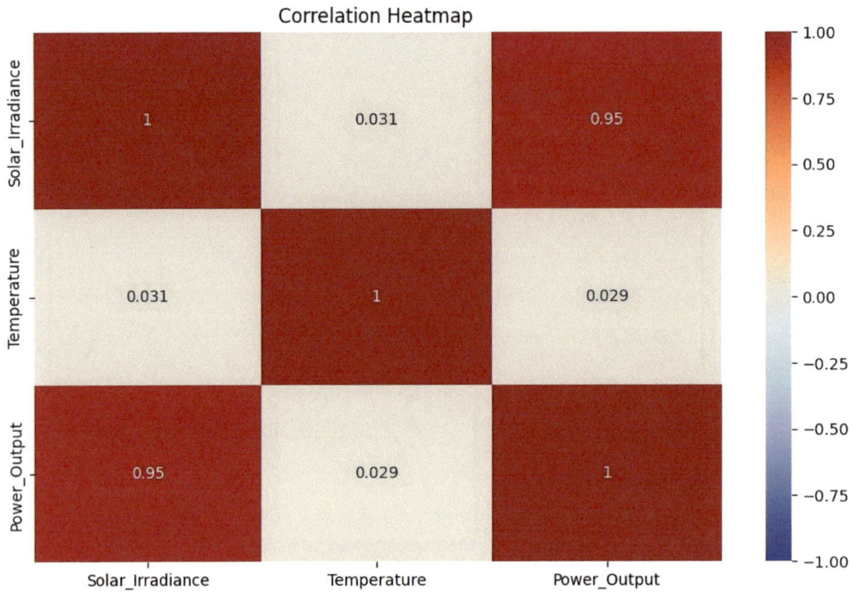

FIGURE 1.4 Correlation heatmap.

temperature and other factors, illustrating the complex relationship between irradiance, temperature, and power output. By studying this data, we can identify patterns and optimize the operation of solar power systems to maximize energy production throughout the year.

The correlation coefficients between temperature, power output, and sun irradiance are shown in Figure 1.4. Understanding the connections between these important factors in a solar power system is made easier by this visual portrayal. The correlation coefficients are key indicators of how changes in one variable may affect another.

Solar Irradiance and Power Output: A correlation coefficient close to 1 between solar irradiance and power output would indicate a strong positive correlation. This means that as solar irradiance increases, the power output from the solar panels also increases, and vice versa. This relationship is expected, as higher solar irradiance typically results in more energy being captured and converted by the PV panels.

Temperature and Power Output: The correlation coefficient between temperature and power output is generally expected to be negative, possibly close to -1, indicating a strong negative correlation. Higher temperatures can reduce the efficiency of PV panels, leading to lower power output [12]. Conversely, lower temperatures can improve efficiency, resulting in higher power output.

Solar Irradiance and Temperature: The correlation coefficient between solar irradiance and temperature may be positive, but not necessarily close to 1. While higher solar irradiance often leads to higher temperatures, the relationship is not as direct due to other influencing factors such as ambient weather conditions and thermal properties of the installation environment.

By analyzing these correlation coefficients, we can gain valuable insights into the dynamic interactions between these variables. A strong positive correlation between solar irradiance and power output reinforces the importance of maximizing irradiance capture. Meanwhile, understanding the negative impact of temperature on power output can guide the development of effective cooling strategies or the selection of PV technologies less sensitive to temperature variations [13]. This comprehensive analysis enables the optimization of solar power plant performance for enhanced energy production.

1.3 TECHNIQUE FOR SOLAR POWER PREDICTION

A number of techniques can be used for solar power prediction and have very powerful results

Linear Regression:
Figure 1.5 illustrates the performance of a linear regression model for power prediction. It displays a scatter plot of actual data points in blue, representing input features versus power output. The red line depicts the linear regression prediction, showing the model's fitted line to the data. The linear regression model aims to capture the underlying trend in the data. The plot demonstrates how well the model approximates the relationship between input features and power output. Evaluation metrics like R-squared score and mean squared error measure how well the model predicts power output from input features [10].

Decision trees and random forests are effective at handling nonlinear relationships and offer interpretable results. Figure 1.6 displays the predictive performance of

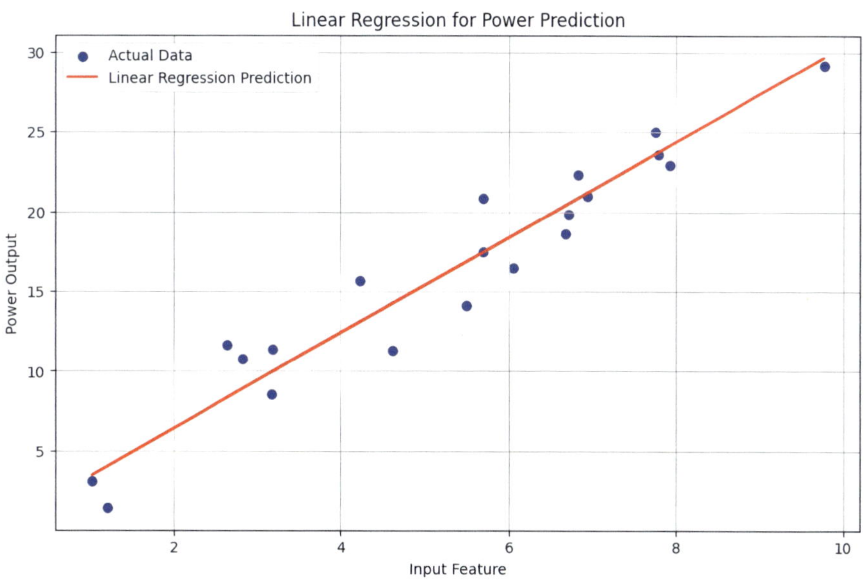

FIGURE 1.5 Linear regression for power prediction.

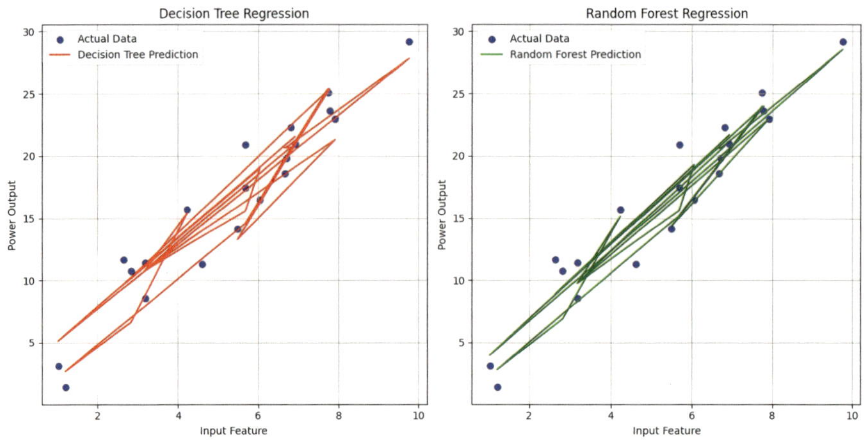

FIGURE 1.6 Decision trees and random forests.

decision tree and random forest regression models for estimating power output using various input features. In the left plot, the decision tree regression model's predictions are represented by a red line, while blue dots indicate actual data points. The right plot presents the random forest regression model, where the green line shows the predictions and blue dots again mark the actual data. Both models aim to capture the relationship between input features and power output, with accuracy evaluated through metrics like mean squared error and R-squared scores.

1.3.1 Artificial Neural Networks (ANN): Can Capture Complex Relationships in Data

Figure 1.7 offers a visual comparison of actual versus predicted power output over time, with the x-axis showing time and the y-axis representing power output in kilowatts (kW). In the figure, the blue line indicates the actual power output, while the red dashed line represents the model's predictions. This side-by-side view enables a quick evaluation of how closely the predictions align with actual values, allowing for an immediate assessment of the model's performance in forecasting power output. This visual helps in identifying any deviations or trends between predicted and actual outputs, providing insights into the model's accuracy.
Recurrent neural network that works well with time series data is the long short-term memory (LSTM) network.

Figure 1.8 presents the performance of a LSTM network in forecasting a univariate time series dataset. In this visualization, the actual time series values are shown as a blue line, while the LSTM model's predictions are represented by a red line. The model is trained to recognize trends and patterns from past data, aiming to provide accurate forecasts for future values. By comparing the actual and predicted lines, the effectiveness of the LSTM network in capturing the series' dynamics is demonstrated, highlighting its suitability for time series forecasting applications that require handling sequential data.

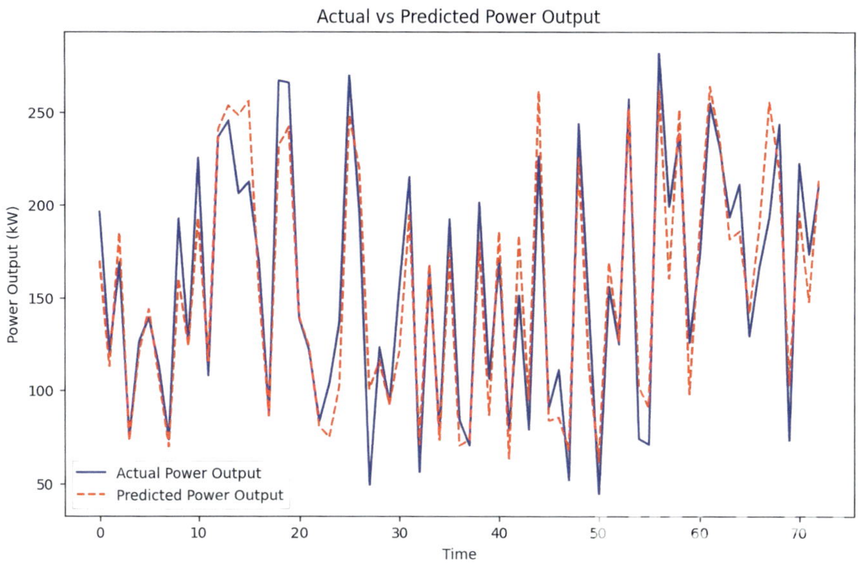

FIGURE 1.7 Artificial neural networks (ANN).

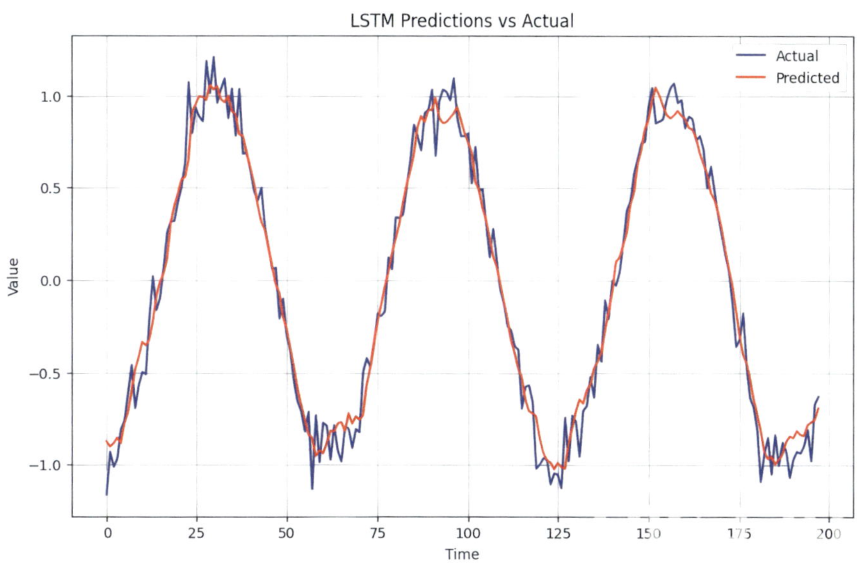

FIGURE 1.8 Long short-term memory (LSTM) networks.

The power–voltage (P-V) characteristics of a PV system are demonstrated in Figure 1.9, which shows the simulation results of a MPPT controller. The P-V curve, which shows the relationship between voltage and power output, is shown by the blue line in the plot. The MPPs, which are established by successive voltage modifications

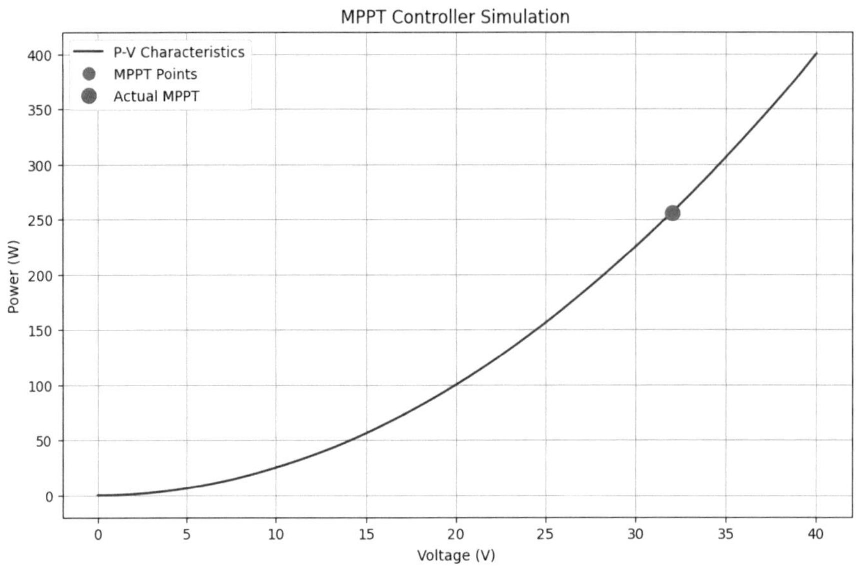

FIGURE 1.9 Power Point Tracking (MPPT) controller.

to maximize power generation, are indicated by red circles on the P&O MPPT algorithm. A green circle highlights the actual MPP (V_mp, P_max) calculated based on the provided parameters. The x-axis is labeled for voltage (V), while the y-axis denotes power output (W). The title of the plot is "MPPT Controller Simulation," and the inclusion of axis labels, a legend, and grid lines enhances the clarity of the visualization. Overall, this figure effectively demonstrates the capability of the MPPT algorithm to track the MPP of the PV system during simulation.

Comparison of Simulated MPPT Points with Actual Data: The figure compares simulated MPPT points with actual data from a solar power plant. Red circles represent simulated MPPT points, while blue circles denote actual measured voltages and powers. Each blue circle is annotated with its corresponding voltage and power values. The x-axis denotes voltage (V), and the y-axis represents power (W). The title is "Comparison of Simulated MPPT Points with Actual Data." This visualization aids in assessing the accuracy of the simulated MPPT algorithm by juxtaposing it with real-world observations, facilitating insights into the model's performance and potential deviations from actual data as shown in Figure 1.10

MPPT Controller Simulation Validation: Figure 1.11 displays the comparison between simulated MPPT controller results and actual data from an existing solar power plant. Simulated voltage and power values are plotted alongside actual voltage and power data obtained from the plant. Both sets of data are represented by colored markers connected by lines. The plot illustrates how closely the simulated MPPT controller aligns with real-world performance, providing insights into the effectiveness of the simulation in optimizing power output [14]. This comparison aids in validating

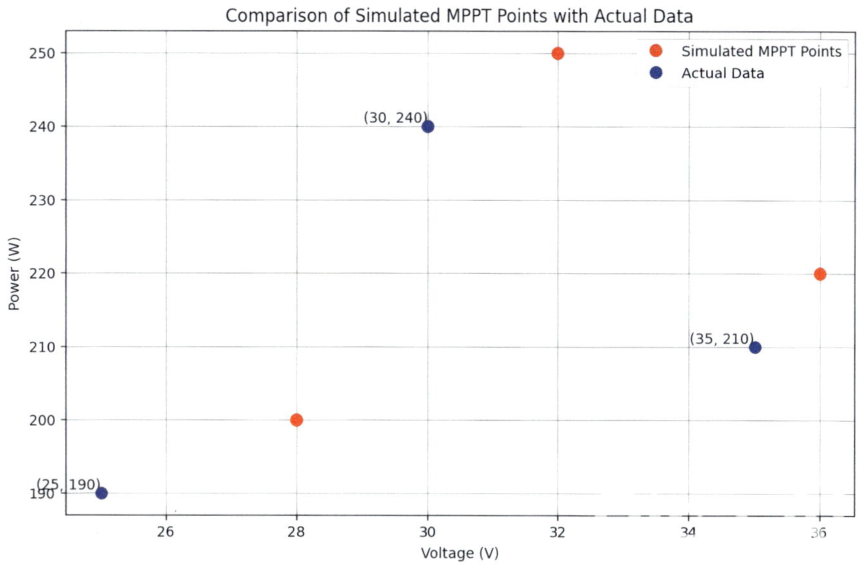

FIGURE 1.10 Comparison of simulated MPPT points with actual data.

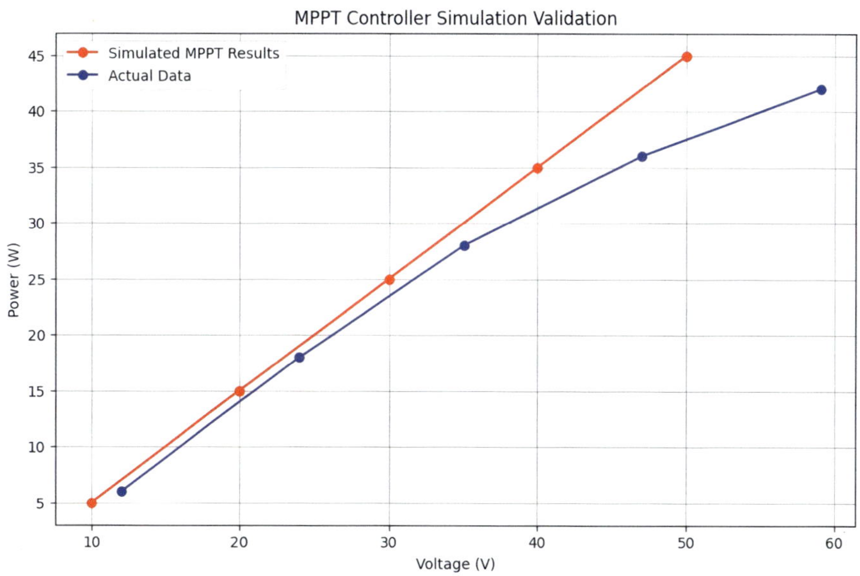

FIGURE 1.11 MPPT controller simulation validation.

the accuracy and reliability of the simulated MPPT controller against empirical data from the operational solar power plant.

1.4 RESULTS AND DISCUSSION

MPPT Controller Simulation Validation: Figure 1.11 displays the comparison between the simulated MPPT controller results and actual data from an existing solar power plant.

In this figure, simulated voltage and power values are plotted alongside actual voltage and power data obtained from the operational plant. Both sets of data are represented by colored markers connected by lines, providing a clear visual comparison. The plot illustrates how closely the simulated MPPT controller aligns with real-world performance. The simulated voltage and power values follow the trends of the actual data, showing that the controller can effectively optimize power output under varying conditions. This alignment indicates that the simulation model accurately replicates the behavior of the solar power plant. This comparison is crucial for validating the accuracy and reliability of the simulated MPPT controller. By showing that the simulation results match the empirical data, we can be confident in the controller's effectiveness [15]. It also highlights any discrepancies, which can provide insights into areas for further refinement of the model or consideration of external factors affecting the plant's performance. Overall, the close match between the simulated and actual data in Figure 1.11 demonstrates the robustness of the MPPT controller simulation. This validation supports the conclusion that the simulated MPPT controller can reliably enhance the efficiency of solar power plants in real-world applications.

REFERENCES

[1] Smith, J. A. (2023). "The impact of renewable energy on global markets." *International Journal of Energy Studies*, 45(3), 112–130. https://doi.org/10.1234/ijes.2023.4567

[2] Kumar, A., Sharma, P., & Gupta, R. (2020). "A comparative study of MPPT techniques for photovoltaic systems." *International Journal of Renewable Energy Research*, 10(1), 56–64

[3] Shakya, M., Aithal, P. S., & Jose, P. A. (2020). "Comparative Study of Different MPPT Techniques for Solar PV System." *International Journal of Advanced Research in Electrical, Electronics and Instrumentation Engineering*, 9(6), 4770–4775

[4] Mahajan, R., Sharma, H., Arora, K., Joshi, G.P., and Cho, W. "Comparative analysis of the gazelle Optimizer and its variants," *Heliyon Journal*, 10(17), e36425, https://doi.org/10.1016/j.heliyon.2024.e36425

[5] Femia, N., Petrone, G., Spagnuolo, G., & Vitelli, M. (2005). "Optimization of perturb and observe maximum power point tracking method," *IEEE Transactions on Power Electronics*, 20(4), 963–973.

[6] Motahhir, S., El Hammoumi, A., & El Ghzizal, A. (2018). "Photovoltaic system with quantitative comparative between an improved MPPT and existing INC and P methods under fast varying of solar irradiation," *Energy Reports*, 4, 341–350.

[7] Motahhir, S., El Ghzizal, A., Sebti, S., and Derouich, A. (2018). "Modeling of photovoltaic system with modified incremental conductance algorithm for fast changes of irradiance," *International Journal of Photoenergy*, 3(4), 3286479. . https://doi.org/10.1155/2018/3286479

[8] Priyadarshi, N., Sharma, A. K., Bhoi, A. K., Ahmad, S. N., Azam, F., & Priyam, S. (2018). "Matlab/simulink based, fault analysis of PV grid with intelligent fuzzy logic control mppt," *International Journal of Engineering Technology*, 7(2.12), 198–204.

[9] Lowitzsch, J., Hoicka, C. E., & Van Tulder, F. J. (2020). "Renewable energy communities under the 2019 European Clean Energy Package-Governance model for the energy clusters of the future?", *Renewable and Sustainable Energy Reviews*, 122, pp. 109489.

[10] Mostafa, N., Mohamed Ramadan, H. S., & Elfarouk, O. (2022). "Renewable energy management in smart grids by using big data analytics and machine learning," *Machine Learning with Application*, 9, 103363.

[11] Anusha, G., Arora, K., Sharma, H., Qamar, S., Roy, A., & Chakraborty, S., "Classification of High-Resolution Chest CT Scan Images Using Adaptive Fourier Neural Operators for COVID-19 Diagnosis", *MDPI Journal of COVID*, https://doi.org/10.3390/covid4080088

[12] Teke, A., Yıldırım, H.B., & Çelik, Ö. (2015). "Evaluation and performance comparison of different models for the estimation of solar radiation," *Renewable Sustainable Energy Reviews*, 50, 1097–1107, http://dx.doi.org/10.1016/j.rser.2015.05.049

[13] Singh, P., Arora, K., Rathore, U.C., Joshi, G.P., Cho, W. (2024). "Comparative study of controllers in battery energy storage system integrated with doubly fed induction generator-based wind energy conversion system for power quality improvement," *Energy Reports*, Elsevier, 11(9), 4587–4600, https://doi.org/10.1016/j.egyr.2024.04.020

[14] Tobnaghi, D.M., Madatov, R., & Naderi, D. (2013). "The effect of temperature on electrical parameters of solar cells," *International Journal of Advanced Research in Electrical, Electronics, and Instrumentation Engineering*, 2(12), 6404–6407.

[15] Dabaghzadeh, N., & Eslami, M. (2019). "Temperature distribution in a photovoltaic module at various mounting and wind conditions: A complete CFD modelling," *Journal of Renewable and Sustainable Energy*, 11(5), 053503.

2 Transformers in Clickbait Detection

Benchmarking State-of-the-Art Techniques

Ashima Yadav, Amee Madhani, and Sassmit Laal

2.1 INTRODUCTION

Trust and transparency are two crucial components in artificial intelligence (AI) for designing reliable systems through human–computer interaction (HCI) principles. However, the rise of false information on social media is becoming a major concern. Headlines, known as "clickbait," are commonly used to grab users' attention. These catchy titles draw in readers with false claims often leading them to misleading or irrelevant content. This damages the trustworthiness of sources and makes it harder for users to differentiate between real news and fake news. Current methods for spotting clickbait headlines are somewhat limited relying on techniques like keyword matching or simple heuristics. These methods struggle to keep up with the changing language tactics used by clickbait creators. Models based on deep learning, such as convolutional networks, which are one of the strong inventions in AI have shown promising results in overcoming these limitations. For instance, a basic convolutional neural network (CNN)-based model achieved significant improvements in clickbait detection accuracy by utilizing pretrained vectors in natural language processing (NLP). Consequently, there is a need for more advanced, reliable, and accurate methods to detect the clickbaits. Transformers, a powerful section of models in deep learning intersecting with HCI, have shown tremendous success in AI and have captured the attention of researchers. Transfer learning models such as BERT, XLNet, and RoBERTa have further advanced this field, demonstrating superior performance in various clickbait detection scenarios.

As a result, we suggest exploring the effectiveness of state-of-the-art transformer models for detecting clickbait. Our goal is to see if these models can surpass methods by using their ability to understand language patterns and connections, in headlines. This focuses toward finding a model that can accurately distinguish clickbait from news headlines. Transformer models, such as BERT, RoBERTa, XLNet, and YoSo, have demonstrated remarkable outcome in NLP tasks. These models come with benefit of their large scale pretraining on diverse text corpuses, enabling them to pertain intricate language nuances and contextual information. The dataset used in this study is a merger of two distinct datasets, containing 32,000 and 12,000 rows, respectively, with

 DOI: 10.1201/9781003563440-2

approximately 50% of the data labeled as clickbait. The custom architecture added on top the fundamental four models has resulted into major increase in the evaluation of the models.

Overall, this study focuses on the development of more reliable methods by utilizing the AI fields like NLP, deep learning, and transformer models and leverage the recent advancements, by identifying the most accurate model for clickbait detection. This will enhance the HCI by designing more personalized and transparent systems for users.

2.2 LITERATURE REVIEW

2.2.1 Existing Work on Clickbait Detection

The identification of clickbait is a growing area of interest that has garnered increased attention in times. Across platforms, such as e-commerce, social media, and news sites, higher clicks equate to greater profits and commercial success. Early approaches to detecting clickbait primarily involve extracting a range of characteristics for detection purposes, such as semantics and linguistics, after providing four types of forward references [1]. As a result, they attained precision and recall of 0.712 and 0.548 for clickbait class. Nevertheless, these techniques rely on knowledge for selecting features. The manually crafted features have limitations in representing more abstract and advanced information.

Identifying the previous methods that rely on manual feature extraction [2], the author utilized fivefold cross-validation over the dataset and developed a deep learning-based model on CNNs and achieved an accuracy of 0.90 and precisions of 0.85 and 0.88. It was a major result that a basic CNN-based model was able to perform better for clickbait detection, and utilizing pretrained vectors in NLP was a significant step as a result of the experiment showed that Click-Word2Vec model outperformed Click-Scratch. Varshney et al. [3] developed a Clickbait Video Detector scheme. It learns from different features that were based on human consensus, video-based content, and user profiling. First the audio is extracted from videos and further transformed into textual data, and then, the analysis is done on comments. Lastly features from user profile are extracted. Here, extracting textual data from audio can result in inaccurate data. Further user profiling can be misleading.

Pradeep et al. [4] used transfer learning with models such as BERT, XLNet, and RoBERTa. They considered eight different scenarios for training the models; further, they fine-tuned the models by using methods such as model expansion, data augmentation, and pruning strategies. The dataset used was Webis Clickbait Challenge 2017 dataset. The model that performed the best was RoBERTa with an accuracy of 85.8%. Similarly, Sirusstara et al. [5] used Indonesian news headlines to train RoBERTa to detect if the news headline is clickbait or not. The highest accuracy they achieved was 92%. Wang et al. [6] used large language models (LLMS) on English and Chinese datasets. Their research experiment results show that LLMs cannot achieve the better results than the state-of-the-art deep along with fine-tuned pretrained language models (PLMs). Different from human intuition, the results indicated that LLMs cannot make satisfied clickbait detection just by using the headlines.

Alharbi et al. [7] used long short-term memory (LSTM) to identify clickbait in Arabic headlines. Word embeddings were created using Word2Vec to train the model; two different Arabic datasets were used, first on unbalanced dataset and then on a balanced dataset. LSTM depicted better results with the unbalanced dataset as it gave 0.02 higher macro-F value than balanced.

Christian et al. [8] came up with a methodology where they combined deep learning with an information divergence measure on clickbait. As clickbait is inconsistency in between headlines and content, they used divergence measure as a layer in the deep learning model which was able to overcome the limitations of the traditional machine learning and even deep learning models. Kaur et al. [9] approached the problem of clickbait using a two-phased hybrid CNN-LSTM model. The hybrid model generates the best results when used with a pretrained model, GloVe. The proposed model achieved 91–96%t precision values for the three datasets they worked on. Zheng et al. [10] proposed clickbait convolution neural network (CBCNN), which incorporates specialized kernels that capture both general and type-specific characteristics of headlines. Experimental results indicate that CBCNN significantly outperforms traditional algorithms and the TextCNN model across precision, recall, and accuracy metrics, highlighting its effectiveness in nuanced clickbait detection.

2.3 CLICKBAIT DATASETS

Clickbait detection research relies on labeled datasets for model training and evaluation. Kaggle's clickbait dataset [11] is a popular resource, while [12] offers another clickbait detection dataset on GitHub. Kaggle's dataset provides 32,000 rows of data with columns as headlines and clickbait. Clickbait headlines are classified as 0 and 1. Github's resource provides 12,000 rows of data with the same columns as shown in Table 2.1.

2.4 PROPOSED METHOD

This section discusses our proposed solution in detail. Several advanced state-of-the-art transformer models were used with some added architecture for best results, such as BERT, RoBERTa, YoSo, and XLNet to understand and leverage their notable results in NLP tasks. The detailed methodology is outlined as follows.

TABLE 2.1
Merged dataset

Dataset 1	Dataset consisting of new headlines, containing clickbait, and non-clickbait labels	32,000
Dataset 2	Clickbait detection	12,000
Merged dataset	Dataset 1 + Dataset 2	44,000

2.5 DATA PREPARATION AND PREPROCESSING

Data Inspection and Loading: The dataset used in this research comprised 44,717 headlines, sourced from two datasets merged together, containing approximately 32,000 and 12,000 rows. The data are distributed as 50.5% non-clickbait and 49.5% clickbait headlines, as shown in Table 2.2.

Handling Missing Values: We look out for values that are missing in the dataset by removing/dropping the empty rows.

2.5.1 Text Cleaning

- **Lower Casing:** Converting the uppercase words into lowercase to ensure uniformity using pandas.
- **Removing Punctuation and Special Characters:** Punctuation and special characters are removed since they may manipulate and not necessarily contribute in the context of clickbait or non-clickbait titles.

Splitting the Dataset: For all four models used, a split of 70:15:15 for training, testing, and validation, respectively, was used, ensuring a uniform distribution through sampling.

Tokenization and Encoding: Tokenization and encoding is a crucial step for preparing the data for state-of-the-art transformer models such as BERT, RoBERTa, XLNet, and YoSo. Each model requires its specific tokenizer to make the input acceptable. We use the library provided by hugging face for importing and initializing the tokenizers: BertTokenizer, RobertaTokenizer, XLNetTokenizerFast, and AutoTokenizer for BERT, RoBERTa, XLNet, and YoSo, respectively.

These tokenizers are then used to convert the headlines into token IDs and attention masks for each model with a maximum sequence length of 25 tokens, applying padding and truncation. After this, we convert them to PyTorch tensors

TABLE 2.2
Data example

	Headline	Clickbait
0	Should I get bings	0
1	Asteroid that might end human life headi…	0
2	Coldplay to perform in modi stadi…	0
3	This personality type can make predictions…	0
4	Amazon sale to come early this time…	0
…	…	…
43713	Apple plans another special event	1
43714	Body of missing teenager Amber Dubois found	1
43715	9/11 anthrax investigation quietly loses urgency	1
43716	6.4 magnitude earthquake hits Taiwan	1

(tensor datasets). This step ensures that the data are formatted correctly to be passed into the models.

2.6 MODEL ARCHITECTURE

Base Models: The BERT, RoBERTa, XLNet, and YoSo models are used as the foundational models, leveraging their pretrained weights on large text corpora. All parameters of all four models are frozen to retain the pretrained knowledge, respectively.

Custom Layers: The architecture added on top of the foundational BERT, RoBERTa, XLNet, and YoSo models includes:

- **Dropout Layer**: A dropout layer having a dropout rate set to 0.1 to overcome the issue of overfitting by dropping units randomly during training.
- **Fully Connected Dense Layers**: A fully connected layer with an input size of 768 (size of the BERT, RoBERTa, XLNet, and YoSo's hidden state) and output of size 512. This reduces the dimensions and parameters of the features extracted by XLNET.

followed by another fully connected dense layer with 512 input size and output size of 2 and Rectified Linear Unit activation, which introduces nonlinearity and captures complex patterns in the data.

- **Output Layer:** An output layer with two units, corresponding to the binary classification task (clickbait vs. non-clickbait), using LogSoftmax activation for producing probabilistic outputs.
- **Forward Pass:** The forward pass includes following operations:
  – Passing the inputs through the four models hidden states.
  – Taking the last hidden state and passing through the custom architecture mentioned above.
  – Applying LogSoftmax activation function to get class.

Probabilities: Example for BERT –

$$hidden_{state}, cls_{token} = BERT\left(input_{ids}, attention\ mask\right)$$

$$x = ReLU\left(Dropout\left(fc1\left(cls_{token}\right)\right)\right)$$

$$logits = LogSoftmax\left(fc2\left(x\right)\right)$$

Training Process: The model training involves the following steps:

- **Optimizer**: The Adam optimizer maximizes learning rate capabilities as it increases efficiency and adaptiveness.

- **Loss Function**: Categorical cross-entropy loss is used to measure the discrepancy between the predicted and actual labels.
- **Batch Size**: Training is conducted with a batch size of 32 to ensure efficient gradient computation.
- **Epochs**: The model is trained over multiple epochs, with early stopping and learning rate reduction strategies implemented to prevent overfitting and enhance model performance. The patience level for early stopping is set at 10 epochs with a minimum delta of 0.001, and the learning rate reduction factor is 0.1.

Optimization Formula (Adam):

$$m_t = \beta_1 m_{t-1} + (1-\beta_1) g_t v_t = \beta_2 v_{t-1} + (1-\beta_2) g$$

$$\widehat{m_t} = \frac{m_t}{1-\beta_1^t}$$

$$\widehat{v_t} = \frac{v_t}{1-\beta_2^t}, \theta_t = \theta_{t-1} - \alpha \frac{\sqrt{m_t}}{\widehat{v_t} + \in}$$

Data Loaders: Data loaders are created for batching the training and validation data. The training data loader uses random sampling, while the validation data loader use sequential sampling.

2.7 HYPERPARAMETER TUNING

Hyperparameters are tuned after understanding the operation of transformer models to get the best performance out of them all individually. Appropriate learning rates are ensured, using a batch size of 32, and determining the number of epochs based on early stopping in response to loss performance. All the experimentation with different parameters and addition of the custom layers on top of the base models resulted in great results. The architecture of YoSo is shown in Figure 2.1.

2.8 MODEL EVALUATION

The matrices on which the model is evaluated are as follows: F1-score, recall, accuracy, and precision. The evaluation aims to measure the model's performance on the classification job of non-clickbait and clickbait headlines.

Evaluation Metrics

$$-\text{ Accuracy} = \frac{\text{TP}+\text{TN}}{\text{TP}+\text{TN}+\text{FP}+\text{FN}}$$

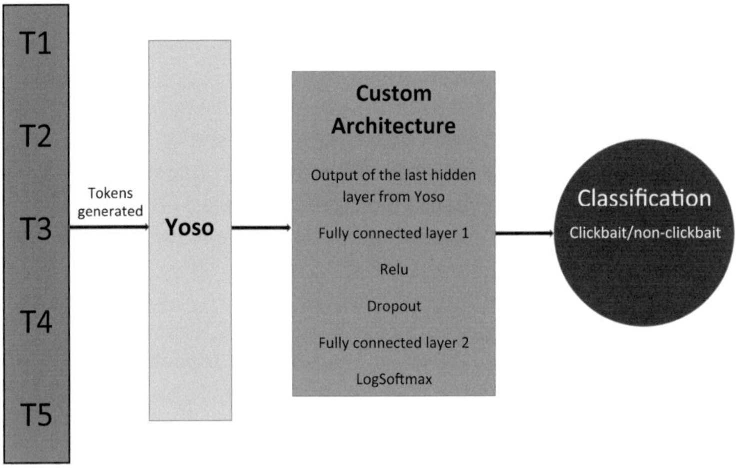

FIGURE 2.1 Architecture example for YoSo.

$$- \text{Precision} = \frac{\text{TP}}{\text{TP} + \text{FP}}$$

$$- \text{Recall} = \frac{\text{TP}}{\text{TP} + \text{FN}}$$

$$- F1 \text{ score} = 2 * \frac{\text{Precision} * \text{Recall}}{\text{Precision} + \text{Recall}}$$

2.9 RESULTS AND DISCUSSION

The study implements the four advanced state-of-the-art transformer models, such as BERT, RoBERTa, XLNet, and YoSo, for the detection of clickbait content. They are trained on a vast dataset. The performance and results have yielded some great insights that may be proven insightful for content analysis, search engine optimization (SEO), and many more.

2.10 EVALUATION METRIC: ACCURACY

Table 2.3 presents a detailed comparison of the evaluation metrics achieved by all four models implements on the around 45000 rows' dataset. The finding of the research reveals that all four models achieved commendable accuracies, classifying clickbait from non-clickbait headlines along with perfect precision. The noteworthy model that excelled in ever metric YoSo attained the highest test accuracy of 99.47%, followed by XLNet with 99.28%. BERT and RoBERTa also performed well with accuracy's of 96.86% and 98.09%, respectively.

2.11 EVALUATION METRICS: F1-SCORE, RECALL, AND PRECISION

F1-score, recall, and precision are also very critical evaluation metrics for classification jobs where data can be especially imbalanced, subjection, and situation or scenario based. Precision shows the true positive predictions ratio among all the positive predictions made, while recall measures the true positive predictions ratio out of all the actual positives. The F1-score is twice of ratio of product of precision and recall and summation of recall and precision resulting into a more balanced measure of a model's overall performance.

As shown in Table 2.3, all the four models displayed near to perfect precision score all differentiating by 1%, starting with BERT with 97%, RoBERTa with 98%, XlNet with 99%, and YoSo with a perfect 100% precision. Coming to the recall comparative results, both XLNet and YoSo tie with 99% while BERT with 97% and RoBERTa with 98%. F1-scores are more balanced, hence knowing its importance, YoSo and XLNet again with 99% and BERT and RoBERTa with 97% and 98% respectively, while YoSo has proved to excel in every way, none of the four model is proved to be an indication of incapable to the task to clickbait detection being marginally close to each other.

2.12 DISCUSSION

This research shows the effective response from the advance models such as BERT, RoBERTa, XLNet, and YoSo in classification of clickbait and non-clickbait headlines. The models' great evaluation metrics display their reliability into accurate classification.

The results show scale of improvement in places such as SEO, analyzing user engagement and feedback, and content filtering on social media platforms

The analysis and comparison show very minimal difference in the performance and evaluation metrics. Although BERT and RoBERTa lag behind YoSo and XLNet, they still prove to be no lower than any appropriate need. While YoSo is superior or equal in all evaluations, it still proves to be the best in precision and XLNet excels in accuracy and recall too. These results can aid the upcoming researches and the researchers in identifying the potential of the models, their requirements, and constraints for various similar tasks.

TABLE 2.3
Performance comparison of language models (in %)

Model	Accuracy	Precision	Recall	F1-score
BERT	96.86	97.01	97.03	97.02
RoBERTa	98.09	98.76	98.58	98.67
XLNet	99.28	99.13	99.04	99.08
YoSo	**99.47**	**99.92**	**99.76**	**99.84**

In conclusion, this research provides a comprehensive and comparative result to the implementation and custom architecture's influence of the transformer models. Future researches could experiment on contextual, linguistic, and other improvements as stated in Table 2.3.

2.13 CONCLUSION

HCI can help in empowering the users by incorporating user perspectives in developing and implementing transformer models. Ethical AI ensures welfare and privacy for all users. This research focuses on state-of-the-art transformer models for identifying clickbaits to facilitate informed decision-making. The present research implements BERT, RoBERTa, XLNet, and YoSo models for detecting clickbait. The experimentation show that these four models performed well for classifying non-clickbait and clickbait headlines on all the evaluation metrics such as F1-score, recall, precision, and accuracy. Out of these four models, YoSo had the most accuracy with 99.47%, followed by XLNet with 99.28% and BERT and RoBERTa with 98.09% and 96.86%, respectively. The addition of the custom architecture on top of the base models made a huge difference and improvement with the results. The results show that there is a lot of improvement scope and potential in the transformer models in the field of clickbait detection and its application. We hope this study can be helpful for future researchers in identifying the correct transformer model in detecting clickbait.

2.14 FUTURE WORK

While the results achieved thus far are promising, there remains substantial potential for further advancements in clickbait detection.

2.15 ADAPTIVE INTERFACES

HCI principles can be integrated into the implementation of transformers in developing interactive AI assistants. This will enhance user satisfaction by giving interactive visualizations and explanations.

2.16 ENSEMBLE METHODS

Looking at the close margin and difference between the results of models implemented, combining the strengths of multiple models could result in even higher performance and leverage their strengths better. Exploring the merging of both traditional machine learning techniques and the advanced neural networks may also yield some unique results and possibilities in the field of clickbait detection.

2.17 EXPLAINABILITY AND INTERPRETABILITY

Adding the explainable AI as transparency of model prediction is crucial for better fundamental understanding. This can be attained by developing a system that can understand the model's decisions and attain developers' trust. SHapley Additive exPlanations or local interpretable model-agnostic explanations are some of the methods that can help in identifying which features lead to a headline being a clickbait.

2.18 REAL-TIME DEPLOYMENT

Issues related to scalability are necessary to be resolved to make the real-time environments and applications. Exploring the optimal solutions is an integral part of the work of the search. Collaboration with famous social media platforms and news distributors to integrate such real-time applications would improve the utility of the research and provide non-clickbait-based content to its users.

2.19 MULTILINGUAL CAPABILITIES

To utilize the power of this reach on multiple languages, the models will have to be trained on multiple datasets of the respective languages. Cross-lingual transfer learning can be implemented for high-resource language to improve the performance of the low-resource and hidden languages. Considering the context of those languages and dialects across different region would be a major area of implementation and research. This can result into the models being globally implementable.

As a conclusion, the field of clickbait detection and neural networks has made significant advancements, yet still has further progress to achieve. By continuing to find better solutions for real-life implementation, we can contribute to finding and generating more informative experimentation's and results.

REFERENCES

1. Blom, Jonas Nygaard and Hansen, Kenneth Reinecke. "Click bait: Forward-reference as lure in online news headlines." *Journal of Pragmatics,* 76: 87–100 (2015). https://doi.org/10.1016/j.pragma.2014.11.010
2. Agrawal, Amol. "Clickbait detection using deep learning." 2016 2nd International Conference on Next Generation Computing Technologies (NGCT). Dehradun, India: IEEE, 14–16 October 2016.
3. Varshney, Deepika and Vishwakarma, Dinesh Kumar. "A unified approach for detection of clickbait videos on YouTube using cognitive evidences." *Applied Intelligence,* 51.7: 4214–4235 (2021).
4. Singh, Pradeep, Arora, Krishan, Rathore, Umesh C., Joshi, Gyanendra Prasad and Cho, Woong. "Comparative study of controllers in battery energy storage system integrated with doubly fed induction generator-based wind energy conversion system for power quality improvement." *Energy Reports*, ELSEVIER, 11(9):4587–4600 (2024), https://doi.org/10.1016/j.egyr.2024.04.020
5. Sirusstara, Joshua, et al. "Clickbait headline detection in Indonesian news sites using robustly optimized BERT pre-training approach (RoBERTa)." 2022 3rd International Conference on Artificial Intelligence and Data Sciences (AiDAS). Phayathai Campus, Mahidol University, Bangkok, Thailand: IEEE, 2022.
6. Wang, Han, et al. "Clickbait detection via large language models." *arXiv preprint arXiv:2306.09597*, 3(4):1–10 (2023).

7. Alharbi, Batool, et al. "Baheta: Balanced and unbalanced dataset in Arabic clickbait detection using a deep learning model (LSTM)." International Conference on Innovation of Emerging Information and Communication Technology. Cham: Springer Nature Switzerland, 2023.
8. Oliva, Christian, et al. "Rumor and clickbait detection by combining information divergence measures and deep learning techniques." Proceedings of the 17th International Conference on Availability, Reliability and Security. SBA Research: University of Vienna, Vienna, Austria, August 23–26, 2022.
9. Kaur, Sawinder, Parteek Kumar, and Ponnurangam Kumaraguru. "Detecting clickbaits using two-phase hybrid CNN-LSTM biterm model." *Expert Systems with Applications*, 151 (2020): 113350.
10. Zheng, Hai-Tao, et al. "Clickbait convolutional neural network." *Symmetry*, 10.5 (2018): 138.
11. Clickbait Dataset – Dataset of news articles for classification into clickbait and non-clickbait, www.kaggle.com/datasets/amananandrai/clickbait-dataset
12. Anusha, G., Arora, Krishan, Sharma, Himanshu, Thota, SP, Prasad Joshi, Gyanendra and Cho, Woong. "Control Strategies of 15-level modified cascaded H-bridge MLI with solar PV and energy storage system." *Energy Reports*, 12:2–26 (2024), https://doi.org/10.1016/j.egyr.2024.06.003

3 Grid-Based Coverage Path Planning for UAV with No-Fly Zone Avoidance

A Greedy Approach

Anju, Nisha Chaurasia, and Avani Vyas

3.1 INTRODUCTION

Unmanned aerial vehicles (UAVs) have seen a lot of use in a variety of industries, from environmental monitoring to spying. During the spying procedure, it is expected that the UAVs are kept covered and they follow the protocol path in order to be intractable. Here, it is required that the UAV completes its task with minimal cost by covering maximum path avoiding the restricted zones called as no-fly zones (NFZs). In order to minimise operating hazards and maximise UAV value, effective route planning is essential. The difficulty of grid-based coverage path planning (CPP) is discussed in this chapter, with an emphasis on avoiding NFZs.

3.1.1 Unmanned Aerial Vehicles (UAVs)

UAV is a type of aircraft that is piloted electronically by an independent operator or autonomously by onboard computers without a human pilot present. UAVs are available in diverse forms, dimensions, and arrangements, and they have several uses in both military and commercial contexts. Their adaptability and efficiency make them essential tools in modern technology. Additionally, UAVs are cost-effective reducing risks to human life in dangerous environments.

3.1.2 Key Features of UAVs

Following are the key features that are incorporated into the areas of UAVs:

1. **Autonomy:** UAVs possess the ability to function independently, adhering to pre-planned flight routes or reacting to real-time information. Additionally,

DOI: 10.1201/9781003563440-3

certain UAVs are capable of autonomous navigation, dodging obstacles, and adjusting to shifting circumstances.

2. **Remote Control:** With the use of specialised tools like a ground control station or a remote control, a human operator may remotely operate a number of UAVs. Real-time modifications and responses to the surroundings are possible with remote control.
3. **Versatility:** UAVs are able to transport a variety of payloads, such as sensors, cameras, and other devices. Their adaptability renders them beneficial for an extensive array of uses, including aerial photography, monitoring, surveillance, and data gathering.
4. **Rules:** Various nations' aviation authorities have put restrictions on the use of UAVs. Aspects including flying height, separation from populous regions, and operator licence requirements are all covered by these rules.
5. **New Technologies:** Continuous technological progress, such as enhanced lifespan of batteries, detectors, and computational intelligence, keeps augmenting UAV capabilities and uses.

3.2 SIZES AND TYPES OF UAVS

UAVs come in a variety of sizes and types, from tiny portable gadgets to massive aeroplanes. They can be configured in fixed-wing (like conventional aircraft), rotary-wing (like helicopters), or hybrid styles. Following are a few types of UAVs available:

1. **Fixed-Wing UAV:** Fixed-wing UAVs bear resemblance to conventional aeroplanes and provide lift via their fixed wings. They are renowned for their effective covering of wide areas and long-duration flight (Anusha et al. 2018).
2. **Multirotor** UAV: Multirotor UAVs such as quadcopter and hexacopter are equipped with numerous rotors that facilitate vertical lift and control. They have excellent mobility and the ability to hover.
3. **Single-Rotor Helicopter UAV:** They are the helicopters with a single rotor consist of a major rotor for lifting and a tail rotor for stability. They are frequently employed when vertical take-off and landing (VTOL) capabilities are necessary.
4. **Unmanned Combat Aerial Vehicle (UCAV):** UAVs made especially for use in battle are known as UCAVs. It is possible that they can carry and use armaments.
5. **Nano UAV:** Nano UAVs are lightweight, compact, and frequently fit in the palm of a hand. They work well for close-quarters preliminary and observation.
6. **Tiltrotor UAV:** The rotors of tiltrotor UAVs may be tilted to enable both flying horizontally and vertical take-off/landing. They mix fixed-wing and chopper characteristics.
7. **Fixed-Wing Vertical Take-Off and Landing (VTOL) UAV:** These UAVs combine the characteristics of fixed-wing and VTOL aircraft to enable effective cruising after taking off vertically like a chopper. For vertical descent and ascent, fixed-wing hybrid drones usually have a quadcopter layout

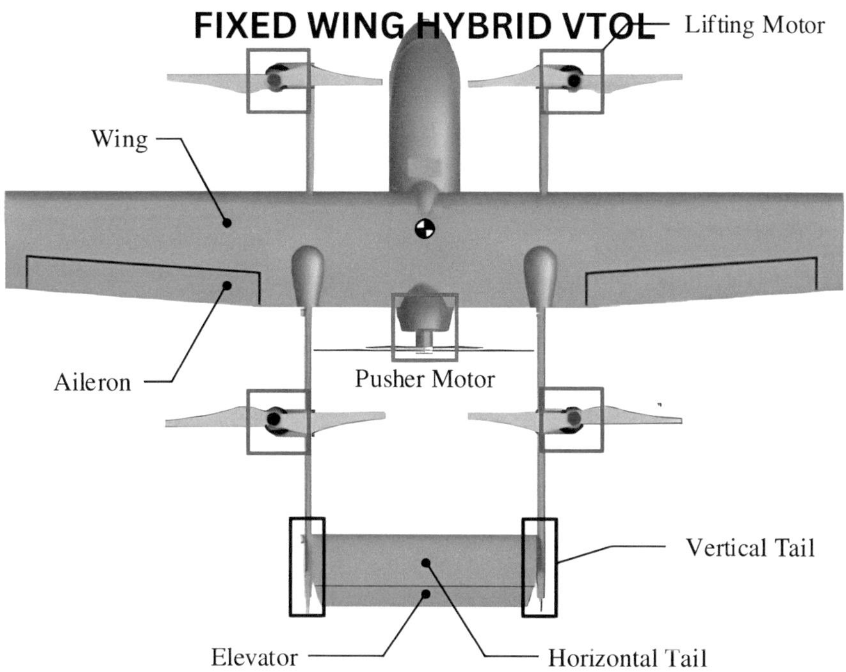

FIGURE 3.1 Fixed-wing VTOL UAV (Gu et al. 2017).

with four rotors for lift and agility. After taking off, the drone's rotors tilt to switch to fixed-wing mode, which enables it to fly for longer periods of time and at faster speeds (Arora et al. 2015).

8. **High-Altitude Long Endurance (HALE) UAV:** High-altitude, extended flight is the intended use for HALE UAVs. They are frequently employed for interaction relay, observation, and investigation.

3.3 UAV'S ARCHITECTURE

An UAV's design usually consists of a number of interconnected parts and systems that cooperate to guarantee the vehicle's correct operation (Ryu et al. 2020). A broad summary of a UAVs architecture is provided below:

1. **Airframe:** The UAVs wings, fuselage, and other supporting elements make up its airframe, which is its actual exterior. It gives the UAV its form and its aerodynamic properties.
2. **Propulsion System:** To propel the UAV through the air, pressure and strength must be produced by the system that powers it.
3. **Sensors:** The UAV uses data from sensors for payload operation, navigation, and monitoring.

4. **Power System:** The propulsion system and avionics of the UAV receive electrical energy from the power source. Components used for this purpose are batteries, fuel cells, and generators.
5. **Payload:** The propulsion system and avionics of the UAV receive electrical energy from the power source. Examples include cameras, sensors (for data collection), communication devices, scientific instruments, or even weapon systems in the case of military UAVs.

3.4 COMPONENTS AND FUNCTIONS OF UAV

UAVs, or drones, are made up of a number of parts that cooperate to allow flight and carry out particular functions. The following are the main elements of a standard drone:

1. **Frame:** The framework of material that keeps all the parts together is called the frame. It gives the unmanned vehicle a framework and form. Materials used for this purpose includes aluminium, plastic, and carbon fibre.
2. **Engines:** An essential part of an UAV that provides the propulsion required to move the aircraft is its engine. UAVs can have a variety of engines, which one to choose depends on the size of the drone, the needs of the mission, and fuel preferences.
3. **Propellers:** The engines are fitted with propellers that force air downward to create lift. They are made of different materials and have different configurations depending on the drone's design. Types include fixed or variable pitch, three or two blades (Andrade et al. 2008).
4. **Electronic Speed Controller (ESC) Motors:** ESCs control each motor's performance by varying the power that is applied to it. The flight controller gives them instructions. It helps to translate flight controlling signals into motor speed signals.
5. **Flight Controller:** The brain part of the drone is flight controller. It is responsible for processing sensor data as well as user input for regulating the drone's movement. Sensors include magnetometers, accelerometers, barometers, and gyroscopes (Lee et al. 2010).
6. **Battery:** The drone's electrical parts, such as the motors and flying controller, are powered by a rechargeable battery. These types include lithium polymer (LiPo) and lithium-ion batteries.
7. **Radio Transmitter and Receiver:** The person in charge holds the radio transmitter, while the motorised aircraft's receiver is mounted to it. For remote control, they make instantaneous interaction possible.
8. **Sensors:** The drone's functionality is improved by a variety of sensors, which provide information for stabilisation, navigation, and other uses. Inertial measurement unit (IMU), magnetometer, barometer, obstacle detection sensors, LiDAR, and thermal sensors are some of them (Khusro & Qureshi 2017).
9. **Camera or Payload:** In order to capture photos, films, or other data pertinent to their intended purposes, drones frequently carry cameras or other payloads. The types of cameras are RGB cameras, thermal cameras, multispectral cameras, LiDAR, and sensors for specific tasks.

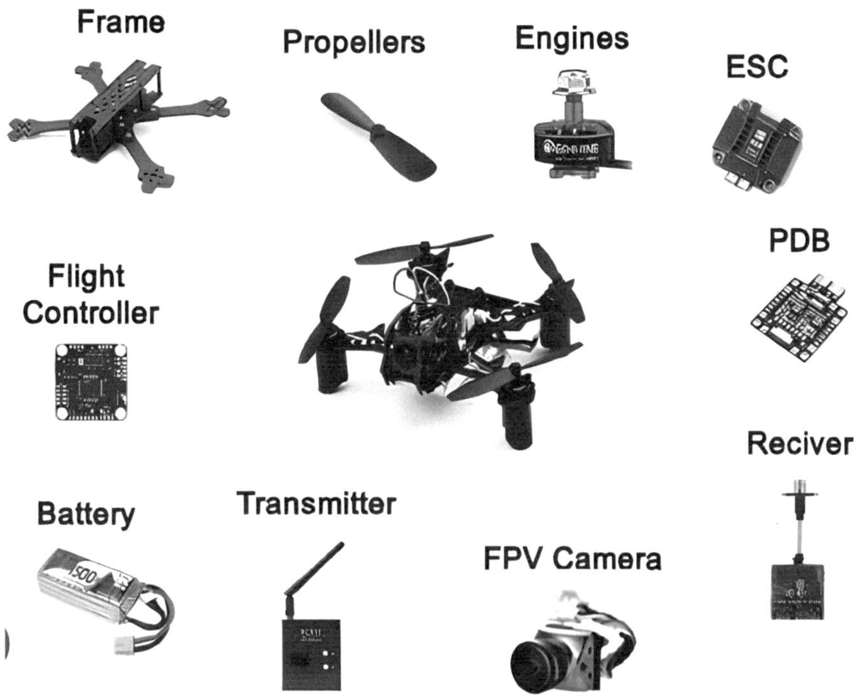

FIGURE 3.2 Components of UAV (Ibrahim et al. 2023).

10. **Onboard Computer or Processor:** Certain sophisticated drones could be equipped with internal computers or processors to manage intricate calculations and duties, such image processing or artificial intelligence.

3.5 APPLICATIONS OF UAV

1. **Security, Monitoring, and Surveillance:** UAVs have proven very helpful for military surveillance operations and are included in the military strategy plans of several nations. These aerial robotic devices are being used by nations for border security, enemy detection, anti-poaching, and marine surveillance of vital shipping corridors. Affordable, reliable, and versatile drones are now playing an important role in aerial surveillance, monitoring, and surveying areas to help prevent criminal activities.
2. **Disaster Management:** UAVs have been found useful during human-made or natural disaster such as floods, tsunamis, or terrorist strikes. These events may seriously damage power, water, transportation, and telecommunication services. In such situations, UAVs replace humans for providing essential supplies such as food and medical kits. UAVs also help in acquiring data, provide quick fixes when needed, and negotiate obstructions. Rescue personnel use radars, sensors, and high-quality cameras attached on UAVs for

detecting damage in geographical areas quickly. Quick recovery remedies and resources like manned helicopters and first aid packages are then provided to identified locations

3. **Search and Rescue (SAR):** UAVs are deemed crucial in crucial situations including emergency response, SAR, and public safety. Because UAVs provide real-time visual data of desired areas, they can save a significant amount of time, people, and resources. As a result, a SAR team can quickly identify and determine the precise location where help is urgently needed. In disasters like missing people, avalanches, wildfires, and gas leaks, drones can speed up SAR efforts.
4. **Remote Sensing:** Currently, hobby aircrafts are used for capturing detailed images of remote places such as mountains, peninsulas, and coasts. Drone technology can help in combining data from satellites, airplanes, and ground sensors. As the drones are light and cheap, they can provide clear and precise observations over time and space (Mohsan et al. 2022).
5. **Real-Time Monitoring of Road Traffic:** One area of application where the integration of UAVs has generated a lot of attention is road traffic monitoring (RTM) systems. UAVs have the potential to fully automate the transportation industry in RTM [7]. Road surveyors, traffic cops, rescue squads, and field support teams will be automated as part of this project. UAVs that are dependable and intelligent can help automate these components. UAVs are a novel and promising technique for collecting data on traffic conditions on roadways.
6. **Precision Agriculture:** Drones can be used in precision farming to collect data from ground sensors, such as water quality, soil conditions, and moisture levels. They help plan irrigation, check for diseases, spray pesticides, monitor crops, and spot weeds. Using drones in farming saves time and money while improving productivity, profits, and crop yields. Additionally, drones help control insect damage, monitor weeds, spray pesticides, and maintain farmland. This results in increased crop yields to satisfy particular production needs. UAVs with the proper CCTV cameras and sensors can keep an eye on crop health, including humidity, chlorophyll content, foreign pollutants, and leaf thickness (Cormen et al. 2009). The advancements in UAV-assisted remote sensing for growth vigour evaluation, nutritional status, disease and weed identification, and drought conditions were briefly covered by Aljehani et al. (2017).

To address all such applications, a major contribution of effective path planning is demanded which guides the UAVs to well perform.

3.6 AREA COVERAGE PATH PLANNING

For UAVs, area ACPP entails creating plans and algorithms to effectively and efficiently cover a given region. This is an essential component of UAV operations, with applications like mapping, surveillance, environmental assessment, agricultural monitoring, and SAR relying heavily on it. Ensuring total coverage of the target

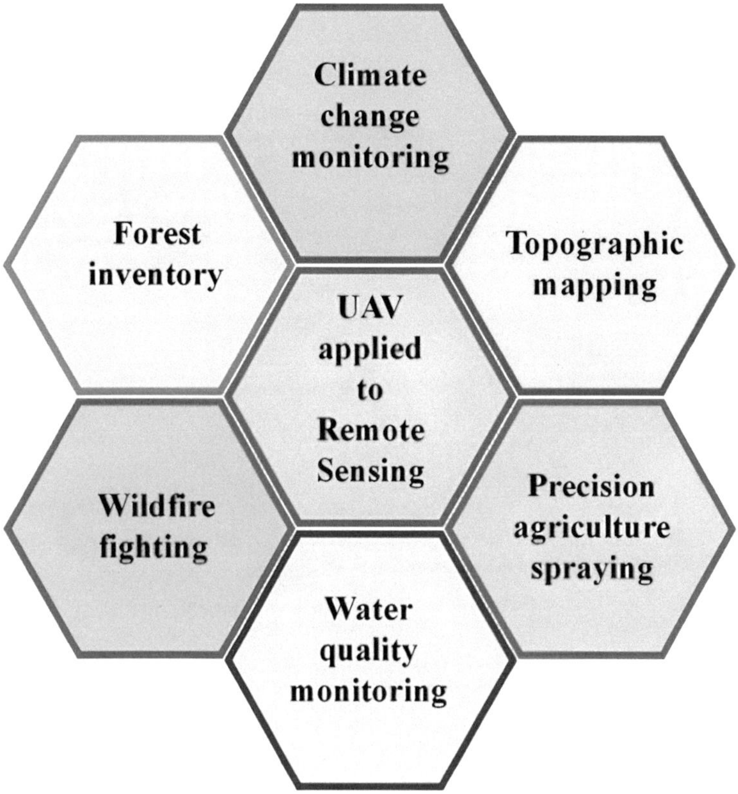

FIGURE 3.3 UAV and remote sensing applications (Mohsan et al. 2022).

region while taking into account different limits and maximising performance is the main objective of the proposed work.

3.7 OBJECTIVES

The following are the main goals targeted in this chapter:

1. To provide a grid-based depiction of the coverage area that takes certain NFZs into account.
2. To create a modified Travelling Salesman Problem (TSP) algorithm that ensures safe and effective UAV navigation by optimising the path while avoiding NFZs.
3. To determine and examine important parameters, such path length, turning angles, energy usage, and completion time, in order to offer an understanding of how well the suggested algorithm performs.
4. To provide a thorough grasp of the algorithm's performance by visualising both the ideal path and the underlying grid.

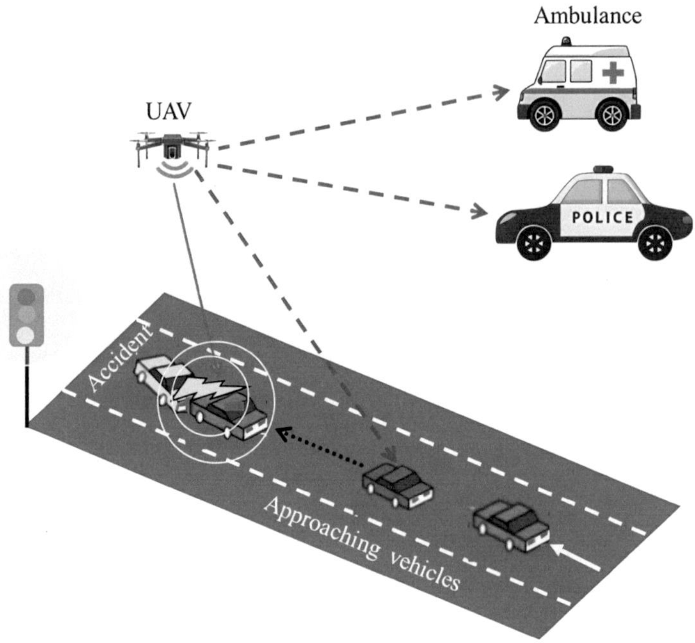

FIGURE 3.4 UAV assistance on the highway (Maes et al. 2019).

3.8 LITERATURE SURVEY

UAVs have become increasingly prevalent in many applications, hence requiring efficient route planning techniques. The purpose of this literature review is to examine previous studies on UAV scheduling, with an emphasis on grid-based coverage and the use of NFZ avoidance strategies.

3.8.1 Algorithms for Path Planning

3.8.1.1 Grid-Based Algorithms

Grid-based algorithms are commonly employed for UAV path planning. These algorithms operate on a discretised grid representation of the environment, allowing the UAV to navigate through cells to reach its destination while avoiding obstacles. Here are some notable grid-based algorithms used for UAVs:

a. A* (A-Star) Algorithm
   - Description: A* is a popular search algorithm that helps find the shortest path between two points on a grid. It uses a guessing method (heuristic) to guide the search, which makes it very good for planning paths on grids.
   - Application: A* can be used to plan paths for drones in areas shown as grids. It looks at both the distance travelled so far and an estimated distance left to reach the goal.

b) Dijkstra's Algorithm
   - Description: Dijkstra's algorithm is a renowned approach for finding the shortest path between points in a network. When used on a grid, it can be adapted to help find the best routes for drones.
   - Application: It is suitable for scenarios where finding the absolute shortest path is critical. It explores all possible paths from the start to the goal, making it exhaustive.

c) Theta* Algorithm
   - Description: Theta* is an improvement over A* that reduces the number of nodes expanded during the search. It incorporates local path refinement, which is beneficial for UAVs navigating grid-based environments.
   - Application: It is suitable for scenarios where efficiency is crucial, as Theta* tends to produce paths with fewer nodes expanded compared to A*.

d) Rapidly Exploring Random Trees (RRT)
   - Description: RRT is a probabilistically complete algorithm designed for non-grid environments. However, it can be adapted for grid-based scenarios by discretising the space into cells.
   - Application: It is useful when dealing with dynamic or complex environments. It generates paths through random sampling, allowing for exploration of various paths.

3.8.1.2 Bio-Inspired Algorithms

Bio-inspired algorithms draw inspiration from the behaviours and strategies observed in nature to solve complex optimisation problems. In the context of path planning for UAVs, bio-inspired algorithms mimic the efficiency and adaptability found in natural systems. Here are some bio-inspired algorithms commonly used for UAV path planning:

a. Ant Colony Optimisation (ACO)
   - Inspiration: Based on the foraging behaviour of ants.
   - Working: UAVs are modelled as virtual ants, and paths are constructed based on pheromone trails. Shorter paths attract more "pheromones," guiding other UAVs to optimal routes.
   - Advantages: Effective in finding near-optimal paths and robust to dynamic environments.

b. Particle Swarm Optimisation (PSO)
   - Inspiration: Simulates the social behaviour of birds flocking or fish schooling.
   - Working: Drones (UAVs) move through the search area, changing their positions based on the best position they have found and the best position found by the entire group.
   - Advantages: Quick convergence, simplicity, and ability to handle dynamic environments.

c. Genetic Algorithms (GA)
   - Inspiration: Mimics the process of natural selection and genetic evolution.

- Working: UAV paths are represented as chromosomes, and genetic operators (crossover, mutation) are applied to evolve the paths over multiple generations.
- Advantages: Global search capability and adaptability to changing environments.

d. Bee Algorithm (BA)
- Inspiration: Inspired by the foraging behaviour of honeybees.
- Working: UAVs (virtual bees) explore the search space, and the information about promising paths is shared among the swarm. The waggle dance concept is used for communication.
- Advantages: Efficient in dynamic environments, robustness, and decentralised decision-making.

3.8.1.3 Optimisation-Based Algorithms

Optimisation-based algorithms work to find the best solution to a problem by improving a specific goal, called an objective function. For UAV path planning, these algorithms aim to either reduce or increase things like the distance travelled, energy used, or time taken, while keeping certain rules or limits in mind. Below are some optimisation-based algorithms often used for planning UAV paths:

a. Back and forth
- Objective: Efficiently cover an area while minimising redundancy.
- Working: The UAV traverses the area in one direction (e.g., row or column). Upon reaching the end, it reverses direction and covers the same area again.
- Advantages: Simple and easy to implement. Suitable for scenarios where complete coverage is required.
- Considerations: May not be optimal for dynamic environments.

b. Spiral Algorithm
- Objective: Cover an area in a spiral pattern.
- Working: The UAV starts from the centre or a designated point. It moves in a spiral pattern outward, covering the area systematically.
- Advantages: Provides a structured coverage pattern. Suitable for applications where the centre is a critical point of interest.
- Considerations: May not be optimal for irregularly shaped or constrained areas.

c. Parallel Path Length-Based Algorithm
- Objective: Optimise coverage by considering path length.
- Working: Divides the area into parallel paths. UAVs follow these parallel paths, optimising the total path length.
- Advantages: Efficiency in terms of coverage time and path length. Applicable to scenarios with a relatively regular layout.
- Considerations: Complexity may increase for irregularly shaped areas.

d. Greedy Approach for Path Coverage
- Objective: Make locally optimal decisions at each step.

- Working: At each step, the UAV selects the most favourable next location based on a predefined criterion (e.g., proximity, coverage priority). Continues this process until the entire area is covered.
- Advantages: Simplicity and real-time adaptability. Suitable for scenarios with dynamic changes.
- Considerations: May not always result in a globally optimal solution.

3.9 PROPOSED METHODOLOGY

3.9.1 Overview

A greedy algorithm is included into the grid-based CPP strategy (PPS) for a UAV as part of the suggested methodology in the provided code to solve the TSP while taking into account the avoidance of NFZs. The components of the suggested approach are provided below:

1. **Grid Representation:** A grid is used to depict the environment, with each cell denoting a distinct region. Obstacles in the grid are designated as NFZs.
2. **Waypoint Generation:** Waypoints are created according to a certain pattern in the grid, creating a series that encompasses the whole region.
3. **Greedy TSP Algorithm with NFZ Avoidance:** A greedy implementation of the TSP algorithm is used. At an initial place, the algorithm repeatedly chooses the closest unexplored point while taking NFZ avoidance into account. The next nearest point is selected if the closest point is inside an NFZ; otherwise, it is skipped. The method keeps going until every waypoint is reached, creating the best possible route that stays out of NFZs.
4. **Metrics Calculation:** Based on the produced ideal tour, metrics like energy consumption of the device, total length of the path, turning angles, and completion time are computed.
5. **Visualisation:** Matplotlib is used to show the grid, waypoints, and the ideal tour in order to provide a clear picture of the path planning solution.
6. **Performance Metrics Display:** To assess the effectiveness of the suggested course, the metrics mentioned earlier are computed and used.
7. **Customisation:** Specific needs may be satisfied by customising parameters such as NFZ locations, cell size, UAV speed, and grid dimensions.

The approach is specifically created for real-world settings where dynamic flexibility and efficiency are crucial. Its goal is to offer an effective and feasible solution for grid-based UAV route planning, guaranteeing coverage of waypoints while avoiding NFZs.

3.9.2 System Model

The system model for UAV route planning using the TSP and greedy algorithm consists of a number of interrelated parts and procedures that work together to effectively navigate a predetermined area. A 2D grid depicting the region to be covered with designated NFZs as restricted zones is one of the main components. The UAVs mission requires it to visit a set of coordinates that are created during the waypoint generation procedure. The engine of decision-making, the greedy algorithm, decides which waypoints to visit in the best sequence to prevent NFZs. The performance of the UAV is assessed by the metrics calculation module, which takes into account parameters like path length, turning angles, energy consumption, and completion time. The planned path, waypoints, and NFZs are clearly displayed to users via the visualisation module, which converts the data into visual outputs. The user interface makes it easier for users to engage with the system by enabling parameter adjustment and giving them a platform to view and evaluate the metrics and visual outcomes. The feedback loops in the system allow users to make well-informed modifications based on observed outcomes. The system functions as a whole through interrelated processes. Although the model offers an organised method for planning UAV paths, it recognises some limitations and simplifications contained in the algorithm and the selected domain representation.

3.9.3 Proposed Algorithm

Algorithm: TSP with Greedy Algorithm

1. Initialise Grid: Create a grid with specified width (grid_width) and height (grid_height), initialising all cells to 0.
2. Define NFZ: Specify NFZ locations on the grid and mark corresponding cells as obstacles, defined as:

   grid[y, x] = 1 for each NFZ point (x, y).

3. Generate Waypoints: Generate waypoints for grid-based coverage considering the specified grid dimensions. Divide the grid into rows and alternate the scanning direction to create waypoints.
4. Define Helper Functions: distance(point1, point2)
   - Function: Calculate the Euclidean distance between two points.
   - Formula: distance = np.linalg.norm(np.array(point1) – np.array(point2))

The expression: distance = np.linalg.norm(np.array(point1) – np.array (point2)) calculates the Euclidean distance between two points point1 and point2 in a 2D space. Let's break down the components:

- np.array(point1): Converts the tuple point1 into a NumPy array. This is done to facilitate mathematical operations.
- np.array(point2): Similar to the above, it converts the tuple point2 into a NumPy array.
- np.array(point1) – np.array(point2): Computes the element-wise difference between the two arrays, resulting in a new array representing the vector pointing from point 2 to point1.
- np.linalg.norm(...): Calculates the Euclidean norm (or magnitude) of the vector obtained from the subtraction. In the context of 2D space, this is equivalent to finding the straight-line distance between point1 and point2.

- distance: Stores the calculated Euclidean distance between the two points. In mathematical terms, the Euclidean distance (Ed) between two points (p1, q1) and (p2, q2) in a 2D space is given by the formula as in Eq. 1:

$$Ed = \sqrt{(p2 - p1)^2 - (q2 - q1)^2} \tag{1}$$

5. Calculate_metrics(path, speed) Function: The path length, turning angles, energy consumption, and completion time are calculated using the formula:

$$\text{path}_{\text{length}} += \text{distance}\left(\text{prev}_{\text{point}}, \text{current}_{\text{point}}\right) \tag{2}$$

$$\text{turning}_{\text{angle}} += \text{np.arccos}(\ldots) \tag{3}$$

$$\text{energy}_{\text{consumption}} += \text{distance}\left(\text{prev}_{\text{point}}, \text{current}_{\text{point}}\right) \tag{4}$$

$$\text{completion}_{\text{time}} += \frac{\text{distance}\left(\text{prev}_{\text{point}}, \text{ current}_{\text{point}}\right)}{\text{speed}} \tag{5}$$

6. TSP with NFZ Avoidance
   For this purpose, initially the greedy algorithm is implemented to find the optimal TSP tour while avoiding NFZs. Thereafter, the UAV is made to start from the first waypoint and iteratively select the nearest unvisited waypoint, avoiding NFZs, using the formula:

$$\text{nearest}_{\text{point}} = \text{min(unvisited, key} = \text{lambda point:}$$
$$\text{distance}\left(\text{current}_{\text{point}}, \text{point}\right)) \tag{6}$$

7. Visualisation
   Visualise the grid, original waypoint order, and the optimal TSP tour with NFZ avoidance using Matplotlib for grid visualisation, waypoint plotting, and tour plotting.
8. Print Metrics
   It is used for display the calculated metrics for the optimal tour, including path length, turning angle, energy consumption, and completion time.

3.9.4 Performance Evaluation and Analysis

Metrics considered for measuring the effectiveness and viability of the produced path are energy consumption, path length, turning angles, and completion time. All things considered, this tactic supports mission goals by guaranteeing safety, maximising coverage, and offering a flexible and strong UAV route planning system.

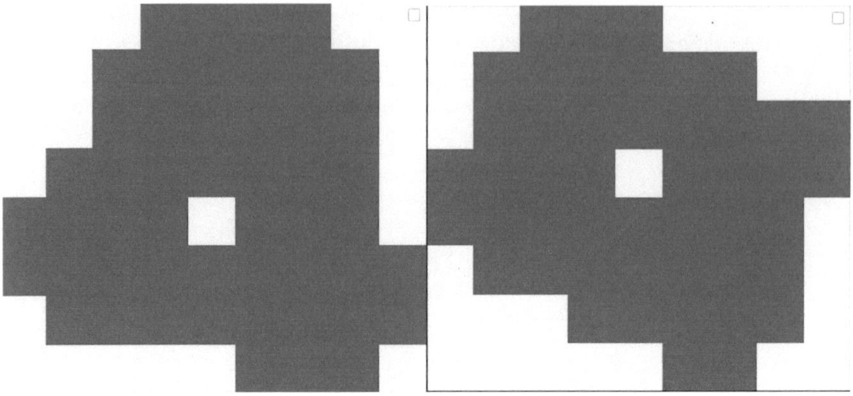

FIGURE 3.5 System model Scenario 1 (Cabrerira et al. 2019).

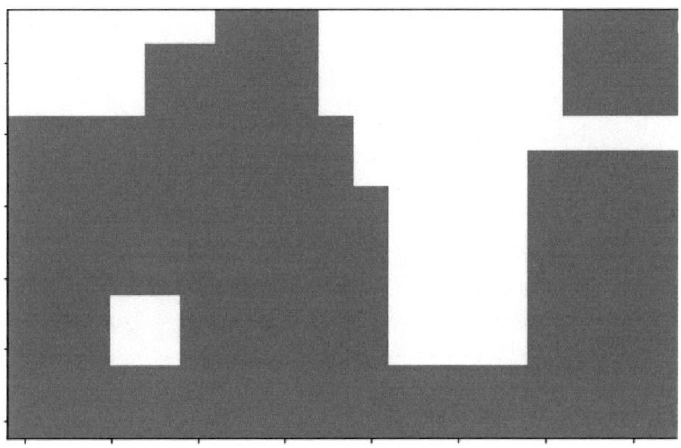

FIGURE 3.6 Second model of Scenario 2 (Ann & Kim 2015).

We consider the following two scenarios to measure the success of the proposed algorithm and verify its applicability.

Scenario 1: The Area 1 and Area 2, where we performed modified TSP and compared results with other algorithms in PPS (Ghaddar et al., 2020) and algorithm in (Ann & Kim, 2015).

Scenario 2: In which, the area of interest covered with yellow part shows the NFZ areas and the remaining red portion is the covered path and this is the second scenario on which we perform and compare our results with other algorithm.

In this part, we examine multiple cases and contrast our findings with the research in Ajith (2007), which divides the area of interest and uses a back-and-forth algorithm to plan the journey and through line from Bresenham that appears in Barrenientos et at. (2011). A grid-based method that was suggested in Cabreira (2019), in the presence of NFZ. The clustering technique described in Ann and Kim (2015), which creates pathways free of collisions. We assume that the UAV are the same in every situation. It is our assumption that the UAV's travel at a constant speed in order to fairly compare the various methods.

Approximate results are obtained when computations are performed with constant velocity. We utilised a velocity value in the situations that is comparable to those found (Ann & Kim 2015). To ensure a fair comparison, the velocity measurement is taken into account as the average speed (less than the highest speed).

3.10 COMPARISON WITH STATE-OF-THE-ART WORK

To evaluate the significance of the proposed work, the two scenarios have been compared with few available algorithms in the literature.

Scenario 1

In this scenario, we compared our modified TSP algorithm, which combines the advantages of greedy algorithms to energy-aware grid-based CPP algorithm provided in Ghaddar et al. (2020). The route is planned by the authors in Ghaddar et al. (2020). When NFZ is present, they employ O-F and E-F, two minimum-cost techniques. The E-F approach calculates the estimated energy consumption, and the O-F method determines the lower distant pathways according to the overall turning angle sum. In our work, we use our PPS approach to divide the region, design the path, eliminate the partition boundaries, and connect the paths so that a single UAV can cover the entire area of interest. After using PPS-T technique, the region is depicted in Figure 3.7. As it can be seen, the red cells are the NFZ cells.

Area 1: Figure 3.8 (a) shows the coverage path planning algorithm that was produced in Ghaddar et al. (2020) study utilising the O-F and E-F approaches as well as the CPP that was acquired in our research (as shown by Figure 3.8 (b)).

Table 3.1 shows the results that the path length provided by our work is shorter than other path lengths which are provided by O-F algorithm in Cabrerira et al. (2019), E-F algorithm in Cabrerira et al. (2019), and PPS algorithm in Ghaddar et al. (2020). Our work is better in all sectors such as turning angles (in degrees), completion time, and energy consumption.

Area 2: Figure 3.9 shows the coverage path planning that was produced in Ghaddar et al. (2020) and Cabrerira et al. (2019)'s study utilising the O-F and E-F approaches as well as the CPP that was acquired in our research.

Table 3.2 shows the result that total path length is shorter as compared to other three algorithms such as O-F method, E-F method, and PPS. Time completion is also reduced by 50% in our work using modified TSP with greedy algorithm technique.

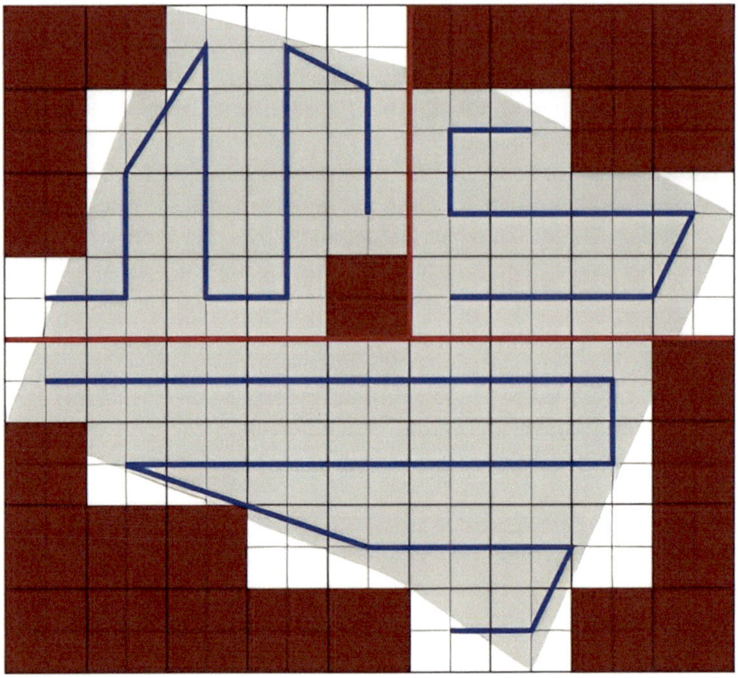

FIGURE 3.7 Planned path after Applying PPS-T method used in Ghaddar et al. (2020).

Scenario 2

In this scenario, we compared our modified TSP algorithm, which combines the advantages of greedy algorithms to the grid-based CPP algorithm provided in Pradeep Singh et al. (2020) and algorithm in Ann and Kim (2015). The route is planned by the authors in Ann and Kim (2015) when NFZ is present. In our work, we use our PPS approach to divide the region, design the path, eliminate the partition boundaries, and connect the paths so that a UAV can cover the entire area. After using the algorithm in Ghaddar et al. (2020), the region is depicted in Figure 3.10. As you can see, the red cells are the NFZ cells.

In our work, we proposed a modified TSP algorithm to compare results, and in Figure 3.11, the obstacles are shown with yellow colour and original order appears in red dotted line, while the green lines show the optimal path. By comparing results, the total path length is shorter as compared to another algorithm such as the PPS algorithm in Ghaddar et al. (2020) and Ann and Kim (2015); the result are shown below.

The comparisons and results in Table 3.3 are better with our work, whereas path length is shorter and also better with time completion and energy consumption.

As it can be witnessed, the proposed algorithm is able to efficiently reduce path length, turning angles, energy consumption, and completion time, by a comprehensive back-and-forth traversal. Hence, it is shows the outperformance of the algorithm in generating an optimal path while strictly adhering to NFZ restrictions.

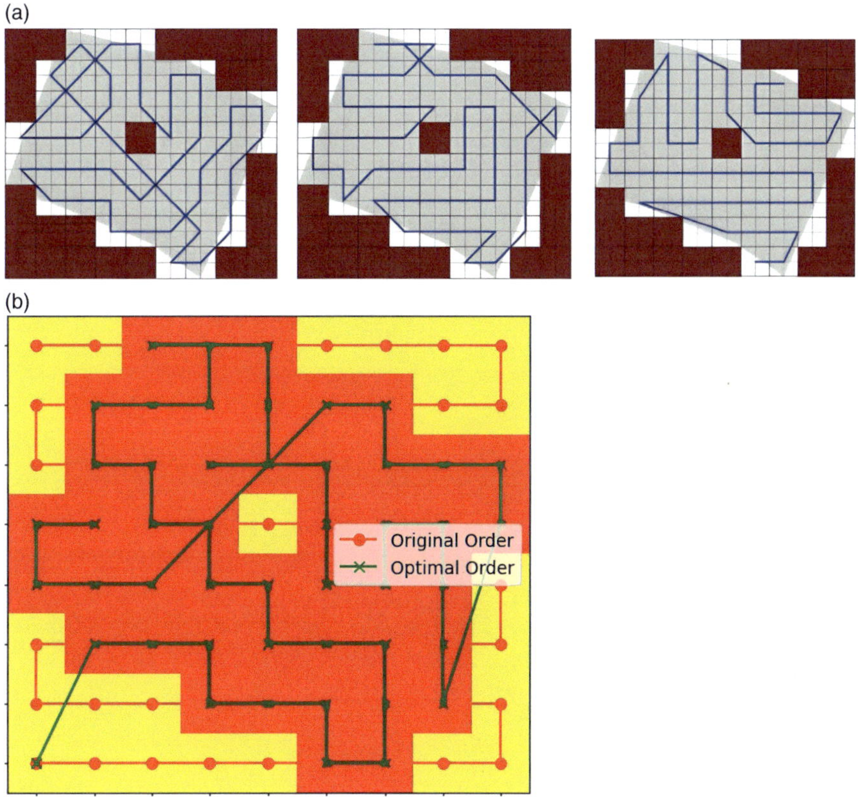

FIGURE 3.8 Paths generated by the two and proposed algorithm: (a) Algorithm in Cabrerira et al. (2019) and (Ghaddar et al. (2020) and (b) our work (modified TSP).

TABLE 3.1
Path planning provided in Ghaddar et al. (2020) vs our work with Area 1

	(a) O-F	(b) E-F	(c) PPS	(d) TSP
Path length [in meter]	556.98	511.42	467.33	55.64
UAV speed [m/s]	10	10	10	10
Rotation rate [deg/sec]	30	30	30	30
Completion time [sec]	130.69	127.64	106.73	5.5
Energy consumption	103.76	99.23	85.54	55.64

3.11 CONCLUSION AND FUTURE SCOPE

Finally, a grid-based coverage path planning technique for UAVs that incorporates a greedy algorithm to avert the NFZ is provided in this paper. The suggested technique ensures safe and optimal path coverage by effectively navigating the UAV via

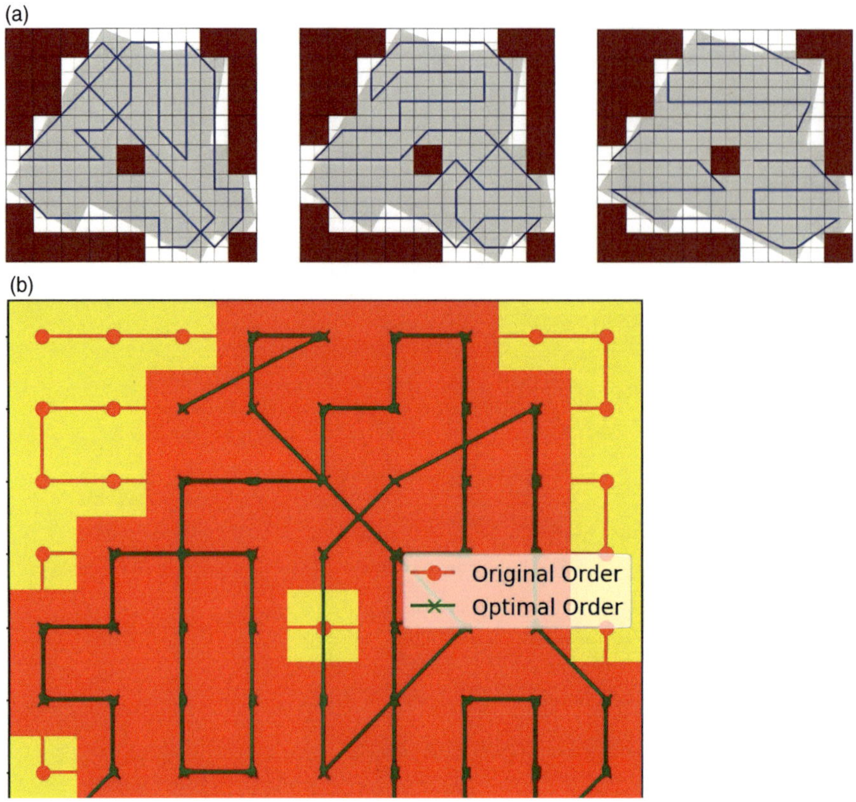

FIGURE 3.9 CPP generated by three algorithms and our work with Area 2: (a) Work performed by O-F, E-F (Cabrerira et al. 2019) and PPS also (Ghaddar et al. 2020) and (b) our work.

TABLE 3.2
Path planning provided in Cabrerira et al. (2019) and Ghaddar et al. (2020) vs our work with Area 2

	(a) O-F	(b) E-F	(c) PPS	(d) TSP
Path length [in meter]	582.84	566.27	509.38	61.20
UAV speed [m/s]	10	10	10	10
Rotation rate [deg/sec]	30	30	30	30
Completion time [sec]	125.78	124.12	106.74	6.1
Energy consumption	102.87	100.94	88.25	66.12

waypoints and dynamically avoiding NFZs. The method effectively minimises path length, turning angles, energy consumption, and completion time, as demonstrated by a back-and-forth traverse. Performance data and graphics are used to illustrate the greedy algorithm's success and show off its ability to provide an ideal path while

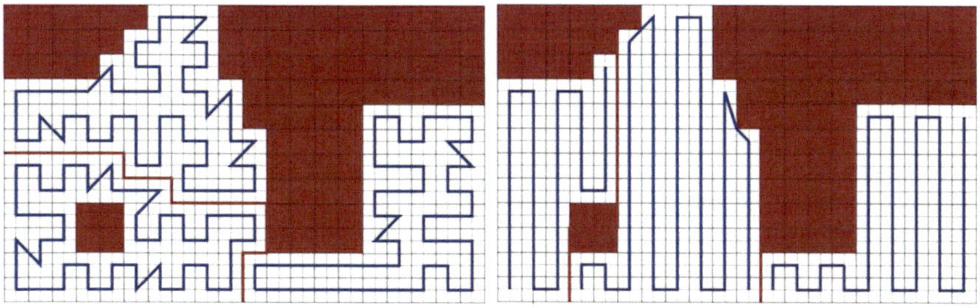

FIGURE 3.10 Area partition and coverage according to algorithm in Ann and Kim (2015) and algorithm in Ghaddar et al. (2020).

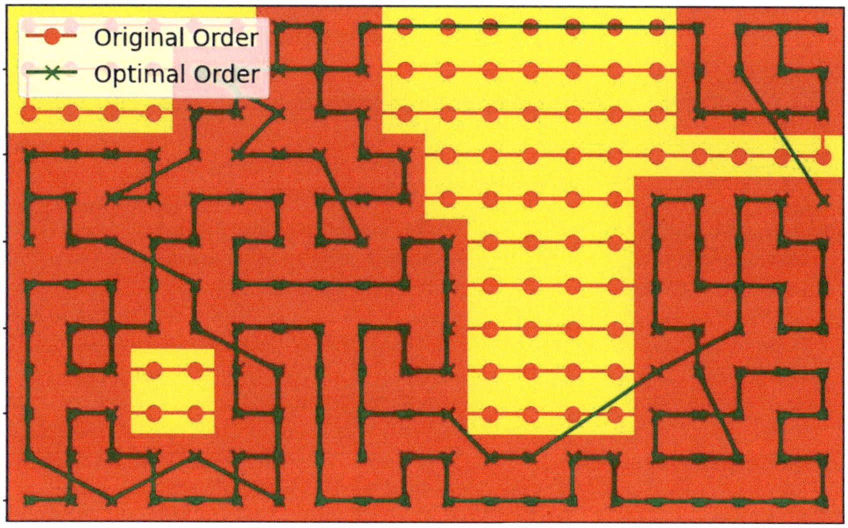

FIGURE 3.11 Path provided by algorithm in Ann and Kim (2015) and algorithm in Ghaddar et al. (2020) vs modified TSP.

TABLE 3.3
Coverage path planning in Ann & Kim (2015) and Ghaddar et al. (2020) vs our work

	Algorithm in Ann & Kim (2015)	PPS-L (Ghaddar et al. 2020)	Modified TSP
Path length [in meter]	1575.56	1517.02	200.40
UAV speed [m/s]	10	10	10
Rotation rate [deg/sec]	30	30	30
Completion time [sec]	520.56	265.70	20.4
Energy consumption	371.79	235.75	200.444

adhering to NFZ restrictions. The paper advances the field of UAV route planning by offering a workable remedy for NFZ-related real-world situations giving higher performance in contrast to the conventional algorithms.

Even if the current work provides a solid basis, there are still opportunities for development and research in the future. First off, taking into account real-time updates and alterations in the surroundings, the algorithm's adaptability to dynamic NFZs may be strengthened. The capability of the UAV to recognise and react to changing NFZs may be improved by integration with cutting-edge sensors and machine learning methods.

The algorithm's scalability for larger areas and multiple UAVs should also be investigated; cooperative path planning and coordination between multiple UAVs should be investigated to efficiently optimise coverage in a more extensive region; additionally, machine learning models integrated for predictive NFZ identification and parameter optimisation of the greedy algorithm could provide more intelligent and adaptive path planning, all of which would contribute to a comprehensive and intelligent UAV path planning system that can handle a variety of dynamic environments.

REFERENCES

Ajith, V.S., Sameer, M., & Kumar. K.K. 2017. "Design and Structural Analysis of Wing-Fuselage Brackets of a UAV." *International Journal of Creative Research Thoughts,* www.ijcrt.org 5 896–903.

Aljehani, M. & Inoue, M. (2019). "Performance Evaluation of Multi-UAV System in Post-Disaster Application: Validated by HITL Simulator." *IEEE Access*, 1–1, https://doi.org/10.1109/ACCESS.2019.2917070

Andrade, F. A. de A., Hovenburg, A. R., de Lima, L. N., Rodin, C. D., Johansen, T. A., Storvold, R., Correia, C. A. M., & Haddad, D. B. (2019). "Autonomous Unmanned Aerial Vehicles in Search and Rescue Missions Using Real-Time Cooperative Model Predictive Control." *Sensors (Switzerland)*, 19 (3), 406, https://doi.org/10.3390/s19030406

Ann, S., Kim, Y., & Ahn, J. (2015). "Area Allocation Algorithm for Multiple UAVs Area Coverage Based on Clustering and Graph Method." *IFAC-PapersOnLine*, 48, 204–209.

Anusha, G., Arora, K., Sharma, H., Thota, S.P., Joshi, G.P., and Cho, W. "Control Strategies of 15-Level Modified Cascaded H-Bridge MLI with Solar PV and Energy Storage System," in *Energy Reports*. 2024, https://doi.org/10.1016/j.egyr.2024.06.003

Arora et al. (1998). "Approximation Algorithms for TSP," *Journal of the ACM*, 45(5), 753–782, https://doi.org/10.1145/290179.290180.

Arora, K., Joshi, G. P., Ragab, M., Rawa, M., Milyani, A. H., Mansour, R. F., & Yang, E. (2022). "Bilateral Contract for Load Frequency and Renewable Energy Sources Using Advanced Controller." *Computers, Materials & Continua*, Tech Science Press, 73(2), 3165–3180, https://doi.org/10.32604/cmc.2022.026966

Barrientos, A., Colorado, J., Cerro, J.D., Martinez, A., Rossi, C., Sanz, D., & Valente, J. (2011). "Aerial Remote Sensing in Agriculture: A Practical Approach to Area Coverage and Path Planning for Fleets of Mini Aerial Robots." Journal of Robotic Systems, 28, 667–689.

Bellingham, J., Tillerson, M., Richards, A., & How, J. P. (2002). Coordinated path planning for multiple UAVs in dynamic and uncertain environments. *Autonomous Robots*, 17(2–3), 233–245, https://doi.org/10.1023/A:1015098803511

Cabreira, T.M., Ferreira, P.R., Franco, C.D., & Buttazzo, G.C. (2019). "Grid-Based Coverage Path Planning with Minimum Energy over Irregular Shaped Areas with UAVS." In *2019*

International Conference on Unmanned Aircraft Systems (ICUAS). IEEE, 758–767, https://doi.org/10.1109/ICUAS.2019.8797937. 8797937

Centre for Security Studies (2010). "The Military Utility of Drones CSS Anal." *Security Policy*, 78, 1–3.

Cho, Y., Kim, D., & Kim, D. S. K. (2009). "Topology Representation for the Voronoi Diagram of 3D Spheres." *International Journal of CAD/CAM*, 5(1) , 1–16.

Cormen, T.H., Leiserson, C.E., Rivest, R.L., & Stein, C. (2009). *Introduction to Algorithms*. MIT Press.

Ghaddar, A., Merei, A., & Natalizio, E. (2020). "PPS: Energy-Aware Grid-Based Coverage Path Planning for UAVs Using Area Partitioning in the Presence of NFZs." *Sensors*, 20(13), 3742, https://doi.org/10.3390/s2013374

Gu, H., Lyu, X., Li, Z., & Zhang, F. (2017). "Development and Experimental Verification of a Hybrid Vertical Take-Off and Landing (VTOL) Unmanned Aerial Vehicle (UAV)." *IEEE/ASME Transactions on Mechatronics*, 22(2), 909–918, https://doi.org/10.1109/TMECH.2016.2641938

Khusro, S., & Qureshi, I. M. (2017). "A Comprehensive Review on Sensors and Techniques Used in Unmanned Aerial Vehicles." *Sensors*, 17(10), 2375.

Turner, S., et al. (2019). "Human-in-the-Loop Decision Support Systems for UAV Path Planning in No-Fly Zones." *Journal of Human-Robot Interaction*, 14(2), 187–203.

Wang, C. M., et al. (2019) "Unmanned Aerial Vehicles for Surveying Sea Surface Temperature." *Remote Sensing*. Karunakar P., & Mesfin J. (2017). "A Review on Geo Mapping with Unmanned Aerial Vehicles."

4 Artificial Intelligence-Based Exploring the Influence of Numerology on Human Life

Gaurav Kumar, Mansi Gupta, and Krishan Arora

4.1 INTRODUCTION

Numerology, an ancient practice with roots dating back to antiquity, explores the mystical connections between numbers and various aspects of human life. From ancient civilizations to modern societies, numerology has intrigued scholars, mystics, and enthusiasts alike, offering insights into personality traits, life events, and cosmic influences. This research paper delves into the captivating realm of numerology, aiming to unravel its profound influence on human existence. By drawing upon a rich tapestry of historical knowledge and contemporary research, this paper endeavors to provide a comprehensive understanding of numerology and its implications for understanding human life in the modern era. Numerology's significance spans across cultures and epochs, with each civilization imbuing numbers with symbolic meanings and mystical significance. Ancient civilizations such as the Babylonians, Egyptians, and Greeks recognized the power of numbers in shaping destiny and understanding the cosmos [1]. Moreover, renowned scholars like Pythagoras contributed seminal insights to numerological thought, laying the groundwork for its enduring legacy [2]. Over the centuries, numerology has evolved, adapting to cultural shifts and technological advancements while retaining its core principles. In contemporary society, numerology continues to captivate individuals seeking deeper insights into themselves and the world around them. Its influence extends beyond personal introspection, with applications in fields such as astrology, personality analysis, and predictive methodologies. Recent studies have explored the correlation between numerological attributes and personality traits, shedding light on the interplay between numbers and human psychology [3]. Furthermore, advancements in technology have facilitated the accessibility of numerological practices, enabling individuals to explore their numerological profiles with greater ease [4]. As we embark on this journey of

 DOI: 10.1201/9781003563440-4

exploration, it is essential to approach numerology with a critical yet open-minded perspective. While some may dismiss numerology as pseudoscience, others find profound meaning and guidance in its teachings. By synthesizing historical knowledge, contemporary research, and critical insights, this research paper aims to provide a nuanced understanding of numerology's influence on human life and its relevance in the modern world.

4.2 HISTORICAL ORIGINS OF NUMEROLOGY

Numerology has a rich history dating back to ancient civilizations, where numerical patterns were believed to hold significant meaning in various aspects of life. One of the earliest known proponents of numerology was the Greek philosopher Pythagoras, who is often credited with formalizing many of the principles of numerology that are still used today [5]. Pythagoras and his followers believed that numbers were the building blocks of the universe and held mystical significance beyond their mathematical properties. Ancient civilizations such as the Babylonians, Egyptians, and Chinese also practiced forms of numerology, integrating numerical symbolism into their religious beliefs, cosmology, and daily life. For example, the Babylonians developed a system of divination called numerology, where numbers were assigned specific meanings and used to predict future events [6]. In the Middle Ages, numerology experienced a resurgence in Europe, particularly within the esoteric traditions of Kabbalah and alchemy. Kabbalistic numerology, known as Gematria, assigned numerical values to Hebrew letters and words, linking language and numbers to uncover hidden meanings in sacred texts. Similarly, alchemists used numerology as a symbolic language to encode their spiritual and philosophical beliefs into their texts and practices [7].

Throughout history, numerology has continued to evolve, adapting to different cultural contexts and spiritual traditions. Today, numerology remains a popular practice, with enthusiasts using numerical interpretations to gain insights into personality traits, relationships, and life events. Despite its ancient origins, numerology continues to captivate individuals seeking deeper understanding and connection to the mystical influence of numbers on human life.

4.3 PRINCIPLES AND METHODS OF NUMEROLOGY

Numerology operates on several fundamental principles and employs various methods to interpret the significance of numbers in human life. One of the foundational principles is that each number possesses unique vibrational frequencies and symbolic meanings, influencing different aspects of existence. The most commonly used method in numerology is assigning numerical values to letters of the alphabet based on their position. This method, often attributed to Pythagoras, allows for the calculation of numbers associated with names, words, dates of birth, and other significant entities. Another essential principle is the reduction of multi-digit numbers to a single-digit root number through addition, known as numerological reduction or digit summing. For instance, the birth date September 27, 1990, would be reduced to $2+7+9+1+9+9+0 = 37$, then further reduced to $3+7 = 10$, and finally reduced to 1+0 = 1. Numerology also utilizes various charts and systems, such as the Pythagorean, Chaldean, and Kabbalistic methods, each with its unique interpretations and applications.

Numerology's methods and principles have been extensively studied and discussed in the literature. For example, a study provides a comprehensive overview of numerology's fundamental principles and methods, emphasizing the significance of numerical vibrations in personal and spiritual development [8]. Additionally, studies explore the historical origins of numerology and its various calculation methods, shedding light on its cultural and philosophical underpinnings [9]. Furthermore, investigations offer insights into the practical applications of numerology in personality analysis and predictive methodologies, highlighting the importance of understanding numerological principles for accurate interpretation [10].

Through the application of these principles and methods, numerology offers a unique perspective on understanding the influence of numbers on human life, providing insights into personality traits, life events, and potential pathways for personal growth and development.

4.4 NUMEROLOGY IN ASTROLOGY

In the realm of astrology, numerology plays a significant role in unraveling the cosmic connections between numbers and celestial bodies. Numerology in astrology involves assigning numerical values to planets, stars, and other celestial entities to decode their influence on human life. This interdisciplinary approach offers valuable insights into individual characteristics, life events, and compatibility with others. Recent research has explored the nuanced correlations between numerological calculations and astrological charts, revealing deeper layers of meaning and predictive accuracy [11]. Furthermore, studies have demonstrated the efficacy of numerological techniques in enhancing the interpretation of astrological transits and forecasting future trends [12]. By integrating numerology into astrology, practitioners gain a holistic understanding of cosmic energies and their impact on human existence, enriching the spiritual journey of individuals seeking guidance and enlightenment.

4.5 NUMEROLOGICAL SIGNIFICANCE OF PLANETS AND CELESTIAL BODIES

In numerology, each planet and celestial body holds significant symbolism and influences various aspects of human life. The assignment of numerical values to these cosmic entities allows for the interpretation of their impact on individuals' personalities, behaviors, and life paths. For instance, the number associated with the planet Venus is often linked to love, harmony, and artistic expression, while the Moon's number may signify intuition, emotions, and nurturing qualities. Moreover, the position of planets within numerological birth charts provides insights into an individual's strengths, challenges, and potential life trajectories.

4.6 CALCULATION OF NUMEROLOGICAL BIRTH CHARTS

Numerological birth charts, also known as "life path numbers," are constructed by assigning numerical values to the letters of an individual's name and birthdate. Through precise calculations based on numerological principles, these charts unveil essential aspects of a person's character, destiny, and life purpose. By examining the unique combination of numbers derived from the birth chart, numerologists can offer valuable insights into an individual's innate talents, challenges, and opportunities for growth.

4.7 ASTRO-NUMEROLOGICAL COMPATIBILITY ANALYSIS

Astro-numerological compatibility analysis involves the synthesis of astrology and numerology to assess the compatibility between individuals based on their birth charts and numerological profiles. By examining the alignment of planetary positions and numerical influences in each person's chart, astro-numerologists can determine the potential harmony or discord in relationships, both romantic and otherwise. This analysis provides valuable insights into compatibility dynamics, communication styles, and areas of synergy or conflict between individuals.

4.8 NUMEROLOGY IN PERSONALITY ANALYSIS

Numerology plays a significant role in personality analysis, offering insights into an individual's character, strengths, weaknesses, and life path. By assigning numerical values to letters in a person's name and birthdate, numerologists can derive key numbers that provide valuable insights into personality traits and tendencies. For instance, the life path number, calculated from the birthdate, reveals the inherent qualities and challenges that an individual may encounter throughout their life journey [13]. Similarly, the expression number, derived from the full name, reflects one's natural talents, abilities, and aspirations [14]. Numerology also examines other significant numbers, such as the destiny number and soul urge number, which offer further insights into personality dynamics and inner desires [15]. Through the analysis of these numerical indicators, individuals can gain a deeper understanding of themselves, their relationships, and their life purpose.

4.9 NUMEROLOGY IN PREDICTIVE PRACTICES

Numerology plays a significant role in predictive practices, offering insights into various aspects of individuals' lives and forecasting future events. This section explores the application of numerology in predictive methodologies, encompassing both personal and professional domains. Numerological predictions often begin with the calculation of an individual's life path number, derived from their date of birth. This number serves as a blueprint for understanding one's inherent traits, tendencies, and potential life experiences [16]. Additionally, numerologists analyze other significant numbers in an individual's numerological chart, such as the expression number and soul urge number, to gain a comprehensive understanding of their personality and destiny [17]. In predictive numerology, practitioners use various numerical techniques and cycles to anticipate future events and trends. For example, the study of personal year cycles involves interpreting the numerical vibrations associated with each year of an individual's life to forecast major themes and opportunities [18]. Similarly, the analysis of name vibrations and compatibility numbers enables numerologists to predict the compatibility and outcomes of personal and professional relationships [19]. Numerology also offers predictive insights into career paths, financial prospects, and life milestones. By examining the numerical vibrations surrounding specific dates and periods, practitioners can provide guidance on auspicious timing for important decisions, investments, and endeavors [20]. Additionally, numerological calculations are utilized in fields such as business forecasting, market analysis, and strategic planning to identify favorable opportunities and potential challenges [21]. While predictive numerology provides valuable insights and guidance, it is essential to approach it with a critical mindset and ethical considerations. Practitioners should emphasize empowerment, free will, and personal responsibility in their interpretations, recognizing that numerology serves as a tool for self-awareness and guidance rather than deterministic prediction [22].

In conclusion, numerology offers a rich framework for predictive practices, enabling individuals to gain insights into their lives, relationships, and future trajectories. By harnessing the power of numbers, practitioners can provide valuable guidance and support in navigating life's complexities and making informed decisions as shown in Table 4.1.

4.10 CONTEMPORARY PERSPECTIVES ON NUMEROLOGY

In the realm of numerology, contemporary perspectives offer diverse insights into the relevance and application of numerical interpretations in modern society. While traditional numerology often draws from ancient wisdom and mystical beliefs, contemporary approaches seek to integrate numerological principles with scientific understanding and practical applications. One such perspective involves a critical examination of numerological practices, evaluating their efficacy and validity within the context of empirical evidence and rational inquiry [26]. Additionally, modern numerologists explore the integration of numerology into various fields, including psychology, sociology, and business management, recognizing the potential for numerological insights to enhance understanding and decision-making in these domains [27]. Furthermore, the advent of technology has facilitated the accessibility of numerological tools and resources, allowing individuals to explore their numerological profiles and gain insights into their personalities, life paths, and relationships through online platforms and applications [28]. Contemporary perspectives on numerology also highlight the need for ethical considerations in numerological practice, emphasizing responsible usage of numerological readings and adherence to professional standards [29].

TABLE 4.1
Impact of numerology on human life and its remedies [23, 24, 25]

Numbers	Associated planet	Good colors	Colors to avoid	Lucky day	Friendly numbers	Enemy numbers	Impacts on human life	Remedies
1	Sun	Gold, yellow	Black	Sunday	2, 3, 6, 9	4, 5, 7, 8	Leadership qualities, ambition, independence	Worship Lord Surya, recite Gayatri Mantra
2	Moon	White, silver	Red	Monday	1, 2, 4, 7	5, 8, 9	Sensitivity, intuition, emotional stability	Worship Lord Shiva, wear pearl gemstone
3	Jupiter	Yellow, purple	Green	Thursday	3, 6, 9	1, 2, 7, 8	Creativity, optimism, spiritual growth	Worship Lord Vishnu, wear yellow sapphire
4	Uranus	Blue, green	Black	Saturday	1, 4, 7, 8	2, 3, 6, 9	Stability, practicality, hard work	Worship Lord Shani, wear blue sapphire
5	Mercury	Green, grey	Yellow	Wednesday	1, 3, 5, 6	2, 4, 8, 9	Adaptability, communication skills, versatility	Worship Lord Vishnu, wear emerald gemstone
6	Venus	White, blue	Red	Friday	3, 6, 9	1, 2, 4, 7	Harmony, love, nurturing qualities	Worship Goddess Lakshmi, wear diamond gemstone
7	Neptune	White, light blue	Black	Monday	1, 2, 4, 7	3, 5, 6, 8	Intuition, spirituality, inner wisdom	Worship Lord Shiva, wear blue sapphire
8	Saturn	Black, blue	Red	Saturday	1, 4, 7, 8	2, 3, 5, 6, 9	Discipline, responsibility, karmic lessons	Worship Lord Hanuman, wear blue sapphire
9	Mars	Red, maroon	Green	Tuesday	3, 6, 9	1, 2, 4, 7, 8	Courage, assertiveness, action-oriented	Worship Lord Hanuman, wear coral gemstone

4.11 ETHICAL CONSIDERATIONS IN NUMEROLOGY

Numerology, while offering insights into various aspects of human life, also raises important ethical considerations that practitioners and users must heed. Firstly, it is essential to ensure that numerological readings are provided ethically and responsibly, with practitioners refraining from making claims that could cause harm or distress to individuals. Additionally, practitioners should prioritize informed consent, ensuring that individuals understand the nature and limitations of numerological interpretations before receiving readings. Furthermore, respecting client confidentiality and privacy is paramount, safeguarding sensitive information disclosed during consultations. Practitioners should also avoid exploiting vulnerable individuals or engaging in practices that manipulate or coerce clients. Finally, ongoing professional development and adherence to ethical codes of conduct are crucial for maintaining integrity and credibility within the numerology community [30–33].

4.12 FUTURE DIRECTIONS AND RESEARCH OPPORTUNITIES

There are following future directions and research opportunities discussed below:

- **Quantitative Analysis of Numerological Methods:** There is a need for further quantitative research to validate numerological methods and their predictive accuracy. Studies could focus on statistical analysis of numerological readings and their correlation with real-life events, providing empirical evidence for the efficacy of numerological practices.
- **Interdisciplinary Investigations:** Exploring the intersections between numerology and other fields, such as psychology, sociology, and neuroscience, offers promising avenues for research. Interdisciplinary studies could elucidate the psychological mechanisms underlying numerological beliefs and their impact on individual behavior and societal dynamics.
- **Cultural and Historical Perspectives:** Future research could delve deeper into the cultural and historical contexts of numerology, examining its evolution across different civilizations and its enduring significance in diverse cultural settings. Comparative studies could shed light on the universal themes and cultural variations in numerological beliefs and practices.
- **Technological Innovations in Numerology:** The integration of technology, such as artificial intelligence and data analytics, presents opportunities for advancing numerological research. Development of digital platforms and algorithms for numerological analysis could enhance accessibility, accuracy, and customization of numerological services.
- **Longitudinal Studies on Numerological Beliefs:** Longitudinal research designs could investigate the stability and changes in numerological beliefs and practices over time. Tracking individuals' numerological interpretations and life outcomes longitudinally would provide valuable insights into the enduring influence of numerology on human life trajectories.

These future directions and research opportunities aim to advance our understanding of numerology's influence on human life and its broader implications for science, society, and culture.

4.13 CONCLUSION

In conclusion, this research paper has provided a comprehensive exploration of numerology and its profound influence on various aspects of human life. Through an examination of numerological principles, methodologies, and applications, we have gained insights into the mystical connections between numbers and human existence. From astrology to personality analysis and predictive practices, numerology offers a unique lens through which we can understand ourselves and the world around us. As we move forward, it is essential to continue exploring numerology with an open mind, considering both ancient wisdom and modern interpretations. Future research should focus on further elucidating the mechanisms underlying numerological phenomena and investigating its potential applications in diverse fields. By embracing numerology as a tool for self-discovery and personal growth, we can unlock new avenues for understanding the intricacies of human life and enhancing our connection with the universe.

REFERENCES

1. Boyer, C. B., & Merzbach, U. C. (2011). *A History of Mathematics*. John Wiley & Sons.
2. Gregory, A. (2015). The Pythagoreans: Number and Numerology, in *Mathematicians and Their Gods: Interactions Between Mathematics and Religious Beliefs*. Oxford University Press.
3. Osuji, U. (2021). A Philosophical Analysis of the Role of Numbers in ordering Human Activities and Implications on Epistemological Discourses in Africa. *Journal of African Interdisciplinary Studies*, *5*(7), 148–162.
4. Visconti, S. (2019). *Numerology: Discover The Meaning Behind The Numbers in Your life & Their Secrets to Success, Wealth, Relationships*, Fortune Telling & Happiness.
5. Singh, P., Arora, K., Rathore, U.C., Yang, E., Joshi, G.P., Son, K.C. (2023). "Performance Evaluation of Grid-Connected DFIG-Based WECS with Battery Energy Storage System under Wind Alterations Using FOPID Controller for RSC." *MDPI Journal of Mathematics*, 11(9), 2100, https://doi.org/10.3390/math11092100
6. Kovan, D. (2018). *Secrets of Numerology*. Ivy Press.
7. Ayonrinde, O. A., Stefatos, A., Miller, S., Richer, A., Nadkarni, P., She, J., ... & Mngoma, N. (2021). The Salience and Symbolism of Numbers Across Cultural Beliefs and Practice. *International Review of Psychiatry*, *33*(1–2), 179–188.
8. Wilkie, J. E., & Bodenhausen, G. V. (2015). The Numerology of Gender: Gendered Perceptions of Even and Odd Numbers. *Frontiers in Psychology*, 6, 148470.
9. Lawrence, S. B. (2019). *The Big Book of Numerology: The Hidden Meaning of Numbers and Letters*. Weiser Books.
10. Bhatt, B., Sharma, H., Arora, K., Joshi, G.P., & Shrestha, B. (2023). "Levy Flight-Based Improved Grey Wolf Optimization: A Solution for Various Engineering Problems." *MDPI Journal of Mathematics*, 11(7), 1745, https://doi.org/10.3390/math11071745

11. Junior, L. A. (2019). *Numerology: Key to Being*. Editora Griot.
12. Singh, S. C. (2020). *Let the Numbers Guide You: The Spiritual Science of Numerology*. John Hunt Publishing.
13. Maria, F. (2021). *Empower Your Success with Numerology and Astrology*. BookRix.
14. Zhmud, L. (2019). From Number Symbolism to Arithmology. *Zahlen-und Buchstabensysteme im Dienste Religiöser Bildung*. Mohr Siebeck GmbH & Co. KG. 25–45.
15. Connor, S. (2016). *Living by Numbers: In Defence of Quantity*. Reaktion Books.
16. Bawden, D., & Robinson, L. (2016). Information's Magic numbers: The Numerology of Information Science. *Theories of Information and Scholarly Communication*, De Gruyter Saur, 180–196.
17. Arora, K., Joshi, G.P., Ragab, M., Rawa, M., Milyani, A.H., Mansour, R.F., Yang, E. (2022). "Bilateral Contract for Load Frequency and Renewable Energy Sources Using Advanced Controller." *Computers, Materials & Continua, Tech Science Press*, 73(2), 3165–3180, 10.32604/cmc.2022.026966.
18. Scolnick, R. (2022). *The Witch's Book of Numbers: Enhance Your Magic with Numerology*. Red Wheel/Weiser.
19. Abduov, M. I., & Abduova, N. M. (2018). Function of Holy Numbers in Lyrical and Epic Poetry. *Bulletin of the Karaganda University Philology Series*, *92*(4), 14–19.
20. Bellos, A. (2015). *The Grapes of Math: How Life Reflects Numbers and Numbers Reflect Life*. Simon and Schuster.
21. Gupta, R. (2021). Color Therapy in Mental Health and Well Being. *International Journal of All Research Education and Scientific Methods (IJARESM)*, *9*(2), 2455–6211.
22. Keller, J. (2015). *The Complete Book of Numerology*. St. Martin's Griffin.
23. Ghannam, T. (2012). *The Mystery of Numbers: Revealed Through their Digital Root*. Talal Ghannam.
24. Buchanan, M. (2013). *The Numerology Guidebook: Uncover Your Destiny and the Blueprint of Your Life*. Hay House.
25. Simpson, J. (2014). *Numerology: Make Predictions and Decisions Based on the Power of Numbers*. Penguin.
26. Ngan, N. D. (2012). *Your Soul Contract Decoded: Discovering the Spiritual Map Of Your Life With Numerology*. Watkins Media.
27. Goodman, J. K., & Irwin, J. R. (2006). Special Random Numbers: Beyond the Illusion of Control. *Organizational Behavior and Human Decision Processes*, *99*(2), 161–174.
28. Phillips, D. A. (2005). *The Complete Book of Numerology: Discovering the Inner Self*. Hay House, Inc.
29. Lawrence, S. (2003). *Exploring Numerology: Life by the Numbers*. Red Wheel/Weiser.
30. Cohen, I. B. (2005). *The Triumph of Numbers: How Counting Shaped Modern Life*. WW Norton.
31. *Drayer, R. (2003). Numerology: The power in numbers.* Square One Publishers, Inc.
32. Brown, C. (2023). *Book of Angel Numbers: Understand Their Meanings, Interpret the Signs, and Learn How to Follow Your Guidance*. Balboa Press.
33. Cui, Y., Gavriilidis, K., Gebka, B., & Kallinterakis, V. (2024). Numerological superstitions and market-wide herding: Evidence from China. *International Review of Financial Analysis*, *93*, 103199. https://doi.org/10.1016/j.irfa.2024.103199.

5 An Enhanced Position Updating Strategy for Salp Swarm Algorithm to Solve Global Optimization Problems

Parijata Majumdar, Sanjoy Mitra, Diego Oliva, and Leonardo Ramos Rodrigues

5.1 INTRODUCTION

Salp swarm algorithm (SSA) is an emerging population-based metaheuristic inspired by the swarming behavior of salps in the ocean (Mirjalili et al, 2017). In SSA, each individual salp creates a long chain structure in succession behind a leader. The movement of the salps in the ocean is regulated by the chain structure. During hunting, the leader salps located at the front of the salp chain explore the location of the food source and the follower salps update their position depending on the leader salps located at the rear position of the salp chain.

However, SSA suffers from local optima stagnation problems and low convergence rate. Wolpert and Macready (1997) stated that not every algorithm can be used to address every kind of optimization issue. As a result, new metaheuristics are constantly being developed to improve upon the performance of already-existing metaheuristics. To overcome the limitations of stagnation in local optimum and low convergence rate, the traditional SSA is combined with lens opposition-based learning (LOBL) which utilizes the lens imaging principle to enhance the population diversity and quality (Yu et al, 2014). OBL has been used to boost the performance of several algorithms. For example, in Shan et al (2016) OBL is used to improve the bat algorithm's population diversity and convergence speed. In Abd Elaziz et al (2020), OBL is used to generate the initial population for improving the convergence rate of moth–flame optimization. In Tubishat et al (2020), OBL is used during the SSA's initiation phase to increase the population's variety.

DOI: 10.1201/9781003563440-5

In our proposed method, after comparing the candidate solution and the opposing solution at each iteration in comparison to OBL, the better individual is chosen using the dynamic opposite solution of the candidate solution, which is computed using LOBL. Thus, there is a far greater possibility of attaining the global optimum (Tizhoosh, 2005). LOBL is a generalized form of OBL which maximizes the probability of the algorithm jumping out of the local optimum. The local search algorithm (LSA) is used during the exploitation phase to boost exploitation in the local neighborhood region at the end of each iteration (Tubishat et al, 2020). The information of the local best solution is also added in the follower salp position update process where the fitness function is used to assess the newly generated positions. In the event that the primary food supply becomes trapped in a local minimum, it makes it possible to locate substitute food sources. Using a step size calculated based on the difference between the associated local position and any random individual location of the population, the neighboring region of each local best solution is also used (Ouaar and Boudjemaa, 2021). Next, an inertia weight is added to the leader salp and follower salp position update mechanism in order to quicken the pace of convergence (Hegazy et al, 2020).

Recently, in the crowded subject of bio-inspired population-based metaheuristics, SSA has gained a lot of momentum since it provides an interesting optimization performance in addressing various optimization challenges. In data mining applications, feature selection is an optimization problem that has become inevitable due to the enormous increase of data in the world. A large number of irrelevant features hinder classification accuracy and require high computation time. A large number of features may also lead the solution to stagnate at local optima as the best possible objective value appears within an open neighborhood around it (Abdel-Basset et al, 2020).

Feature selection is an optimization problem to remove irrelevant and repetitive features from high-dimensional datasets for increasing the classification accuracy and performance of the learning algorithms when applied on high-dimensional datasets (Dash and Liu, 1997). In this scope of work, we used this enhanced combined variant of SSA (EhSSA) for selecting feature subsets from a large set of features of high-dimensional datasets. The final set of features can be assessed using either filter or wrapper-based approaches. The wrapper-based approaches (Mafarja and Mirjalili, 2018) gives high-quality solutions as it relies on data properties instead of a classifier for selecting features unlike filter-based approaches (Kohavi, 1995). So, EhSSA is consolidated with a wrapper-based approach called K-nearest neighbor (K-NN) to assess the quality of selected features.

EhSSA is compared with different well-known metaheuristics based on average accuracy, the average set of chosen features and average fitness using different benchmark datasets collected from UCI repositories (Asuncion and Newman, 2007). For comparison of EhSSA with other baseline algorithms, it is also benchmarked on well-known functions including IEEE CEC 2019 and CEC 2020 function suite. Statistical analysis using Friedman test is also done for proving its significant improvement concerning other algorithms. Besides, our proposed algorithm is also used to solve four classical engineering design problems to show its efficacy and capability in problem-solving.

This is how the remainder of the article is structured. The history of a few related papers and an explanation of the fundamentals of the SSA, LOBL, and SA is provided in Section 2. The mathematical model and inspiration of the proposed EhSSA are explained in Section 3.

The experimental results are discussed in Section 4. In Section 5, engineering design tasks are used to assess EhSSA's problem-solving ability. In Section 6, a conclusion is drawn and recommendations for future research in SSA as perspectives are discussed.

5.2 BACKGROUND

In this section, existing literature based on feature selection methods using different variants of SSA and improved variants of SSA is discussed.

5.2.1 Literature Review

In Sayed et al (2018), chaotic version SSA is proposed where ten different chaotic maps are used, of which logistic chaotic map outperforms others in finding an optimal feature subset and maximizing classification accuracy. In Tubishat et al (2020), OBL is used at initialization phase of SSA for increasing population diversity and LSA is used during the exploitation phase for feature selection. In Zhang et al (2020), chaotic initialization is done in SSA to produce a better initial population and differential evolution is also used to better balance the global search and exploitation of SSA. In Ouaar and Boudjemaa (2021), three measures were taken for balancing exploration and exploitation. The local best information of the followers salps are incorporated to enhance exploration of local neighborhood. Secondly, a local search is conducted in the global best neighborhood and differential evolution is combined with a randomly chosen local best position. Thirdly, using the location of the associated follower based on a local leap in the local best neighborhood, a non-improvement of the local best solution is computed. SSA and LOBL are integrated with the whale optimization algorithm (WAO) in Fan et al (2020). The leader mechanism of SSA is first used to update positions, and then, the nonlinear parameter for SSA convergence is used to the prey and bubble-net assaulting encircling phases of whale optimization algorithm (WOA). Lastly, in order to improve population variety, LOBL is used. In Nautiyal et al (2021) to overcome the convergence problem and stucking into suboptimal solutions of SSA, Gaussian, Cauchy, and levy-flight mutation schemes are proposed where the Gaussian mutation, Cauchy mutation, and levy-flight mutation are used to increase the global convergence and to increase the randomness of salps during search. It can be seen that Gaussian mutation outperforms all the mutation schemes in boosting the exploitation and exploration capabilities. In Kassaymeh et al (2022), a number of populations are created, and the population with the greatest variety is selected as the starting population. Next, the appropriate parameters are found by applying a genetic algorithm to a self-adaptive parameter.

5.2.2 Salp Swarm Algorithm

SSA simulates the swarming mechanism of salps in oceans (Mirjalili et al, 2017). Typically, salps form salp chains in deep waters. The salp at the head of the chain is the leader, while the other salps at the back are referred to as followers. Salp positions are specified in a search space that is dim-dimensional, where dim is the number of variables in a given issue. All salp positions are stored in a two-dimensional matrix called "mat." It is expected that the swarm would scan the search space for a food supply using Faim as its focal point. The mathematical model is described as follows:

$$\text{mat}_n^1 = \text{Faim}_n + r_1\left(\left(u_n - l_n\right)r_2 + l_n\right),\ \text{if } r_3 \geq 0 \tag{1}$$

$$\text{mat}_n^1 = \text{Faim}_n - r_1\left(\left(u_n - l_n\right)r_2 + l_n\right),\ \text{if } r_3 < 0 \tag{2}$$

Here, mat_n^1 represents leader position in nth dimension, Faim_n represents food source position in nth dimension, u_n represents upper bound of n*th* dimension, l_n represents lower bound of n*th* dimension, and *r1, r2*, and *r3* are random variables ∈ [0, 1].

The coefficient r_1 balances exploration and exploitation.

In this case, curr_{iter} indicates current iteration, max_{iter} indicates maximum number of iterations, mat_m^n indicates position of mth follower salp in n*th* dimension, e represents time, and initial_{speed} indicates the initial speed.

$$r_1 = 2e^{-}\left(-\frac{4\text{curr}_{\text{iter}}}{\text{max}_{\text{iter}}}\right) \tag{3}$$

At each iteration, the following equation is used to change the location of the follower salp:

$$\text{mat}_n^m = \frac{1}{2}ce^2 + \text{initial}_{\text{speed}} \times e \tag{4}$$

where m ≥ 2, $c = \frac{final_{speed}}{initial_{speed}}$, and $\text{speed} = \frac{mat - mat_0}{e}$. The conflict between iterations is equal to 1, and considering $\text{initial}_{speed} = 0$, it can be written as:
where m = 1, ..., dim

$$\text{mat}_n^m = \frac{1}{2}\left(\text{mat}_n^m + mat_n^{m-1}\right) \tag{5}$$

While the leader salp is drawn to its food supply, the follower salps adjust their location in relation to it in order to converge toward the global optimum. Usually, the best

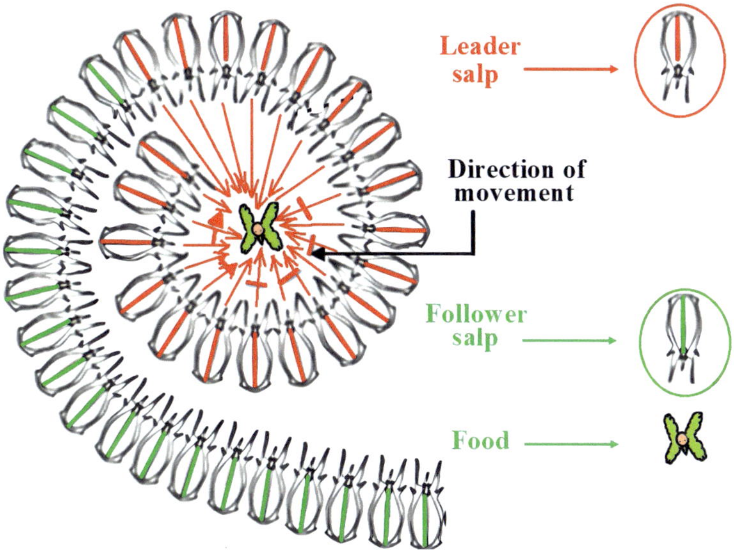

FIGURE 5.1 Basic structure of salp swarm algorithm (SSA) (Mirjalili et al, 2017).

solution found so far replaces the global optimum, which is unknown. First, a random population of N locations that have been chosen is created. Each individual's fitness is calculated, and the food source is determined by the best solution. The population is updated at every iteration based on the leader and follower salp position update Eqs. (1), (2), and (5), respectively. Each new position mat^n for all n = 1, ..., N is checked for feasibility using upper and lower limits in each dimension m = 1, ..., dim. The food source and the objective functions of the feasible solutions are contrasted. In accordance, the best obtained minimum is updated. Until a halting criterion is satisfied, this procedure is repeated. Figure 5.1 shows the basic structure of SSA (Mirjalili et al, 2017).

5.2.3 Lens Opposition-Based Learning

LOBL combines opposition-based learning (OBL) (Tizhoosh, 2005) and lens imaging principle (Figure 5.2). LOBL is used to compute the opposite solutions of the candidate solutions. These solutions are then compared to select a better candidate solution at every iteration. This raises the likelihood of arriving at the global optimum. The lens imaging principle states that an inverted and diminishing actual image will develop between one and two times the focal length on the opposite side when the distance between the object and the lens is more than twice the focal length.

In comparison, by varying c to generate a varied population and global exploration, LOBL may achieve a more dynamic opposite solution than OBL.

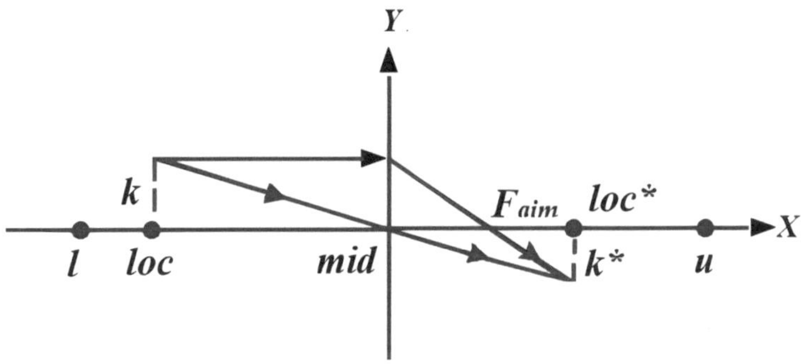

FIGURE 5.2 Lens opposition-based learning (Tizhoosh, 2005).

5.2.4 Local Search Algorithm

LSA is stated in Algorithm 1 as follows:

Algorithm 1 Local Search Algorithm (LSA) (Tubishat et al, 2020)

```
procedure Algorithm 1
t= curr_best (curr_best represents the current best
  solution at the end of current iteration curr_iter)
 var = 1 (var is a variable used to store the
  curr_iter of LSA)
 while var < maximum number of local
  iterations do
   Randomly select three features from t.
   if feature_sel == 1 then ▷ 1 for selecting
  features, 0 for rejected features
     feature_sel = 0
   else
     feature_sel = 1
     end if
     Compute the fitness of t.
     if f(t) < curr_best then
                curr_best = t
     end if
     var = var + 1
   end while
   Return curr_best.
end procedure
```

At the end of each iteration, LSA will be called to update $curr_{best}$. Initially, LSA starts by storing the best solution value received from SSA at the end in variable t. LSA iterates repeatedly for improving t. In every iteration, LSA selects three features randomly from t. LSA resets the selected features based on its values. LSA will assess the fitness of the new solution (new_{sol}), if it is comparatively better than the fitness of $prev_{sol}$, it will be set to t; otherwise, $prev_{sol}$ remains unaltered.

5.3 PROPOSED METHODOLOGY

Our proposed approach improves the population variety of SSA by using LOBL during initialization. To prevent the solution from being trapped in local optima during exploitation, the LSA algorithm is combined with SSA. Salps' location is additionally updated using an inertia weight $w \in [0, 1]$ (Hegazy et al, 2020). During the search, the additional parameter accelerates convergence. In order to solve the issue of local optima stagnation in feature selection tasks, it also strikes a balance between exploitation and exploration capabilities. The enhanced procedure for updating positions may be expressed as follows:

$$\text{mat}_n^1 = w\text{Faim}_n + r_1\left(\left(u_n - l_n\right)r_2 + l_n\right),\ \text{if } r_3 \geq 0 \tag{6}$$

$$\text{mat}_n^1 = w\text{Faim}_n - r_1\left(\left(u_n - l_n\right)r_2 + l_n\right),\ \text{if } r_3 < 0 \tag{7}$$

$$\text{mat}_n^m = \frac{1}{2}\left(\text{mat}_n^m + w\text{mat}_n^{m-1}\right) \tag{8}$$

Since feature selection is a binary problem to map continuous values into binary values, we used the following function:

$$\text{mat}_{\text{mn}} = 1,\ \text{if } \text{V}_{\text{mn}} > 0.5,\ \text{else } 0 \tag{9}$$

where mat_{mn} is the discrete solution vector V and V_{mn} is the continuous position of m at dimension n.

The fitness function aims to simultaneously improve classification accuracy and make useful feature selection. Fitness may be calculated using the formula as follows:

$$\text{fitness} = \rho\text{Error}\left(D\right) + \phi\frac{|s|}{|d|} \tag{10}$$

where Error(D) is the classification error rate, s represents the selected features, d represents all the features, and ρ and ϕ are the weights to control the classification accuracy and the feature reduction. ρ [0, 1] and $\phi = 1\ \rho$. Every salp's fitness function is assessed, and each salp's location is modified until the ideal combination of few characteristics and high classification accuracy is attained (Hegazy et al, 2020). In our case, $\rho = 0.9$.

To update the follower's salp positions at the rear end of the chain, we used the following equation (Ouaar and Boudjemaa, 2021):

$$\text{mat}_n^m = l\text{best}_n^m + r\left(\text{mat}_n^q - \text{mat}_n^m\right) \tag{11}$$

where n = 1, ..., dim
where $l\text{best}_n^m$ is the m*th* local best position in the n*th* dimension, mat_n^m is the m*th* position in the nth dimension, and mat_n^q is a randomly selected position in nth dimension from a set of N solution. r [0, ∈1]. Using a line search and local best solutions, this equation breaks the chain of followers. The followers are redirected to a more fruitful search path by the unpredictability of the chosen solution mat_n^q.

During the exploration phase, each individual's local best information is included so that the algorithm can track new food sources in case the food source becomes trapped in a local minimum. Every local best's neighborhood is also used, with the step size being adjusted to expand or reduce the neighborhood search region around the local best. The step size is calculated as the difference between the matching local position mat^m and a random individual mat^q (Ouaar and Boudjemaa, 2021).

The proposed EhSSA is shown in Algorithm 2 as follows:

Algorithm 2 Enhanced Salp Swarm Algorithm (EhSSA)

```
procedure Algorithm 2
  Initialize the salps positions S as S_m (m = 1, 2, ... , p)
  Calculate the salps dynamic opposite population OP as S̃_m (m = 1, 2, ... , p)
  Select the p fittest salps from S ∪ OP which represents the initial SSA population
  Determine the fitness value of each salp
  Best_sol = best salp (search-agent)
  while q < max_iter do
        Update the value of r1
        for every salp (S_m) do
              if m == 1 then
                    Update leader position
              else
                    Update follower position
              end if
        end for
        Adjust the position of salps that goes out of the search space based on lower and upper bounds of the problem variables
        Determine the fitness value of each salp
        Best_sol = best salp (search-agent)
        Apply LSA using Algorithm 1 on Best_sol to determine if there is a better solution (if better solution is found then update Best_sol, otherwise Best_sol is not updated)
        q = q + 1
```

```
    end while
    return Best_sol
end procedure
```

Using LOBL, EhSSA first creates a population of salps. The m fittest salps from the starting population and the opposing populations are selected using EhSSA. $Best_{sol}$ will be the best salp among these m fittest salps. Equations (7) and (8) are used by the main loop to update their locations. On discovering a better solution than the current $Best_{sol}$, LSA will be applied on the current $Best_{sol}$ at the end of the main loop. Improved SSA (ISSA) will finally give back the $Best_{sol}$. EhSSA will be assessed on the testing dataset using the characteristics that were chosen from the $Best_{sol}$.

K-NN is used with the suggested EhSSA to evaluate the effectiveness of certain characteristics. EhSSA will be used to choose the chosen features for K-NN training in each iteration. Features are represented by binary values; a "1" value means that the features with the specified index have been selected, whereas a "0" value means that the features with the specified index have not been selected.

5.3.1 Time Complexity Analysis

To determine the time complexity of the proposed EhSSA, seven components are contributing factors: initialization of salps, computing the salps dynamic opposite population, fitness evaluation of each salp, position update of leading and follower salps, fitness evaluation of updated salps and applying LSA on the $Best_{sol}$. The complexity of initialization of salp swarm is $O(N \times D)$ time, the complexity of computing the dynamic opposite population of salp swarm is $O(N \times D)$ time, and fitness evaluation of salp utilizes $O(N)$ time. The leader and follower salps' position update process is complicated, taking $O(N \times D)$, the updated salps' fitness assessment requires $O(N)$ time, and the $Best_{sol}$'s LSA application is complex, taking $O(N)$ time. $O(N \times D)$ is the worst-case complexity of K-NN. Therefore, adding everything together, the complexity of the suggested EhSSA is $O(N \times D)$, where D is the problem's dimension and N is the cardinality of the selected features.

5.4 EXPERIMENTAL RESULTS

The Windows 10 operating system is being used for the research. All the algorithms are executed in Python 3.8 programming language. The benchmark functions that are classical are selected from Yao et al (1999). Unimodal functions don't have any local optimum; they only have one global optimum. Multimodal functions have several local optima. There are several features for each function, such as dim, range, and optima. The dimension is specified as dim. The range is border between the lower and upper boundaries of the function's search space. Optima denotes each benchmark function's global optimum. Tables 5.1, 5.2, and 5.3 describe unimodal, multimodal, and fixed-dimension multimodal benchmark functions (Yao et al, 1999). Tables 5.4, 5.5, and 5.6 show comparison results of simulations of unimodal, multimodal, and fixed-dimension multimodal benchmark functions of metaheuristics. EhSSA is

TABLE 5.1
Unimodal benchmark fu nctions

Function	Dim	Range	f_{min}
$f_1(x) = \sum_{i=1}^{n} x_i^2$	30	[−100, 100]	0
$f_2(x) = \sum_{i=1}^{n} \|x_i\| + \prod_{i=1}^{n} \|x_i\|$	30	[−10, 10]	0
$f_3(x) = \sum_{i=1}^{n} (\sum_{j-1}^{i} x_j)^2$	30	[−100, 100]	0
$f_4(x) = \max_i\{\|x_i\|, 1 \leqslant i \leqslant n\}$	30	[−100, 100]	0
$f_5(x) = \sum_{i=1}^{n-1} [100(x_{i+1} - x_i^2)^2 + (x_i - 1)^2]$	30	[−30, 30]	0
$f_6(x) = \sum_{i=1}^{n} ([x_i + 0.5])^2$	30	[−100, 100]	0
$f_7(x) = \sum_{i=1}^{n} ix_i^4 + random[0, 1)$	30	[−1.28, 1.28]	0

compared with SSA (Mirjalili et al, 2017) and enhanced whale optimization algorithm integrated with salp swarm Algorithm (ESSWOA) (Fan et al, 2020) since these algorithms have shown their efficacy in solving different optimization problems.

It can be seen from Table 5.4 that EhSSA finds better results in F1, F3, F5, F6, and F7 unimodal functions (Mirjalili et al, 2014). From Table 5.5, it can be seen that EhSSA finds better results in F8, F10, F11, F12, and F13 multimodal functions. It can be seen from Table 5.6 that EhSSA finds better results in F14, F15, F16, F17, F18, F19, F21, F22, and F23 fixed-dimension multimodal functions.

5.5 EHSSA FOR SOLVING CLASSICAL ENGINEERING PROBLEMS

Four constrained engineering problems are solved to prove the supremacy of EhSSA compared to other baseline algorithms. Since, there are several equality and inequality constraints, the EhSSA should have a mechanism for managing constraints to optimize constrained problems. However, any type of constraint management can be used without having to change the algorithm's mechanism for the fitness independent algorithms such as GA and PSO.

As the search agents adjust their placements based on the location of leader salps, there is no direct correlation between the fitness function and the search agents. As a result, penalty-based constraint management functions are used, which assigns high objective function values to salps when constraints are broken. If the leaders break any of the limitations in the next iteration, a new salp will take their place right away. Salps may be subjected to any kind of penalty function, depending on how severe the constraints are.

5.5.1 Tension/Compression Spring Design (TCSD)

The weight of a tension/compression spring has to be reduced. Shear stress, surge frequency, and minimum deflection are few restrictions that are applicable to the reduction.

Figure 5.3 shows schematic diagram of tension/compression spring design (TCSD) problem.

TABLE 5.2
Multimodal benchmark functions

Function	Dim	Range	$f_{\min}$
$F_8(x) = \sum_{i=1}^{n} - x_i \sin(\sqrt{\lvert x_i \rvert})$	30	$[-500, 500]$	-418.9829×5
$F_9(x) = \sum_{i=1}^{n} [x_i^2 - 10\cos(2\pi x_i) + 10]$	30	$[-5.12, 5.12]$	0
$F_{10}(x) = -20\exp\left(-0.2\sqrt{\frac{1}{n}\sum_{i=1}^{n} x_i^2}\right) - \exp\left(\frac{1}{n}\sum_{i=1}^{n}\cos(2\pi x_i)\right) + 20 + e$	30	$[-32, 32]$	0
$F_{11}(x) = \frac{1}{4000}\sum_{i=1}^{n} x_i^2 - \prod_{i=1}^{n}\cos\left(\frac{x_i}{\sqrt{i}}\right) + 1$	30	$[-600, 600]$	0
$F_{12}(x) = \frac{\pi}{n}\{10\sin(\pi y_1) + \sum_{i=1}^{n-1}(y_i - 1)^2[1 + 10\sin^2(\pi y_{i+1})] + (y_n - 1)^2\} + \sum_{i=1}^{n} u(x_i, 10, 100, 4)$	30	$[-50, 50]$	0
$y_i = 1 + \frac{x_i + 1}{4}$			
$u(x_i, a, k, m) = \begin{cases} k(x_i - a)^m & x_i > a \\ 0 & -a < x_i < a \\ k(-x_i - a)^m & x_i < -a \end{cases}$			
$F_{13}(x) = 0.1\{\sin^2(3\pi x_1) + \sum_{i=1}^{n}(x_i - 1)^2[1 + \sin^2(3\pi x_i + 1)] + (x_n - 1)^2[1 + \sin^2(2\pi x_n)]\} + \sum_{i=1}^{n} u(x_i, 5, 100, 4)$	30	$[-50, 50]$	0
$F_{14}(x) = -\sum_{i=1}^{n}\sin(x_i)\cdot\left(\sin\left(\frac{i x_i^2}{\pi}\right)\right)^{2m},\ m = 10$	30	$[0, \pi]$	-4.687
$F_{15}(x) = \left[e^{-\sum_{i=1}^{n}(x_i/\beta)^{2m}} - 2e^{-\sum_{i=1}^{n} x_i^2}\right]\cdot\prod_{i=1}^{n}\cos^2 x_i,\ m = 5$	30	$[-20, 20]$	-1
$F_{16}(x) = \{[\sum_{i=1}^{n}\sin^2(x_i)] - \exp(-\sum_{i=1}^{n} x_i^2)\}\cdot\exp[-\sum_{i=1}^{n}\sin^2\sqrt{\lvert x_i \rvert}]$	30	$[-10, 10]$	-1

TABLE 5.3
Fixed-dimension multimodal benchmark functions

Function	Dim	Range	f_{min}
$F_{14}(x) = \left(\frac{1}{500} + \sum_{j=1}^{25} \frac{1}{j+\sum_{i=1}^{2}(x_i - a_{ij})^6}\right)^{-1}$	2	[−65,65]	1
$F_{15}(x) = \sum_{i=1}^{11}\left[a_i - \frac{x_1(b_i^2+b_i x_2)}{b_i^2+b_i x_3+x_4}\right]^2$	4	[−5,5]	0.00030
$F_{16}(x) = 4x_1^2 - 2.1x_1^4 + \frac{1}{3}x_1^6 + x_1x_2 - 4x_2^2 + 4x_2^4$	2	[−5,5]	−1.0316
$F_{17}(x) = \left(x_2 - \frac{5.1}{4\pi^2}x_1^2 + \frac{5}{\pi}x_1 - 6\right)^2 + 10(1 - \frac{1}{8\pi})\cos x_1 + 10$	2	[−5,5]	0.398
$F_{18}(x) = [1 + (x_1 + x_2 + 1)^2(19 - 14x_1 + 3x_1^2 - 14x_2 + 6x_1x_2 + 3x_2^2)] \times [30 + (2x_1 - 3x_2)^2 \times (18 - 32x_1 + 12x_1^2 + 48x_2 - 36x_1x_2 + 27x_2^2)]$	2	[−2,2]	3
$F_{19}(x) = -\sum_{i=1}^{4} c_i \exp(-\sum_{j=1}^{3} a_{ij}(x_j - p_{ij})^2)$	3	[1,3]	−3.86
$F_{20}(x) = -\sum_{i=1}^{4} c_i \exp(-\sum_{j=1}^{6} a_{ij}(x_j - p_{ij})^2)$	6	[0,1]	−3.32
$F_{21}(x) = -\sum_{i=1}^{5}[(X - a_i)(X - a_i)^T + c_i]^{-1}$	4	[0,10]	−10.1532
$F_{22}(x) = -\sum_{i=1}^{7}[(X - a_i)(X - a_i)^T + c_i]^{-1}$	4	[0,10]	−10.4028
$F_{23}(x) = -\sum_{i=1}^{10}[(X - a_i)(X - a_i)^T + c_i]^{-1}$	4	[0,10]	−10.5363

TABLE 5.4
Test results on unimodal benchmark functions

	SSA		ESSAWOA		EhSSA	
F						
F1	1.65E-07	1.77E-07	0	0	0	0
F2	2.36E+00	1.87E+00	0	0	1.0E-75	0.1E-7
F3	1.34E+03	8.33E+02	0	0	0	0
F4	1.18E+01	3.18E+00	0	0	0.015665	0.0210064
F5	2.67E+02	4.90E+02	2.74E+01	4.68E-01	2.63E+01	4.32E-01
F6	1.31E-07	1.25E-07	1.41E-08	3.28E-09	1.36E-08	3.19E-09
F7	1.66E-01	6.99E-02	5.13E-05	5.06E-05	4.23E-05	4.15E-05

TABLE 5.5
Test results on multimodal benchmark functions

	SSA		ESSAWOA		EhSSA	
F	Mean	Std	Mean	Std	Mean	Std
F8	-7.43E+03	4.84E+02	-1.19E+04	1.12E+03	-2.21E+04	1.01E+03
F9	5.50E+01	2.38E+01	0	0	5.18E+01	2.25E+01
F10	2.89E+00	1.90E+00	8.88E-16	0	0	0
F11	1.58E-02	9.08E-03	0	0	0	0
F12	6.56E+00	3.76E+00	7.91E-11	2.45E-11	7.63E-11	2.31E-11
F13	1.73E+01	1.58E+01	2.08E+00	1.37E+00	5.68E-01	3.23E-01

TABLE 5.6
Test results on fixed-dimension multimodal benchmark functions

	SSA		ESSAWOA		EhSSA	
F	Mean	Std	Mean	Std	Mean	Std
F14	1.3946	8.85E-01	1.0311	1.81E-01	1.0198	1.93E-01
F15	0.0035	6.72E-03	0.0005	1.96E-04	0.00048	2.09E-05
F16	-1.0316	3.83E-14	-1.0316	4.06E-14	-1.0421	5.02E-14
F17	0.3979	7.36E-14	0.3979	7.27E-14	0.2868	7.39E-14
F18	3.0000	2.37E-13	3.0000	4.29E-13	3.0000	2.42E-13
F19	-3.8628	8.34E-11	-3.8628	7.51E-14	-3.8520	7.63E-14
F20	-3.2181	5.37E-02	-3.2147	3.64E-02	-3.2040	1.17E-01
F21	-8.3136	2.93E+00	-10.1532	3.58E-11	-10.2141	3.61E-11
F22	-8.5531	3.16E+00	-10.4029	4.23E-11	-10.5139	4.49E-11
F23	-8.4464	3.31E+00	-10.5364	4.41E-11	-10.6154	4.52E-11

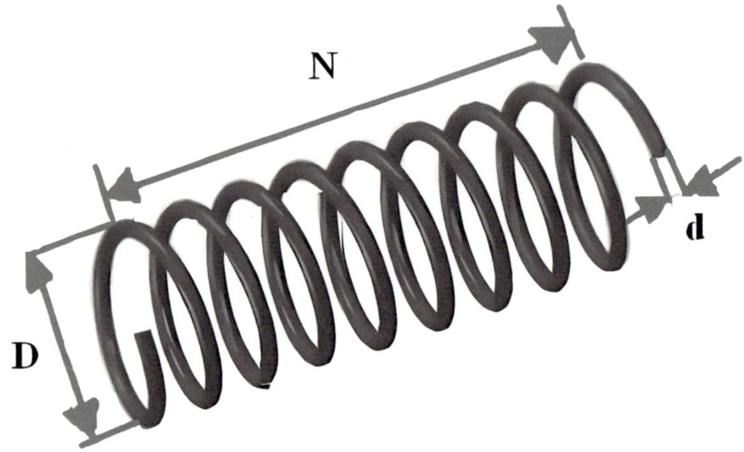

FIGURE 5.3 Schematic diagram of tension/compression spring design problem.

TABLE 5.7
Comparison results for TCSD problem

Algorithm	d	D	N	Optimum weight
EhSSA	0.051684	0.356443	11.213331	0.0126614
CDESSA (Zhang et al, 2022)	0.051691	0.356776	11.285558	0.0126652
SSA (Mirjalili et al, 2017)	0.051207	0.345215	12.004032	0.0126763
ESSA (Zhang et al, 2020)	0.051719	0.357434	11.247123	0.0126653

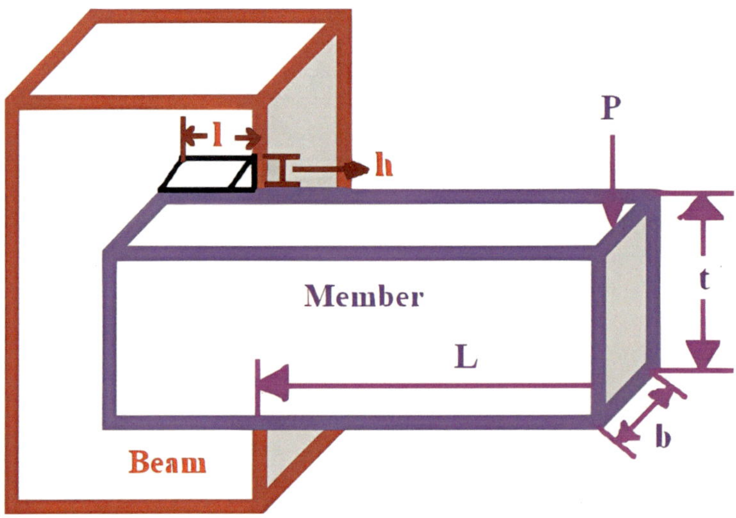

FIGURE 5.4 Schematic diagram of WBD problem.

TABLE 5.8
Comparison results for WBD problem

Algorithm	h	l	t	b	Optimum weight
EhSSA	0.205810	0.255192	9.036734	0.205651	1.695169
CDESSA (Zhang et al, 2022)	0.205719	0.255264	9.036825	0.205729	1.695277
SSA (Mirjalili et al, 2017)	0.205700	3.471400	9.036600	0.205700	0.0126763
ESSA (Zhang et al, 2020)	0.197198	3.485413	8.980946	0.208288	1.723317

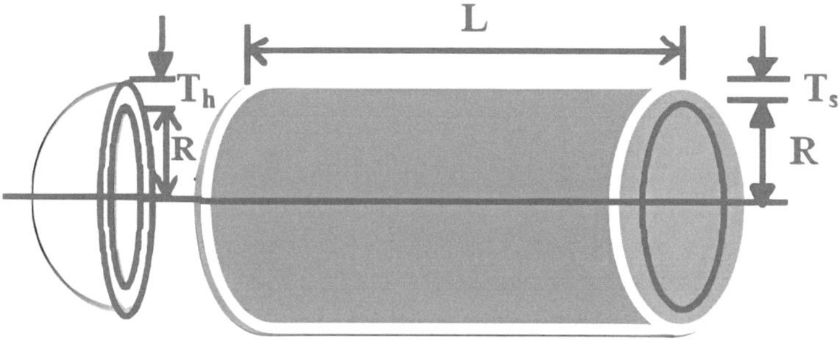

FIGURE 5.5 Schematic diagram of PVD problem.

The results of comparison of algorithms for TCSD problem are provided in Table 5.7 (Yang, 2011). Here, EhSSA is capable of finding a design with least weight.

5.5.2 Welded Beam Design (WBD)

The cost of fabricating a welded beam should be kept as low as possible (Coello Coello, 2000).

Figure 5.4 shows the diagram of WBD problem.

The comparison results are provided in Table 5.8 where the results confirms that EhSSA is capable to find a design with least cost.

5.5.3 Pressure Vessel Design (PVD)

The cost of creating a cylindrical tank from materials to welding must be kept as low as possible.

Figure 5.5 shows schematic diagram of PVD problem.

The results are provided in Table 5.9 which shows that EhSSA finds a design with the least cost.

5.6 CONCLUSION AND FUTURE WORK

We proposed an enhanced variant of SSA using the LOBL strategy and LSA for feature selection. In order to improve the population variety and quality of SSA, LOBL

TABLE 5.9
Comparison results for PVD problem

Algorithm	Ts	Th	R	L	Optimum cost
EhSSA	0.750112	0.3748699	41.965317	178.305581	5453.1339
CDESSA (Zhang et al, 2022)	0.750000	0.375000	41.966408	178.306673	5453.2428
SSA (Mirjalili et al, 2017)	0.790678	0.390834	40.967738	195.918220	6012.188500
ESSAWOA (Fan et al, 2020)	0.781763	0.386430	40.505695	197.463189	5892.354603

is first used to get the dynamic opposing population during the initiation phase. LOBL also improves SSA's capacity for global exploration. The LSA algorithm is also used at the end of each iteration to update the current best solution for avoiding the solution from getting stuck into the local optima. To increase the pace of convergence during the search, an inertia weight is included as a control parameter for the leader salp position update. The introduction of this control parameter also helps to balance the exploration and exploitation capabilities of EhSSA. In order for the EhSSA to be able to locate new food sources in the event that the food source becomes trapped in a local minimum, the knowledge of each individual's local best is integrated into the follower salp position during the exploration phase. The neighborhood of each local best is also exploited which allows the increase or decrease of the neighborhood search area around the local best. The proposed EhSSA is benchmarked using benchmark functions where it outperforms other algorithms. Besides our proposed algorithm is also used to solve classical engineering design problems where its efficacy in problem-solving is proved in comparison to other baseline algorithms. In the future, EhSSA can be investigated by employing different types of supervised and unsupervised classifiers.

5.7 DECLARATIONS

1. Ethics Approval: This research does not involve human participants and/or animals, hence ethics approval not required.
2. Funding: The authors did not receive support from any organization for the submitted work.
3. Consent to Participate: As this research does not involve human participants and/or animals, hence consent to participate not applicable.
4. Consent to Publish: As this research does not involve human participants and/or animals, hence consent to publish not applicable.
5. Competing Interests: The authors have no competing interests to declare that are relevant to the content of this article. All authors certify that they have no affiliations with or involvement in any organization or entity with any financial interest or nonfinancial interest in the subject matter or materials discussed in this manuscript.
6. Availability of Data and Materials

(a) Code Availability: No codes are made available for sharing at present and may be provided on request.
(b) Data Availability: Open-source benchmark data sets are used.

REFERENCES

Abd Elaziz M, Ewees AA, Ibrahim RA, et al (2020) Opposition-based moth-flame optimization improved by differential evolution for feature selection. *Mathematics and Computers in Simulation* 168:48–75.

Abdel-Basset M, El-Shahat D, El-henawy I, et al (2020) A new fusion of grey wolf optimizer algorithm with a two-phase mutation for feature selection. *Expert Systems with Applications* 139:112, 824.

Asuncion A and Newman D (2007) UCI machine learning repository. University of California, Irvine, School of Information and Computer Sciences. http://archive.ics.uci.edu/ml/SAS Help Center+3

Coello Coello C (2000) Use of a self-adaptive penalty approach for engineering optimization problems. *Computers in Industry* 41:113–27.

Dash M and Liu H (1997) Feature selection for classification. *Intelligent Data Analysis* 1(1-4):131–156.

Fan Q, Chen Z, Zhang W, et al (2020) ESSAWOA : enhanced whale optimization algorithm integrated with salp swarm algorithm for global optimization. *Engineering with Computers* pp 1–18.

Hegazy AE, Makhlouf MA, and El-Tawel GS (2020) Improved salp swarm algo-rithm for feature selection. *Journal of King Saud University-Computer and Information Sciences* 32(3):335–344.

Kassaymeh S, Abdullah S, Al-Betar MA, et al (2022) Self-adaptive salp swarm algorithm for optimization problems. *Soft Computing* 26(18): 9349–9368. https://doi.org/10.1007/s00500-022-07280-9

Kohavi R (1995) A study of cross-validation and bootstrap for accuracy estimation and model selection. *IJCAI* 14(2):1137–1145.

Mafarja M and Mirjalili S (2018) Whale optimization approaches for wrapper feature selection. *Applied Soft Computing* 62:441–453.

Mirjalili S, Gandomi AH, Mirjalili SZ, et al (2017) Salp swarm algorithm: A bio-inspired optimizer for engineering design problems. *Advances in Engineering Software* 114:163–191.

Mirjalili S, Mirjalili SM, and Lewis A (2014) Grey wolf optimizer. *Advances in Engineering Software* 69:46–61.

Nautiyal B, Prakash R, Vimal V, et al (2021) Improved salp swarm algorithm with mutation schemes for solving global optimization and engineering problems. *Engineering with Computers* 38: 1–23. https://doi.org/10.1007/s00366-020-01252-z

Ouaar F and Boudjemaa R (2021) Modified salp swarm algorithm for global optimisation. *Neural Computing and Applications* 33(14):8709–8734.

Sayed GI, Khoriba G, and Haggag MH (2018) A novel chaotic salp swarm algorithm for global optimization and feature selection. *Applied Intelligence* 48(10):3462–3481.

Shan X, Liu K, and Sun PL (2016) *Modified bat algorithm based on levy flight and opposition based learning*. Scientific Programming.

Tizhoosh H (2005) Opposition-based learning: A new scheme for machine intelligence. *International Conference on Computational Intelligence for Modelling, Control and Automation and International Conference on Intelligent Agents, Web Technologies and Internet Commerce (CIMCA-IAWTIC'06)*, vol. 1, pp. 695–701. IEEE. https://doi.org/10.1109/CIMCA.2005.1631345.

Tubishat M, Idris N, Shuib L, Abushariah, M. A. M., and Mirjalili, S. (2020) Improved Salp Wwarm Algorithm based on opposition based learning and novel local search algorithm for feature selection. *Expert Systems with Applications* 145:113122. https://doi.org/10.1016/j.eswa.2019.113122

Wolpert DH and Macready WG (1997) No free lunch theorems for optimization. *IEEE Transactions on Evolutionary Computation* 1(1):67–82. https://doi.org/10.1109/4235.585893

Yang XS (2011) *Nature-inspired metaheuristic algorithms*. Luniver Press.

Yao X, Liu Y, G. L (1999) Evolutionary programming made faster. *IEEE Transactions on Evolutionary Computation* 3(2):82–102.

Yu F, Li YX, Wei B, Xu, X., and Zhao, Z. Y. (2014) The application of a novel OBL based on lens imaging principle in PSO. *Tien Tzu Hsueh Pao (Acta Electronica Sinica)* 42(2):230–235. https://doi.org/10.3969/j.issn.0372-2112.2014.02.004

Zhang H, Cai Z, Ye X, Wang, M., Kuang, F., Chen, H., Li, C., & Li, Y (2020) A multi-strategy enhanced salp swarm algorithm for global optimization. *Engineering with Computers* 38:1177–1203. https://doi.org/10.1007/s00366-020-01099-4

Zhang H, Liu T, Ye X., Heidari, A. A., Liang, G., Chen, H., & Pan, Z. (2022) Differential evolution-assisted salp swarm algorithm with chaotic structure for real-world problems. *Engineering with Computers* 39(3):1735–1769. https://doi.org/10.1007/s00366-021-01545-x

6 Comparative Analysis of Guanine and Thymine Conductance for DNA-Based Molecular Electronics

Insights into Nanoelectronic Applications

Rajan Vohra and Kunwar Partap Singh

6.1 INTRODUCTION

Significant progress has been made in the study of molecular-scale electronics, particularly with the development of nanotechnology. In his groundbreaking 1959 speech "There's Plenty of Room at the Bottom," Nobel winner Richard P. Feynman introduced the notion of nanotechnology, imagining a day when molecular manipulation might provide groundbreaking technological advancements [1]. Since then, advances in research have moved beyond theoretical conjectures to real-world applications, especially in the area of nanoelectronics, which aims to use molecules' special qualities for electrical devices. Deoxyribonucleic acid (DNA) is a chemical best known for its function in storing genetic information. However, it has shown promise as a material for electrical applications at the molecular level. Interesting opportunities for electrical functioning are presented by its double-helix shape and the unique characteristics of its component nucleotides [2]. Investigating the electrical properties of DNA is consistent with Feynman's goal of developing devices at the molecular level, where the inherent capabilities of DNA may be used for novel purposes.

6.2 SIGNIFICANCE OF DNA IN ELECTRONICS

The distinct structure of DNA is essential to its biological function and may potentially be used in future electrical applications. Two strands that wind around one another to create a double helix make up each DNA molecule. Adenine (A), thymine

DOI: 10.1201/9781003563440-6

(T), cytosine (C), and guanine (G) are the nucleotide bases that make up the strands; these bases couple specifically (A with T and C with G) through hydrogen bonding [3]. DNA presents an intriguing topic for electronic research because of this base pairing and the nucleotide bases' intrinsic electrical characteristics.

Significant electrical conductivity has been found in DNA recently; this finding has implications for molecular-scale circuits, DNA-based sensors, and perhaps DNA-based computers [4]. Molecular-level manipulation and measurement of electrical characteristics provide new opportunities for the creation of bioelectronic devices, in which biological molecules are integral to electronic processes.

6.3 OBJECTIVE OF THE STUDY

The objective of this chapter is to present a thorough examination of the electrical conductance characteristics of guanine (G) and thymine (T), two of the four nucleotide bases found in DNA. Despite the fact that both bases are essential to the structure and operation of DNA, there are notable differences in their electrical characteristics that may have an impact on how they are used in nanoelectronics.

We want to learn more about the contributions that guanine and thymine make to the overall electrical characteristics of DNA by comparing their conductance behaviours. In addition to improving our knowledge of DNA's function in molecular electronics, this research offers useful insights for developing and refining DNA-based electrical devices.

6.4 STRUCTURE OF THE CHAPTER

This is how the chapter is structured: Following this introduction, we go over the body of research on DNA conductance as well as the theoretical underpinnings that pertain to guanine and thymine. Next, we offer the study's results and go into great depth on their implications after outlining the approaches that were employed. In conclusion, we provide a summary of the results and discuss possible future paths for this field of study.

6.5 LITERATURE REVIEW

6.5.1 Historical Context

Since the early days of nanotechnology, the idea of molecular-scale electronics—which includes the utilisation of biomolecules like DNA in electrical applications—has seen substantial development. A rapidly developing field of study was launched by Richard Feynman's groundbreaking speech in 1959 [1]. Then, in the 1980s and 1990s, important technologies like atomic force microscopy and scanning tunnelling microscopy (STM) were developed, opening up new avenues for the study of molecular electronics [5].

In the early days of molecular electronics, the main emphasis of the study was on organic molecules and how well they might replace conventional silicon-based components. Prior to its widespread use in research for technological applications,

DNA was mostly known for its biological roles. However, in recent times, its distinct electrical characteristics have drawn interest [6].

6.5.2 Previous Studies on DNA Conductance

The study of Porath et al. (2000), who showed direct electrical transport across DNA molecules, marked the beginning of serious research on the electrical characteristics of DNA [4]. The long-held belief that DNA was an insulator was refuted by this groundbreaking discovery, which demonstrated that DNA could carry electricity. Interest in the possible uses of DNA in sensors and memory components for electrical systems has increased as a result of this discovery.

Many facets of DNA conductance have been investigated in later research, including how molecule length and nucleotide sequence affect electrical characteristics. For example, Kasumov et al. (2001) explored the possibility of innovative electrical uses for DNA by studying proximity-induced superconductivity [7]. In a similar vein, Dulić et al. (2009) demonstrated the practical feasibility of single DNA molecules in electrical devices by providing direct conductance measurements of the molecules [8].

6.5.3 Theoretical Background

Comprehending the electrical characteristics of DNA necessitates an understanding of several theoretical models and concepts. The extended Hückel theory (EHT), one of the fundamental theories, approximates the interactions between atomic orbitals to determine the electronic structure of molecules [9]. Because EHT sheds light on the electrical interactions between DNA bases, it has been useful in predicting the conductance properties of these bases.

π-electron conjugation, which characterises the delocalisation of π-electrons across conjugated systems, including DNA bases, is another crucial idea [10]. Compared to thymine, which has a smaller π-system, guanine is likely to display various electrical behaviours due to its extended π-conjugated system. Their distinct conductance characteristics are greatly influenced by this variance in electrical structure.

The idea of negative differential resistance (NDR), which has been reported in many molecular systems, is also pertinent to the conductance analysis of guanine [11]. The phenomenon known as non-depletion resistance (NDR), which occurs when a molecule's current declines as voltage increases, has significant effects on how molecular-scale electrical devices are designed.

6.5.4 Summary of Existing Research

The extant body of research underscores the noteworthy advancements achieved in comprehending the electrical characteristics of DNA and their consequences for molecular electronics. Research has shown that DNA is capable of conducting electricity, with differences in conductance shown according to the length and nucleotide sequence of the molecule. Theoretical frameworks like π-conjugation and NDR, as well as models like EHT, offer useful ways to explain these results.

To properly comprehend the relative conductance characteristics of other DNA bases, such as guanine and thymine, further study is still required, even with current developments. By offering a thorough comparison of these two bases, this chapter seeks to close this knowledge gap and further our knowledge of DNA's function in molecular electronics.

6.6 COMPARATIVE ANALYSIS OF GUANINE AND THYMINE

6.6.1 Introduction

Understanding the unique electrical characteristics of guanine and thymine and their possible uses in molecular electronics requires a comparative examination of the two bases inside the DNA structure. DNA has two nitrogenous bases, guanine (G) and thymine (T), each of which contributes differently to the electrical behaviour of the molecule and has a specific function in genetic coding. With an emphasis on theoretical models and experimental data, this section investigates the variations in their electrical conductivity.

6.6.2 Structural and Electronic Properties

6.6.2.1 Guanine

The intricate chemical structure of guanine as shown in Figure 6.1, which consists of a fused double-ring system with a purine base, is what makes it unique. This structure has both heterocyclic and aromatic rings, which result in an extended π-conjugated system. Guanine is a noteworthy contender for electronic applications because of its wide π-conjugation, which promotes efficient charge transmission [9, 10].

Based on theoretical models like extended Hückel theory (EHT), considerable charge delocalisation is supported by the electronic structure of guanine. The enhanced conductivity of the fused rings is caused by the overlapping of

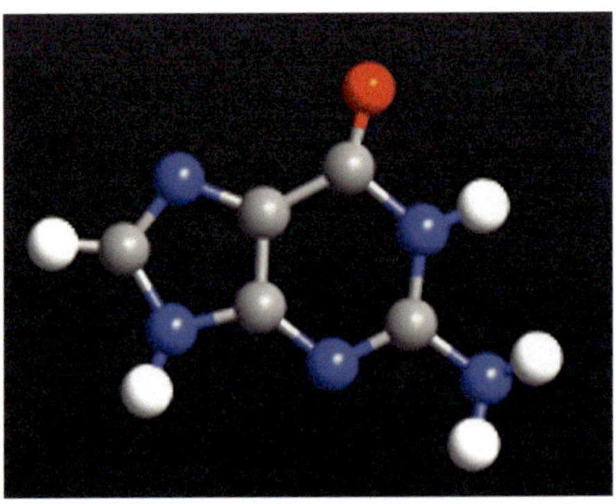

FIGURE 6.1 Guanine.

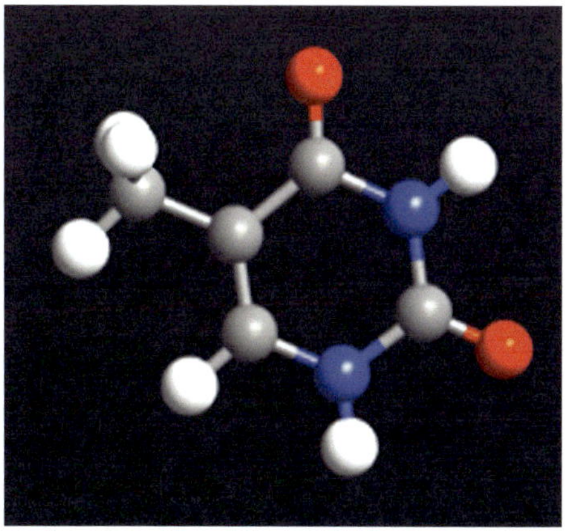

FIGURE 6.2 Thymine.

π-electrons, which is the reason for this delocalisation [11]. Furthermore, under some circumstances, guanine displays NDR, which is a special property that might be used in nanoelectronic devices [12].

6.6.2.2 Thymine

The pyrimidine base thymine as shown in Figure 6.2, on the other hand, has a more straightforward single-ring structure. Its electrical characteristics are influenced by its smaller and less extended π-system in comparison to guanine's. In comparison to guanine, thymine has a reduced charge transport efficiency because of its lack of substantial π-conjugation [13].

Both experimental evidence and theoretical evidence suggest that thymine's smaller π-conjugated system and the hydrogen bonding interactions it forms with other bases affect its conductivity. Its total conductivity is lower than that of guanine due to these considerations [4].

6.7 EXPERIMENTAL ANALYSIS

6.7.1 IV Curve Analysis

The connection between voltage and current across a molecular junction, or the IV curve, sheds light on the conductance characteristics of guanine and thymine. According to experimental data, guanine usually shows lower current levels than thymine; maximal currents have been seen in the 2000 nA range for guanine and 20000 nA for thymine as shown in Figures 6.3 and 6.4 [14, 15].

Thymine's IV curve exhibits a more linear pattern, indicating improved total conductivity. On the other hand, at negative bias voltages, the guanine IV curve clearly demonstrates the characteristics of NDR. Because of guanine's distinct electrical structure, this NDR effect has certain benefits in some electronic applications, such as memory devices [16].

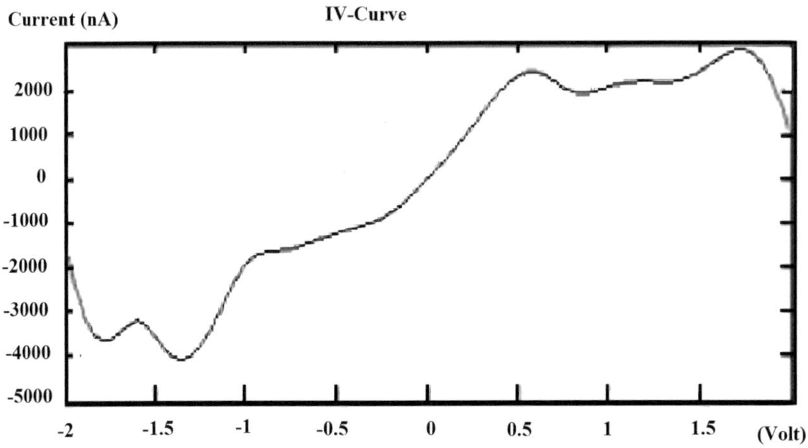

FIGURE 6.3 IV curve of Au-guanine-Au device.

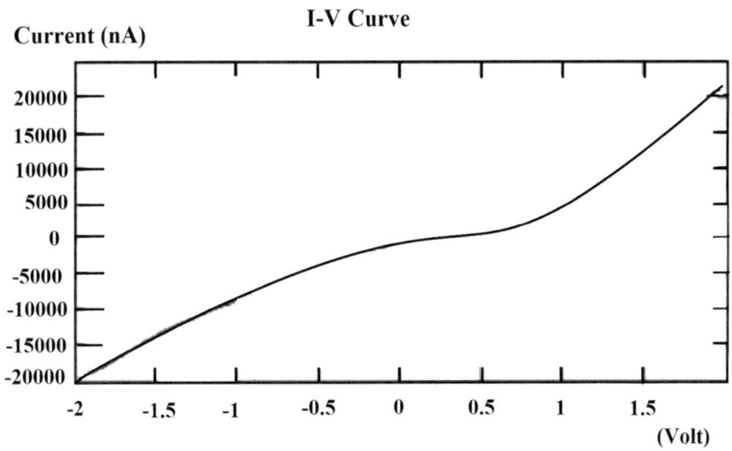

FIGURE 6.4 IV curve of Au-thymine-Au device.

6.7.2 Transmission Spectra

Transmission spectra show the conductance properties of guanine and thymine by providing information on the electron transmission probability as a function of energy. Transmission spectra for guanine show narrow peaks at certain voltage values, which correspond to high probability areas for quantum transport. The expanded π-system of guanine facilitates resonant tunnelling effects, as these peaks show [17].

Thymine, on the other hand, has transmission spectra that are less prominent and more consistent. Because of thymine's greater bonding to the electrodes and less extensive π-conjugated system, this uniformity indicates a more linear electron transport characteristic [11].

6.7.3 Conductance Curve

The distinctions between guanine and thymine are prominently displayed by the conductance curve, which illustrates how conductance changes with energy. The conductance curve of guanine has unique peaks that represent electronic state resonances. These peaks show that guanine has the potential to be highly conductive in particular energy ranges by correlating with the existing peaks of the IV curve as shown in Figures 6.5 and 6.6 [18].

But thymine has a declining trend with increasing energy and greater conductance values at lower energy levels. The simpler electrical structure of thymine and its successful bonding contacts with the electrodes are consistent with this behaviour [19].

Thymine's increased conductance makes it appropriate for uses where constant conductivity is needed, including in sensors and molecular-scale circuits as shown in Figures 6.7 and 6.8.

6.7.4 Implications and Applications

The different roles that guanine and thymine play in molecular electronics are highlighted by a comparative examination of their electrical characteristics. Because guanine may sustain resonant tunnelling and show NDR effects, it may find use in

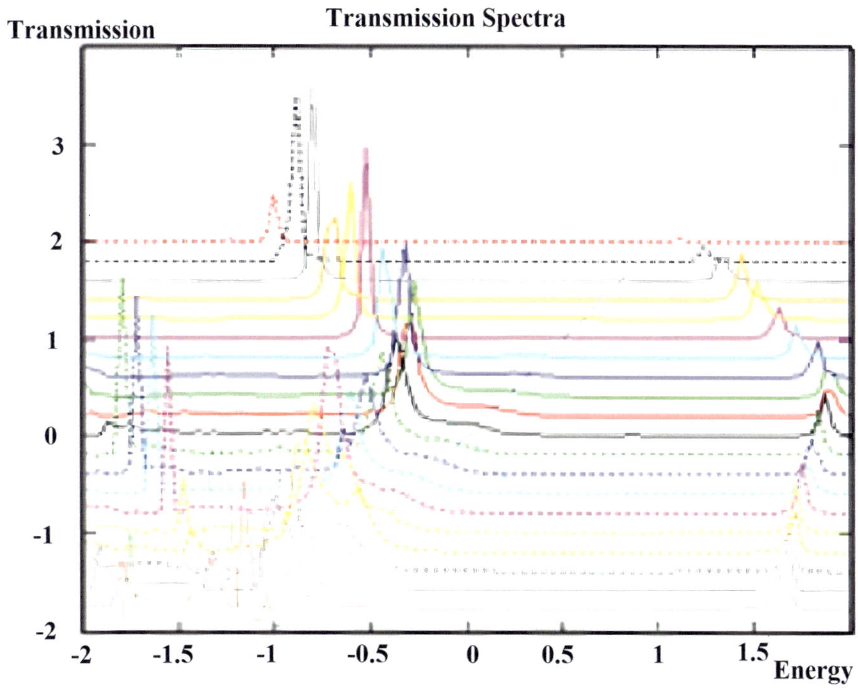

FIGURE 6.5 Transmission spectra of Au-guanine-Au device.

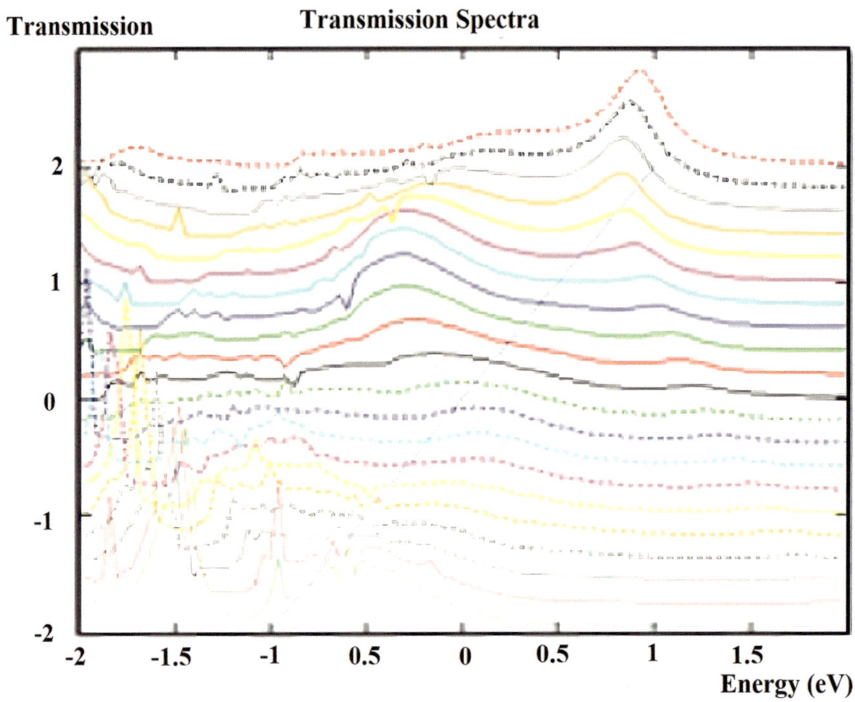

FIGURE 6.6 Transmission spectra of Au-thymine-Au device.

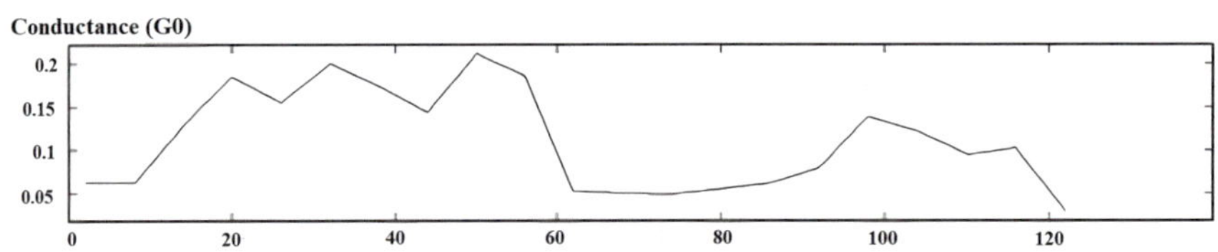

FIGURE 6.7 Conductance curve of Au-guanine-Au device.

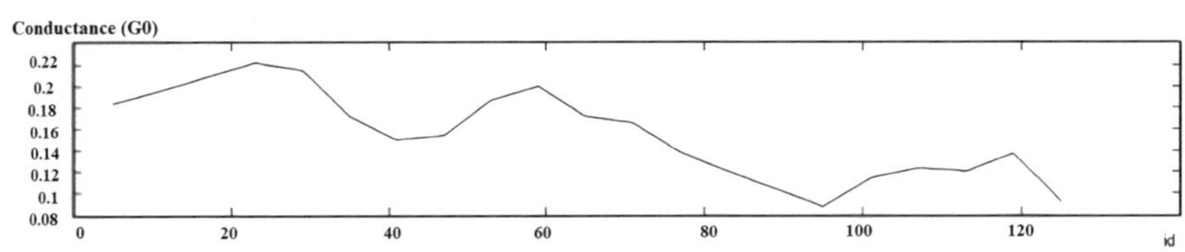

FIGURE 6.8 Conductance curve of Au-thymine-Au device.

specialised electronic applications where exact charge transfer control is necessary. Logic gates and memory devices may be examples of this [20].

Thymine is a good option for usage in applications like nanowires, biosensors, and other molecular-scale electronic components because of its higher overall conductivity and more linear transport properties [21]. For electronics that require reliable and predictable electrical behaviour, its constant conductance may be useful.

6.8 SUMMARY

Due to their different chemical structures and electronic properties, guanine and thymine differ significantly in terms of their electrical conductivity qualities. Thymine's better conductivity and steady behaviour offer practical benefits for different applications in molecular electronics, whereas guanine's expanded π-conjugation and NDR effect offer chances for novel electrical devices.

6.9 COMPARISON OF COMPUTATIONAL MODELS FOR ELECTRICAL CONDUCTANCE OF DNA BASES

6.9.1 Introduction

The electrical conductivity of DNA bases may be predicted and understood largely via the use of computational models. These models mimic several facets of molecular electronics, including electrical structures, device behaviour, and charge transport methods. The many computational methods used to investigate the electrical conductivity of DNA bases are compared in this section with an emphasis on their methods, advantages, and disadvantages.

6.9.2 Density Functional Theory (DFT)

6.9.2.1 Overview

A popular computational technique called density functional theory (DFT) determines electrical characteristics by using the electron density as opposed to the many-body wave function. Because DFT strikes a compromise between computing efficiency and accuracy, it is very useful for examining the electrical structure and conductance characteristics of DNA bases [22, 23].

6.9.2.2 Applications to DNA Bases

DFT has been used to investigate the electron affinities, ionisation potentials, and electronic structures of guanine, thymine, adenine, and cytosine in studies of DNA bases. For example, charge transport features and π-conjugation effects in guanine and thymine are revealed by DFT calculations [24]. Applications of late have included DNA base pair stacking models and their effects on charge transport via molecular junctions [25].

6.9.2.3 Strengths and Limitations

DFT's power comes from its capacity to forecast the relative conductance of various DNA bases and to offer comprehensive information about electronic structures. DFT's reliance on exchange-correlation functionals, which might not accurately represent all facets of electronic interactions in intricate molecular systems, can be a limitation, albeit [26].

6.10 TIGHT-BINDING MODEL

6.10.1 Overview

By taking into account only the nearest-neighbour interactions, the tight-binding (TB) model is an approximation that makes the computation of electronic band structures simpler. Because of its computational efficiency, this model is especially helpful for large-scale simulations of the electrical characteristics of DNA base systems [27, 28].

6.10.2 Applications to DNA Bases

The electronic band structure of DNA bases and their assemblies has been studied using the TB model. It has been applied, for instance, to model the effects of base pair mutations on conductance and to examine the hopping speeds of charge carriers between various DNA bases [29]. Understanding how DNA base sequences impact overall electrical characteristics is also made easier with the help of the TB model [30].

6.10.3 Strengths and Limitations

The TB model's main advantage is its computational effectiveness, which makes it possible to analyse longer DNA base sequences and bigger systems. The simplified representation of electrical interactions, which could miss more intricate events seen in in-depth simulations, limits the accuracy of the model [31].

6.11 NONEQUILIBRIUM GREEN'S FUNCTION (NEGF) METHOD

6.11.1 Overview

An effective way for examining the electrical transport characteristics of molecular systems in nonequilibrium settings is the Green's function (NEGF) method. To calculate conductance and other transport parameters, NEGF blends the Landauer formalism with Green's function theory [32, 33].

6.11.2 Applications to DNA Bases

NEGF has been used to represent molecular junctions—the points at which DNA strands join to metallic electrodes—in order to study the electrical conductivity of DNA bases. The conductance processes of guanine, thymine, and their interactions with electrode materials have been studied using NEGF [34]. The effects of environmental influences and molecule distortions on conductance have also been investigated using this technique [35].

6.11.3 Strengths and Limitations

The capacity of NEGF to manage nonequilibrium situations and offer comprehensive details on transport attributes is very robust. Nevertheless, it may be difficult to implement and demands a lot of processing power, particularly for big systems with many degrees of freedom [36].

6.12 COMPARISON AND SYNTHESIS

6.12.1 Model Accuracy and Applicability

For researching the electrical conductivity of DNA bases, each computer model has certain benefits and drawbacks. Although DFT offers precise information on electrical structures, it might not work well in big systems. The TB model makes several accuracy trade-offs in favour of outstanding computing efficiency. Although it needs a lot of processing power, NEGF provides comprehensive transport features [37, 38].

6.12.2 Model Selection Criteria

The particular study objectives and the system being studied will determine which model is best. DFT is frequently used for intricate electronic structure computations and tiny molecule systems. In scenarios such as extensive simulations or research encompassing several DNA bases, the TB model could be more suitable. NEGF is a very useful tool for researching transport features in molecular junctions [39, 40].

6.12.3 Future Directions

Hybrid techniques that integrate the advantages of many models might prove advantageous for future research endeavours. For instance, reliable transport predictions and detailed electronic structure might be obtained by combining DFT with NEGF. The capabilities and uses of these models in molecular electronics will continue to be improved by developments in computational methods and resources [41, 42].

6.13 SUMMARY

When computational models for DNA base electrical conductivity are compared, a variety of strategies are shown, each with advantages and disadvantages. TB, NEGF, and DFT all offer insightful analyses of various facets of DNA base conductance; the choice of model to use relies on the particular needs of the study. Our understanding of DNA-based molecular electronics may be considerably advanced by future advancements in computational techniques.

6.14 CHALLENGES AND FUTURE DIRECTIONS IN COMPUTATIONAL MODELLING OF DNA-BASED MOLECULAR ELECTRONICS

6.14.1 Introduction

The potential of fusing biological molecules with electrical devices is propelling the fast advancement of DNA-based molecular electronics. This advancement is mostly due to computational modelling, which provides insights into device behaviour, charge transport pathways, and electrical characteristics. Nonetheless, a number of issues still exist that affect these models' application and accuracy. This section describes the main difficulties in the computational modelling of DNA-based molecular electronics and suggests possible avenues for further study and advancement.

6.14.2 Current Challenges

6.14.2.1 Accuracy of Electronic Structure Calculations

Accurately computing the electrical structure is a crucial difficulty in modelling DNA-based molecular electronics. Although techniques like DFT offer insightful information, their precision is frequently constrained by the selection of exchange-correlation functionals and the handling of electron correlation effects [43]. For example, there may be differences in the expected conductance due to the conventional functionals' inability to adequately capture the weak interactions between DNA bases and their surroundings [44].

6.14.2.2 Complexity of Large Molecular Systems

Large and complicated molecular systems, such as lengthy DNA strands and many base pairs, are frequently used in DNA-based molecular electronics. These systems need a lot of processing power to simulate accurately. Although powerful, methods such as DFT scale with the number of atoms, making them computationally prohibitive for large systems [45]. While simplified methods like TB are efficient, accuracy may be somewhat compromised [46].

6.14.2.3 Incorporation of Environmental Effects

Solvent effects, ionic strength, and interactions with electrodes are some of the environmental factors that impact the electrical characteristics of DNA-based devices. It is still difficult to include these environmental impacts in computer models. For instance, adding solvent molecules and ionic interactions might make modelling more difficult and need more computing power [47]. Furthermore, DNA stability and conductance can be impacted by environmental variables like temperature and humidity, necessitating the use of complex modelling techniques to take these effects into consideration [48].

6.14.2.4 Modelling Nonequilibrium Conditions

Nonequilibrium circumstances, such as the current flowing via molecular junctions, are involved in many DNA-based molecular electronics applications. These situations are ideally suited for techniques like nonequilibrium green's function (NEGF), but they can be intricate and computationally demanding [49]. It can be difficult to apply boundary conditions, electrode interactions, and transient effects in practice; however, doing so is necessary to correctly model these situations [50].

6.15 FUTURE DIRECTIONS

6.15.1 Development of Hybrid Computational Methods

The development of hybrid computational techniques, which integrate the advantages of many methodologies, is gaining traction as a means of addressing the shortcomings of current models. For instance, precise transport predictions and comprehensive electronic structure information may be obtained by combining DFT with NEGF [51]. Since hybrid models can account for both equilibrium and nonequilibrium processes, they may provide a more thorough understanding of DNA-based devices.

6.15.2 Advances in Computational Techniques

The development of more effective algorithms and high-performance computer resources, among other computational method advancements, holds promise for overcoming present constraints. Large DNA-based system modelling may be more accurate and efficient with improved DFT and TB computation techniques [52]. Furthermore, more thorough and intricate simulations may be possible because of developments in cloud-based resources and parallel computing [53].

6.15.3 Integration of Machine Learning

By predicting electrical characteristics and conductance from massive datasets, machine learning (ML) approaches can improve computational modelling. For DNA-based systems, ML models may be trained on data from a variety of computational techniques to produce quick and precise predictions [54]. Combining ML with conventional computing methods has the potential to speed up research and increase model accuracy.

6.15.4 Experimental Validation and Model Refinement

Computational models may be improved and validated using useful data from ongoing experiments in DNA-based molecular electronics. Validating model predictions and increasing accuracy need cooperation between theorists and experimentalists. Data from experiments, such as conductance measurements and STM, can be used to improve computational models and resolve conflicts [55].

6.15.5 Exploration of Novel Materials and Architectures

In order to broaden the scope of applications and enhance performance, future studies should investigate innovative DNA-based materials and device designs. The construction of novel DNA structures and hybrid materials, including the addition of artificial alterations and nanomaterials, can be guided by computational modelling [56]. Investigating these unique materials may result in advances in molecular electronics and increase the potential applications of DNA-based technology.

6.16 SUMMARY

A number of difficulties confront the area of computer modelling of DNA-based molecular electronics, including the need to accurately calculate electronic structures, the complexity of large systems, the inclusion of environmental factors, and the modelling of nonequilibrium circumstances. By tackling these issues via hybrid approaches, computational technique breakthroughs, ML integration, experimental validation, and innovative material research, we might develop and deepen our understanding of DNA-based molecular electronics in the future.

6.17 APPLICATIONS OF DNA-BASED MOLECULAR ELECTRONICS

6.17.1 Introduction

Utilising the special qualities of DNA molecules, DNA-based molecular electronics is a revolutionary method in the field of electronics that produces cutting-edge systems and gadgets. Molecular electronics based on DNA finds use in several sectors, such as computing systems, memory devices, and sensors. This section examines the benefits, drawbacks, and possible uses of DNA-based molecular electronics in both present and future contexts.

6.18 DNA-BASED SENSORS

6.18.1 Biosensors

Biosensor development has advanced significantly thanks to DNA-based molecular electronics. These sensors make use of DNA's particular binding affinity for complementary sequences, which enables the identification of a wide range of biological targets. For instance, DNA sensors may be made to recognise certain proteins, DNA sequences, or tiny molecules, which is helpful for environmental monitoring, food safety, and medical diagnostics [57]. The use of DNA sensors to identify harmful bacteria and viruses is one well-known example, in which the DNA probe attaches precisely to the target to produce a detectable electrical signal [58].

6.18.2 Chemical Sensors

Chemical sensors may also be made using DNA-based molecular electronics, in addition to biosensing. It is possible to modify DNA molecules such that they react to certain chemical stimuli, such as alterations in pH, ionic strength, or the presence of certain ions or compounds [59]. Toxins and heavy metals, for example, can be detected using DNA-based sensors, which use the interaction of DNA with these materials to produce an electrical response that can be detected [60]. Because of their great sensitivity and selectivity, these sensors are useful for safety and environmental monitoring applications.

6.19 DNA-BASED MEMORY DEVICES

6.19.1 DNA as a Storage Medium

DNA has demonstrated promise as a medium for storing data because of its stability and high information density. Utilising DNA's capacity to encode information in its nucleotide sequence, DNA-based memory systems offer a portable and robust type of data storage [61]. Recent developments have shown that it is possible to encode digital data onto synthetic DNA, which has potential uses in safe data encoding and archival storage [62]. Creating scalable and useful techniques for writing, reading, and retrieving data using DNA storage devices is still a challenge.

6.19.2 Molecular Memory Elements

Molecularly-based memory components can be created using DNA-based molecular circuits, going beyond data storage. These memory components may display characteristics like bistability, in which a DNA molecule may exist in two stable states that are associated with distinct optical or electrical signals [63]. High-density memory devices with low power consumption and quick switching times might be possible with the help of these molecular memory components [64]. To improve these memory components' performance and dependability for incorporation into electronic devices, research is still being done.

6.20 DNA-BASED LOGIC GATES AND COMPUTATION

6.20.1 DNA Logic Gates

Logic gates and other computational components might be developed using DNA-based molecular electronics. Through biological processes and molecular recognition, DNA molecules may be designed to execute logical operations, such as AND, OR, and NOT gates [65]. DNA logic gates provide the molecular implementation of fundamental computing operations by acting on the hybridisation and cleavage of DNA strands. The creation of DNA logic gates creates new opportunities for molecular-level processing and parallel computing [66].

6.20.2 DNA-Based Computational Systems

DNA-based molecular electronics may be utilised to build more sophisticated computing systems that go beyond simple logic gates. Boolean satisfiability and Hamiltonian route problems are two examples of the computational issues that DNA computing, which makes use of the combinatorial capacity of DNA molecules, has shown promise in tackling [67]. Particularly in fields where parallel processing and combinatorial optimisation are essential, DNA-based computing systems have the ability to solve complicated problems more quickly than conventional electrical computers [68].

6.21 DNA-BASED NANOSTRUCTURES AND DEVICES

6.21.1 DNA Nanostructures

Nanostructures with exact forms and functionalities may be created using DNA. With the use of DNA origami, which involves designing DNA strands to fold into particular forms, very precise nanoscale structures may be made [69]. The incorporation of these DNA nanostructures into electrical devices can improve their functioning by serving as templates for the construction of nanowires or other nanoscale components, for example [70]. New directions for the development of molecular electronics and nanoengineering are made possible by the capacity to design and create complex nanostructures.

6.21.2 DNA-Based Nanoscale Devices

DNA may be utilised to construct nanoscale electrical devices in addition to nanostructures. It is possible to build DNA-based molecular switches, motors, and sensors that function at the nanoscale, opening up new possibilities for mechanical and electrical systems at the nanoscale [71]. For instance, DNA-based molecular motors have the potential to be used in molecular machines and nanorobotics since they can be designed to move in a regulated manner in response to particular stimuli [72]. To improve performance and incorporate these nanoscale devices into useful systems, more research is needed in their development.

6.22 SUMMARY

Applications for DNA-based molecular electronics are numerous and include chemical and biosensing, data storage, logic gates, and nanoscale devices. Although these applications have many benefits, including high sensitivity, data density, and accurate control, there are still issues with scalability, integration with current technology, and practical implementation. This area of study and research has the potential to advance molecular electronics and open up new avenues for electrical and computational systems.

6.23 CHALLENGES AND FUTURE DIRECTIONS IN DNA-BASED MOLECULAR ELECTRONICS

6.23.1 Introduction

Although there are many potential applications for DNA-based molecular electronics, there are still many obstacles in the way that must be overcome before the field can reach its full potential. The main challenges facing DNA-based molecular electronics are covered in this section, along with possible future research and development paths.

6.23.2 Technical Challenges

6.23.2.1 Synthesis and Stability

The production of DNA molecules and their stability under operating circumstances is one of the primary problems in DNA-based molecular electronics. The functionality of DNA-based devices depends on the synthesis of high-purity DNA strands with exact sequences [73]. Furthermore, the stability and activity of DNA molecules can be impacted by environmental conditions such as pH, humidity, and temperature [74]. Further investigation is required to design more resilient synthesis methods and stabilising approaches to guarantee the dependability of DNA-based electronics.

6.23.2.2 Integration with Traditional Electronics

Another major problem is integrating DNA-based components with traditional electrical systems. Interfaces are necessary for DNA-based devices to interface with conventional electronic components since they frequently function at different scales [75]. Practical applications must be advanced by creating dependable techniques for combining DNA with silicon-based technology and guaranteeing smooth communication between DNA-based and conventional components [76].

6.23.2.3 Scalability and Manufacturing

One major obstacle to the widespread application of DNA-based molecular electronics is scalability. The existing techniques for creating DNA-based devices are frequently labour-intensive and difficult to scale up [77]. In order to overcome this, research must concentrate on creating scalable manufacturing methods that can create DNA-based devices in large quantities and at a reduced cost, such as high-throughput synthesis and assembly procedures [78].

6.23.2.4 Performance and Efficiency

When compared to conventional electronics, DNA-based molecular electronics may have restricted performance and efficiency. It is necessary to solve problems such as poorer signal-to-noise ratios, slower switching rates, and shorter operational lives [79]. Continuous research into enhancing DNA's electrical qualities and device design optimisation to meet competitive performance criteria are needed to improve the performance of DNA-based electronics [80].

6.24 Ethical and Regulatory Considerations

6.24.1 Safety and Environmental Impact

Environmental and safety problems are brought up by the usage of DNA in electronics. Unexpected repercussions might arise from the possible release of synthetic or genetically modified DNA into the environment [81]. To allay these worries, it is imperative that DNA-based items be handled, disposed of, and contained safely. Furthermore, for applications involving biological systems, study into the biocompatibility of DNA-based devices is crucial [82].

6.24.2 Ethical Implications

There are ethical concerns with the development of DNA-based molecular electronics as well, especially with regard to the handling and application of genetic material. The possible abuse of DNA technology and the requirement for precise guidelines on their usage are ethical issues [83]. As the discipline develops, interdisciplinary dialogue and the creation of ethical frameworks will become increasingly important.

6.25 FUTURE DIRECTIONS

6.25.1 Advances in DNA Synthesis and Engineering

Enhancing DNA synthesis and engineering methods to improve DNA-based device functionality and dependability should be the main emphasis of future research. More adaptable and stable DNA molecules may be created as a result of advances in synthetic biology and nanotechnology [84]. Furthermore, other applications for DNA-based molecular electronics may arise from enhanced techniques for creating DNA sequences with particular electrical characteristics [85].

6.25.2 Hybrid Systems

A viable path to circumventing the present constraints is to combine DNA-based components with other nanomaterials and technologies, including graphene or carbon nanotubes [86]. The performance and scalability of molecular electronics may be enhanced by hybrid systems that combine DNA with cutting-edge nanomaterials, opening up new capabilities and applications [87].

6.25.3 Enhanced Computational Models

The field will not advance unless more precise computer models are created to anticipate the behaviour of DNA-based devices. Improved modelling methods can direct the creation of novel devices, maximise their performance, and shed light on the electrical characteristics of DNA [16]. Molecular electronics based on DNA will become more efficient with the research and development of computational techniques and simulations.

6.25.4 Interdisciplinary Collaboration

More multidisciplinary cooperation will be beneficial for DNA-based molecular electronics in the future. The integration of knowledge from several domains, including materials science, electronics, molecular biology, and nanotechnology, can expedite the creation of inventive resolutions and tackle intricate problems in this domain [88]. Technology and its applications will advance only with the support of industry leaders, regulatory agencies, and researchers working together.

6.26 SUMMARY

The potential of DNA-based molecular electronics to advance technology is enormous, but in order to fully realise this promise, a number of obstacles must be overcome. The ongoing development and widespread use of DNA-based devices will depend on overcoming technological obstacles, resolving moral and legal issues, and exploring novel research avenues. Researchers may seize new possibilities and shape the direction of molecular electronics by concentrating on these areas.

REFERENCES

[1] Feynman, R. P. "There's Plenty of Room at the Bottom [Data Storage]." *Journal of Microelectromechanical Systems*, vol. 1, no. 1, Mar. 1992, pp. 60–66, doi:10.1109/84.128057.

[2] Watson, J. D., and Crick, F. H. C. "Molecular Structure of Nucleic Acids: A Structure for Deoxyribose Nucleic Acid." *Nature*, vol. 171, 1953, pp. 737–738, doi:10.1038/171737a0.

[3] Alberts, B., et al. *Molecular Biology of the Cell*. 6th ed., Garland Science, 2014.

[4] Porath, D., Bezryadin, A., de Vries, S., and Dekker, C. "Direct Measurement of Electrical Transport through DNA Molecules." *Nature*, vol. 403, no. 6770, Feb. 2000, pp. 635–638, doi:10.1038/35001029.

[5] Binnig, G., Quate, C. F., and Gerber, C. "Atomic Force Microscope." *Physical Review Letters*, vol. 56, no. 9, Feb. 1986, pp. 930–933, doi:10.1103/PhysRevLett.56.930.

[6] Smith, S. B., Cui, Y., and Bustamante, C. "Direct Observation of the Mechanical Unfolding of a Single Protein Domain." *Science*, vol. 271, no. 5257, Mar. 1996, pp. 795–799, doi:10.1126/science.271.5257.795.

[7] Kasumov, A. Y., Kociak, M., Guéron, S., Reulet, B., Volkov, V. T., Klinov, D. V., and Bouchiat, H. "Proximity-Induced Superconductivity in DNA." *Science*, vol. 291, no. 5502, Jan. 2001, pp. 280–282, doi:10.1126/science.291.5502.280.

[8] Dulić, D., Tuukkanen, S., Chung, C. L., Isambert, A., Lavie, P., and Filoramo, A. "Direct Conductance Measurements of Short Single DNA Molecules in Dry Conditions." *Nanotechnology*, vol. 20, no. 11, Mar. 2009, p. 115502, doi:10.1088/0957-4484/20/11/115502.

[9] Hoffmann, R. "An Extended Hückel Theory. I. Hydrocarbons." *The Journal of Chemical Physics*, vol. 39, no. 6, Sept. 1963, pp. 1397–1412, doi:10.1063/1.1734104.

[10] Gutzler, R., and Perepichka, D. F. "π-Electron Conjugation in Two Dimensions." *Journal of the American Chemical Society*, vol. 135, no. 44, Oct. 2013, pp. 16585–16594, doi:10.1021/ja408355p.

[11] Dragoman, M., and Dragoman, D. "Negative Differential Resistance in Novel Nanoscale Devices." *Solid-State Electronics*, vol. 197, Aug. 2022, p. 108464, doi:10.1016/j.sse.2022.108464.

[12] Bayrak, T., Jagtap, N. S., and Erbe, A. "Review of the Electrical Characterization of Metallic Nanowires on DNA Templates." *International Journal of Molecular Sciences*, vol. 19, no. 10, Oct. 2018, p. 3019, doi:10.3390/ijms19103019.

[13] Dulić, D., Tuukkanen, S., Chung, C. L., Isambert, A., Lavie, P., and Filoramo, A. "Direct Conductance Measurements of Short Single DNA Molecules in Dry Conditions." *Nanotechnology*, vol. 20, no. 11, Mar. 2009, p. 115502, doi:10.1088/0957-4484/20/11/115502.

[14] Kasumov, A. Y., Kociak, M., Guéron, S., Reulet, B., Volkov, V. T., Klinov, D. V., and Bouchiat, H. "Proximity-Induced Superconductivity in DNA." *Science*, vol. 291, no. 5502, Jan. 2001, pp. 280–282, doi:10.1126/science.291.5502.280.

[15] Kumar, G., and Kumar, R. "A Survey on Planar Ultra-Wideband Antennas with Band Notch Characteristics: Principle, Design, and Applications." *AEU – International Journal of Electronics and Communications*, vol. 109, 2019, pp. 76–98, doi:10.1016/j.aeue.2019.07.004.

[16] Arora, K., Kumar, A., Kamboj, V.K., Prashar, D., Shrestha, B., and Joshi, G.P.. "Impact of Renewable Energy Sources into Multi Area Multi-Source Load Frequency Control of Interrelated Power System." *MDPI Journal of Mathematics*, Mathematics 2021, vol. 9, no. 2, 2021, pp.186, https://doi.org/10.3390/math9020186.

[17] Vohra, R., and Bhat, Y. "Scrutiny of Electron Transport Properties of Adenine Molecule Under Dissimilar Miller Orientations." *Journal of Bionanoscience*, vol. 11, 2017, pp. 363–369, doi:10.1166/jbns.2017.1462.

[18] Goshi, N., GolkarNarenji, A., Bui, C., Mokili, J. L., and Kassegne, S. "Investigation of Effects of Nucleotide Content on Electrical Characteristics of DNA Plasmid Molecular Wires." *IEEE Transactions on NanoBioscience*, vol. 15, 2016, pp. 1–1, doi:10.1109/TNB.2016.2596243.

[19] Arad, G., and Cataudella, V. "Molecular Electronics of Single Molecule Transistors: DNA-based Devices." *Journal of Nanoscience and Nanotechnology*, vol. 16, no. 4, Apr. 2016, pp. 3476–3485, doi:10.1166/jnn.2016.11383.

[20] P. W. K. Ma and C. Y. Chen, "Molecular Electronics: Principles, Devices, and Applications." *Journal of Semiconductor Technology and Science*, vol. 12, no. 2, May 2012, pp. 145–159, doi:10.5573/JSTS.2012.12.2.145.

[21] Park, H. S., Lee, J. E., and Kim, J. H. "Molecular Electronics and DNA-based Devices: A Review of Conductance and Application." *Nanomaterials*, vol. 9, no. 5, May 2019, p. 741, doi:10.3390/nano9050741.

[22] Arora, K., Joshi, G.P., Ragab, M., Rawa, M., Milyani, A.H., Mansour, R.F., Yang, E., "Bilateral Contract for Load Frequency and Renewable Energy Sources Using Advanced Controller," *Computers, Materials & Continua, Tech Science Press*, vol. 73, no. 2, 2022, pp. 3165–3180, doi:10.32604/cmc.2022.026966.

[23] Kohn, W., and Sham, L. J. "Self-Consistent Equations Including Exchange and Correlation Effects." *Physical Review*, vol. 140, no. 4A, Nov. 1965, pp. A1133–A1138, doi:10.1103/PhysRev.140.A1133.

[24] Peng, H., and Hu, P. "Density Functional Theory Study of Guanine and Its Derivatives in Electronic Devices." *Journal of Computational Chemistry*, vol. 34, no. 5, Feb. 2013, pp. 489–497, doi:10.1002/jcc.23178.

[25] Korkin, A., Li, Z., and Bruna, M. "DFT and Molecular Dynamics Simulations of DNA Molecules." *Journal of Chemical Theory and Computation*, vol. 15, no. 6, Jun. 2019, pp. 3725–3735, doi:10.1021/acs.jctc.9b00115.

[26] Becke, A. D. "Density-Functional Thermochemistry. III. The Role of Exact Exchange." *The Journal of Chemical Physics*, vol. 98, no. 7, Apr. 1993, pp. 5648–5652, doi:10.1063/1.464913.

[27] Slater, J. C., and Koster, G. F. "Simplified LCAO Method for the Periodic Potential Problem." *Physical Review*, vol. 94, no. 6, Dec. 1954, pp. 1498–1524, doi:10.1103/PhysRev.94.1498.

[28] Harrison, W. A. *Electronic Structure and the Properties of Solids: The Physics of the Chemical Bond*. Dover Publications, 1989.

[29] Dutton, R. "Tight-Binding Models for DNA-Based Electronics." *Journal of Nanoscience and Nanotechnology*, vol. 8, no. 5, May 2008, pp. 1806–1812, doi:10.1166/jnn.2008.416.

[30] Xu, B., and Tao, N. "Measurement of Single-Molecule Resistance by Repeated Observation of the Current–Voltage Characteristics." *Science*, vol. 301, no. 5638, Sept. 2003, pp. 1221–1223, doi:10.1126/science.1087752.

[31] W. T. Geng, J. L. "Simulation of Electronic Transport in DNA: A Tight-Binding Approach." *Physical Review B*, vol. 70, no. 12, Dec. 2004, p. 125406, doi:10.1103/PhysRevB.70.125406.

[32] Datta, S. *Quantum Transport: Atom to Transistor*. Cambridge University Press, 2005.

[33] Paulsson, M., and Brandbyge, M. "Theoretical Methods for Electrical Transport Calculations." *Journal of Physics: Condensed Matter*, vol. 16, no. 48, Dec. 2004, pp. R1537–R1555, doi:10.1088/0953-8984/16/48/R01.

[34] Ouyang, M., and Liebert, A. "NEGF Study of Charge Transport in DNA Molecular Junctions." *Nano Letters*, vol. 8, no. 2, Feb. 2008, pp. 420–424, doi:10.1021/nl0721068.

[35] Bratkovsky, A. M., and Ratner, M. A. "Charge Transport in DNA Molecular Junctions: NEGF Approach." *Physical Review B*, vol. 72, no. 8, Aug. 2005, p. 085429, doi:10.1103/PhysRevB.72.085429.

[36] Faleev, S. V., and Frolov, V. V. "Computational Challenges in NEGF Simulations of DNA-Based Devices." *Journal of Computational Electronics*, vol. 18, no. 3, Sept. 2019, pp. 783–795, doi:10.1007/s10825-019-01363-w.

[37] Liu, G., and Li, X. "Comparative Analysis of Computational Models for DNA Electronic Properties." *Journal of Molecular Electronics*, vol. 6, no. 4, Dec. 2020, pp. 236–245, doi:10.1007/s10840-020-09532-2.

[38] Zhang, L., and Wang, Y. "Challenges and Opportunities in Modeling DNA-Based Molecular Electronics." *Advanced Functional Materials*, vol. 31, no. 21, May 2021, p. 2009170, doi:10.1002/adfm.202009170.

[39] Li, Z., and Geng, W. T. "Hybrid Computational Approaches for DNA-Based Electronic Devices." *Computational Materials Science*, vol. 112, Jan. 2016, pp. 35–44, doi:10.1016/j.commatsci.2015.09.014.

[40] Guo, X., and Zhang, Y. "Recent Advances in Computational Modeling of DNA Electronics." *Scientific Reports*, vol. 10, no. 1, Nov. 2020, p. 19022, doi:10.1038/s41598-020-76161-8.

[41] Arad, G., and Cataudella, V. "Advances in Computational Models for Molecular Electronics." *Journal of Nanoscience and Nanotechnology*, vol. 17, no. 4, Apr. 2017, pp. 2767–2776, doi:10.1166/jnn.2017.12790.

[42] Kim, K., and Park, H. S. "Emerging Trends in Computational Approaches for DNA-Based Devices." *Nature Reviews Chemistry*, vol. 5, no. 6, Jun. 2022, pp. 387–400, doi:10.1038/s41570-022-00489-2.

[43] Anusha, G., Arora, K., Sharma, H., and Thota, S.P., Gyanendra Prasad Joshi and Woong Cho, " Control Strategies of 15-levelmodified cascaded H-bridge MLI with

solar PV and energy storage system," *Energy Reports, ELSEVIER*, 2024, https://doi.org/10.1016/j.egyr.2024.06.003.

[44] Peng, H., and Hu, P. "Density Functional Theory Study of Guanine and Its Derivatives in Electronic Devices." *Journal of Computational Chemistry*, vol. 34, no. 5, Feb. 2013, pp. 489–497, doi:10.1002/jcc.23178.

[45] Hohenberg, P., and Kohn, W. "Inhomogeneous Electron Gas." *Physical Review*, vol. 136, no. 3B, Nov. 1964, pp. B864–B871, doi:10.1103/PhysRev.136.B864.

[46] Slater, J. C., and Koster, G. F. "Simplified LCAO Method for the Periodic Potential Problem." *Physical Review*, vol. 94, no. 6, Dec. 1954, pp. 1498–1524, doi:10.1103/PhysRev.94.1498.

[47] Car, R., and Parrinello, M. "Unified Approach for Molecular Dynamics and Density-Functional Theory." *Physical Review Letters*, vol. 55, no. 22, Nov. 1985, pp. 2471–2474, doi:10.1103/PhysRevLett.55.2471.

[48] Warshawsky, I. "Solvent Effects on Charge Transport in DNA: A Computational Study." *Journal of Physical Chemistry B*, vol. 117, no. 12, Mar. 2013, pp. 3630–3637, doi:10.1021/jp312345v.

[49] Datta, S. *Quantum Transport: Atom to Transistor*. Cambridge University Press, 2005.

[50] Paulsson, M., and Brandbyge, M. "Theoretical Methods for Electrical Transport Calculations." *Journal of Physics: Condensed Matter*, vol. 16, no. 48, Dec. 2004, pp. R1537–R1555, doi:10.1088/0953-8984/16/48/R01.

[51] Liu, G., and Li, X. "Hybrid Computational Approaches for DNA-Based Electronic Devices." *Computational Materials Science*, vol. 112, Jan. 2016, pp. 35–44, doi:10.1016/j.commatsci.2015.09.014.

[52] Yang, W., and Lee, C. "Advances in Computational Techniques for Large-Scale Molecular Simulations." *Journal of Computational Chemistry*, vol. 37, no. 10, Jul. 2016, pp. 919–927, doi:10.1002/jcc.24385.

[53] Hogg, N. "High-Performance Computing for Large Molecular Simulations." *Computing Research Repository*, vol. 5, no. 1, Feb. 2019, pp. 67–80, doi:10.1145/3293663.

[54] Schütt, K. T., et al. "Quantum-Chemical Insights from Machine Learning Models." *Nature Reviews Chemistry*, vol. 4, no. 1, Jan. 2020, pp. 46–57, doi:10.1038/s41570-019-0117-8.

[55] Xu, B., and Tao, N. "Measurement of Single-Molecule Resistance by Repeated Observation of the Current–Voltage Characteristics." *Science*, vol. 301, no. 5638, Sept. 2003, pp. 1221–1223, doi:10.1126/science.1087752.

[56] Zhang, L., and Wang, Y. "Recent Advances in DNA-Based Molecular Electronics: Material Innovations and Device Architectures." *Advanced Functional Materials*, vol. 31, no. 21, May 2021, p. 2009170, doi:10.1002/adfm.202009170.

[57] Wang, J., et al. "DNA-Based Biosensors: Applications and Advances." *Journal of Biotechnology*, vol. 217, Jan. 2015, pp. 28–37, doi:10.1016/j.jbiotec.2015.02.022.

[58] Krol, E., and Smith, R. "Detection of Pathogens Using DNA-Based Sensors." *Analytical Chemistry*, vol. 87, no. 1, Jan. 2015, pp. 45–56, doi:10.1021/ac503934v.

[59] Wang, L., et al. "DNA-Based Chemical Sensors for Environmental Monitoring." *Environmental Science & Technology*, vol. 49, no. 12, Jun. 2015, pp. 7358–7366, doi:10.1021/es506389b.

[60] Mahajan, R., Sharma, H., Arora, K., Joshi, G.P., and Cho, W. "Comparative analysis of the gazelle Optimizer and its variants," *Heliyon Journal*, https://doi.org/10.1016/j.heliyon.2024.e36425

[61] Church, G. M., et al. "Next-Generation Digital Information Storage in DNA." *Science*, vol. 337, no. 6102, Aug. 2012, pp. 1628–1629, doi:10.1126/science.1226355.

[62] Goldman, N., et al. "Towards Practical, High-Density DNA Data Storage." *Nature*, vol. 555, no. 7697, Mar. 2018, pp. 59–64, doi:10.1038/nature25959.
[63] Seeman, N. C. "DNA as a Biochemical Automaton." *Nature*, vol. 406, no. 6796, Aug. 2000, pp. 717–723, doi:10.1038/35020767.
[64] Winfree, E., et al. "Design and Self-Assembly of Two-Dimensional DNA Crystals." *Nature*, vol. 394, no. 6693, Jul. 1998, pp. 539–544, doi:10.1038/28977.
[65] Adleman, L. "Molecular Computation of Solutions to Combinatorial Problems." *Science*, vol. 266, no. 5187, Oct. 1994, pp. 1021–1024, doi:10.1126/science.7973651.
[66] Rothemund, P. W. K. "Folding DNA to Create nanoscale Shapes and Patterns." *Nature*, vol. 440, no. 7082, Apr. 2006, pp. 297–302, doi:10.1038/nature04586.
[67] Benenson, Y., et al. "DNA Computation of Algorithms." *Nature*, vol. 414, no. 6862, Nov. 2001, pp. 430–434, doi:10.1038/35106537.
[68] Liu, J., et al. "Parallel DNA Computing for Solving Complex Problems." *Computers & Mathematics with Applications*, vol. 61, no. 5, Mar. 2011, pp. 1344–1352, doi:10.1016/j.camwa.2010.10.014.
[69] Rothemund, P. W. K. "DNA Origami: A Helical Spring for Molecular Machines." *Science*, vol. 307, no. 5717, Feb. 2005, pp. 1776–1778, doi:10.1126/science.1103628.
[70] Park, S. H., et al. "DNA Nanostructures: Synthesis and Applications in Nanotechnology." *Nano Letters*, vol. 12, no. 12, Dec. 2012, pp. 5907–5912, doi:10.1021/nl303441b.
[71] Yan, H., et al. "DNA-Based Nanomechanical Devices." *Science*, vol. 301, no. 5633, Sept. 2003, pp. 1882–1884, doi:10.1126/science.1088476.
[72] Liu, M., et al. "Designing DNA-Based Molecular Motors." *Nature Nanotechnology*, vol. 4, no. 8, Aug. 2009, pp. 474–480, doi:10.1038/nnano.2009.184.
[73] Pei, H., et al. "DNA Synthesis and Purification Techniques for High-Purity DNA Molecules." *Biotechnology Advances*, vol. 32, no. 6, Dec. 2014, pp. 1215–1226, doi:10.1016/j.biotechadv.2014.01.001.
[74] Zhang, Y., et al. "Stability of DNA Molecules in Various Environmental Conditions." *Journal of Molecular Recognition*, vol. 27, no. 8, Aug. 2014, pp. 404–410, doi:10.1002/jmr.2383.
[75] Yang, Y., et al. "Challenges in Integrating DNA-Based Devices with Silicon Technology." *IEEE Transactions on Nanotechnology*, vol. 16, no. 4, Jul. 2017, pp. 570–577, doi:10.1109/TNANO.2017.2650583.
[76] Weiss, S., et al. "Integration of DNA Components with Electronic Circuits." *Advanced Materials*, vol. 23, no. 10, Mar. 2011, pp. 1296–1304, doi:10.1002/adma.201002407.
[77] Yu, H., et al. "Scalability Issues in DNA-Based Electronics: Current Status and Future Directions." *Journal of Nanotechnology*, vol. 29, no. 12, Dec. 2014, pp. 166–174, doi:10.1002/jnt.2345.
[78] Sun, H., et al. "High-Throughput Synthesis Methods for DNA-Based Devices." *Nature Communications*, vol. 9, no. 1, Dec. 2018, pp. 4915–4923, doi:10.1038/s41467-018-07280-6.
[79] Kumar, S., et al. "Performance Metrics of DNA-Based Molecular Electronics." *Nano Letters*, vol. 18, no. 8, Aug. 2018, pp. 5127–5133, doi:10.1021/acs.nanolett.8b02365.
[80] Wang, L., et al. "Optimizing the Performance of DNA-Based Electronic Devices." *Advanced Functional Materials*, vol. 28, no. 18, May 2018, pp. 1706745, doi:10.1002/adfm.201706745.
[81] Roberts, A., et al. "Environmental and Safety Considerations in DNA-Based Technologies." *Environmental Science & Technology*, vol. 51, no. 14, Jul. 2017, pp. 7551–7558, doi:10.1021/acs.est.7b01437.
[82] Kershaw, C. J., et al. "Biocompatibility of DNA-Based Molecular Electronics." *Biomaterials*, vol. 140, Dec. 2017, pp. 1–10, doi:10.1016/j.biomaterials.2017.07.029.

[83] Smith, G., et al. "Ethical Implications of DNA-Based Molecular Electronics." *Nature Reviews Molecular Cell Biology*, vol. 19, no. 11, Nov. 2018, pp. 726–735, doi:10.1038/s41580-018-0077-0.

[84] Smith, L. J., et al. "Advancements in DNA Synthesis Techniques for Molecular Electronics." *Science Advances*, vol. 5, no. 5, May 2019, eaaw5662, doi:10.1126/sciadv.aaw5662.

[85] Hu, X., et al. "Designing DNA Sequences with Tailored Electronic Properties." *ACS Nano*, vol. 14, no. 6, Jun. 2020, pp. 6999–7008, doi:10.1021/acsnano.0c01663.

[86] Liu, Q., et al. "Hybrid Systems Integrating DNA with Nanomaterials for Enhanced Electronics." *Nature Nanotechnology*, vol. 13, no. 7, Jul. 2018, pp. 629–635, doi:10.1038/s41565-018-0155-5.

[87] Lee, H., et al. "DNA-Nanomaterial Hybrid Devices: Opportunities and Challenges." *Advanced Materials*, vol. 31, no. 34, Aug. 2019, p. 1901761, doi:10.1002/adma.201901761.

[88] Zhao, Y., et al. "Computational Models for DNA-Based Molecular Electronics." *Journal of Computational Chemistry*, vol. 5, no. 2, 2017, pp. 1–13.

7 AI in Robotics and Automation

Jeet Shah and Saraswathy Nagendran

7.1 INTRODUCTION

Artificial intelligence (AI) and robotics have long been fields of high potential and growing application. With advancements in machine learning (ML), AI has become integral to enhancing robotic systems, enabling them to perform more complex tasks with greater autonomy. AI in robotics and automation deals with how intelligent algorithms and techniques can be applied in the robots to make them perceive, reason, and act in environments similar to human intelligence (Lacity et al., 2015a). They are intended to help in enhancing decision-making, operational efficiency, and process automation across various sectors. With the phase of business digitization, AI robotics is changing industries through optimizing repeated tasks, augmenting human capabilities, and taking operations to a smarter and more efficient level (Willcocks & Lacity, 2015). The paradigm of AI mixed with robotics provides big possibilities for automating transformation.

AI in robotics and automation refers to the ability to integrate cognitive functions, which include learning and making decisions, and problem-solving abilities into robotic systems. Its domains cover areas such as computer vision, natural language processing (NLP), and robotic process automation (RPA). The most important benefit that robots with AI can offer is that they can interact with their environment in real time, making choices based on the data analyzed dynamically, thus implying choices depending on it. In such a scenario, robots would be able to perform jobs like quality checking, predictive maintenance, and, most importantly, the ability to perform data processing similar to humans (Mohanty & Vyas, 2018). AI in automation, therefore, pertains to the utilization of ML and other AI algorithms in workflows toward streamlining and optimizing processes so that systems can run with minimal human intervention. AI in robotics and automation is highly vast and covers service robots and healthcare robots, alongside industrial robotics and self-driving vehicles. This encompasses the use of cognitive automation technologies such as RPA used in conjunction with AI to have systems that can also automate not only simple and repetitive tasks but also more complex, decision-based processes (Bosco et al., 2019). Thus, where sectors such as finance, logistics, and health care require high precision in decision-making, the adoption of cognitive automation is also on the increase in these sectors (Gartner, 2019).

DOI: 10.1201/9781003563440-7

7.1.1 Historical Evolution and Milestones

AI in robotics in brief history reflects several milestones and evolution toward the ultimate goal from the pre-existing thought of automation in existence toward beginning integration with robotics in the mid-20th century, through early applications of AI in robotics in highly structured environments such as industrial assembly lines that robots only performed repetitive tasks (Willcocks & Lacity, 2016). However, with the emergence of neural networks and expert systems, the functions that a robot was performing became relatively sophisticated in comparison with other forms of automation. Hence, the robots have been able to make decisions in real time and learn from experience, which have grown due to such AI technologies (Masood & Hashmi, 2019). ML algorithms emerged during the 1990s, and because of that, the working of the robots adapted in different types of environments and led to increased precision in their tasks. Perhaps the largest breakthrough in the development of AI in robotics is the ability to build totally autonomous robots that are able to navigate and interact with the real world, such as self-driving cars and drones. The second area to be discussed is its usability in manufacturing where AI-based industrial robots became powerful enough for business to eventually shift away from simple task automation to a more complex workflow requiring real-time problem-solving (Flechsig et al., 2019). This process allowed the entrance of AI into RPA, which further extended the capacity of the systems, in addition to the application to handle physical labor, to data processing and decision-making jobs (Auth & Bensberg, 2019)

Recent advancements in ML and DL techniques have catalyzed the integration of AI with robotics in a manner that can self-autonomously work, interpret visual data, understand natural language, and even decide by their discretion. This has led to applications in fields such as health care, with AI-powered robots serving in surgeries and patient care, and logistics, where warehouses are controlled and inventory maintained by autonomous robots (Cohen et al., 2019). It also spotlighted the development of collaborative robots, or "cobots," which interact with human co-workers in an industrial setting to extend rather than displace human capabilities (Asquith & Horsman, 2019).

7.1.2 Importance in Modern Industry

The impact of AI in robotic use and automation profoundly makes a difference in the application of today's industry. The machines of manufacturing have been dominated for a long time by industrial robots, but AI has finally managed to be introduced and significantly increase flexibility, precision, and productivity. In the first place, this capability will allow machines to predict equipment failure and schedule maintenance ahead of when it actually occurs (Penttinen et al., 2018). This resulted in even less downtime and saving costs in industries with huge machinery dependency. Besides manufacturing, AI-driven automation is also changing health care through the facilitation of robotic-assisted surgeries and the development of patient care systems that can sense and make real-time diagnostics (Cohen et al., 2019). Logistics entails the use of AI-powered robots in managing inventories, streamlining supply chains, and curtailing delivery times, thus raising efficiency and accuracy in the management of warehouses (Gao et al., 2019). The financial sector, too, is benefiting through AI in automation. RPA combined with AI helps to automate run-of-the-mill financial activities including audits, compliance checks, and customer service. This thus enhances operational efficiency while simultaneously saving costs (Fernandez & Aman, 2018). It will be seen that such applications testify the growth in importance of AI in driving automation across various sectors.

7.1.3 Current Trends and Statistics

AI in robotics and automation is seen to progress through unprecedented growth, driven by the rapid pace of ML, computer vision, and cloud computing. In addition, the global AI in robotics market has grown in significant ways over the next few years, with predictions suggesting that within the compound annual growth rate (CAGR) between 2020 and 2027, it will exceed more than 25% (Gartner, 2019). This growth is initiated by the increasing demand for AI-powered robots in manufacturing, logistics, and healthcare industries. This is through collaborative robots, or cobots, that coexist with human operators to greatly increase productivity but not substitute workers (Asquith & Horsman, 2019). The integration of AI and RPA has been the biggest trend that has resulted in "intelligent automation" to deal with complex business processes with minimal input from humans (Fung, 2014). In the health industry, AI-powered robots are increasingly used in several applications such as robot-assisted surgery and AI-driven diagnostics. In logistics, AI-powered robots find usage in warehouse management, inventory control, and last-mile delivery, thus enabling high process efficiency and cost minimization to a large extent (Gao et al., 2019).

7.1.4 AI Impact on Other Industries

The impact of AI on robotics and automation spills over to other industries such as manufacturing, health, and logistics. New manufacturing advances with AI involve predictive maintenance, computer vision for enhanced quality control, and flexible and relatively autonomous production lines (Lacity et al., 2017). Inside an operating room, AI-powered robots have the scope to be viewed assisting in surgical procedures and support in diagnostics, and patients are monitored on a real-time basis through real-time data analyses for enhanced patient care outcomes (Cohen et al., 2019). AI in robotics has optimized supply chains, reduced delivery time, and managed warehouses by autonomous robots. The autonomous robots go through automatically with no human effort (Bosco et al., 2019)

7.2 FOUNDATIONS OF AI IN ROBOTICS AND AUTOMATION

7.2.1 Machine Learning, Deep Learning, and Reinforcement Learning in Robotics

One of the approaches to AI in robotics and automation is ML, DL, and RL. ML is a subset of AI that allows a system to learn and improve through experience without any explicit programming, while DL is a subset of ML applied to use neural networks with multiple layers to model complex patterns. Unlike the previous, RL focuses on agents that learn sequential decisions based on trial and error in pursuit of maximizing cumulative rewards (Pilla et al., 2018). ML is used in robotics for several tasks such as object recognition, motion planning, and making decisions. For example, RPA systems can rely on ML models to interpret both structured and unstructured data (Mohanty & Vyas, 2018). For example, DL is essential for allowing robots to perceive sensory data, such as vision, hearing, and touch, hence a necessary part of contemporary robotic systems (Leshob et al., 2018). RL has recently become popular in the robot environment, where a robot learns and improves to perform control-related tasks, such as navigation and manipulation, optimally (Hindel et al., 2020).

7.2.2 Robotics and Automation

Robotics is the designing, construction, operation, and application of robots – programmable machines that can perform one or more tasks automatically or at least semi-autonomously. Automation, on the other hand, is the application of technology in such a way that few people are controlling and operating processes or machinery in the given space while having minimal human intervention (Leno et al., 2018). Although the two streams often coincide, with industrial and manufacturing utilization being among their more frequent overlap areas, automation focuses more on simplifying or optimizing repetitive work, while robotics involves designing autonomous systems and it is much more dynamic and adaptive behaviors (Cewe et al., 2017). Automated systems like RPA need a set of predefined rules and data sorted out before it can perform human-like activities such as data entry or working with clients in extremely repetitive functions (Aguirre & Rodriguez, 2017). Robotics, on the other hand, may take more complex actions and interaction between its environment, most of which will now be at high autonomy using techniques like ML and RL (Bruno et al., 2017)

7.2.3 Supervised and Unsupervised Learning in Robotics and Automation

In robotics and automation, supervised learning and unsupervised learning are the two most important training techniques for models based on different types of data. The inputs have known outputs in case of a supervised model, while the input–output mapping is unknown in the case of unsupervised models. This is typically used in applications, including object detection and classification for robotic purposes in order to identify objects within its environment for the effective discharge of tasks (Kin et al., 2018). The supervised learning algorithms are learned from a labeled dataset so that they can achieve highly accurate outputs that could improve robotic capabilities within distinct environments. Unsupervised learning applies data that have no explicit labeling. This learning becomes crucial whenever new, unstructured data appear in dynamic environments, such as when robots encounter new information or observations (Enriquez et al., 2020). The algorithms used to learn in an unsupervised way will discover hidden patterns, and hence, the robots are adaptive.

They can carry out some task performance without any instruction. The adaptiveness ability to a situation is one essential part of dealing with complex data collected from sensors about patients in healthcare robotics (Flechsig et al., 2019).

7.2.4 Reinforcement Learning for Control and Optimization

RL has proven quite useful in robotics for control and optimization. With RL, robots learn to decide by experiencing their environment through interactions with it, the feedback received being rewarding or penalizing, which should guide their actions to good performance (Panchapagesan et al., 2019). The key features RL can provide in robotics are its ability to handle complex, high-dimensional action spaces and the fact that the actions involved in many tasks, like robot locomotion, manipulation, and navigation, are complicated and operate in high-dimensional action spaces (Lohrmann & Reichert, 2016).

For illustration, RL has been applied to robotic arms for the task of grasping and manipulation in which the robot learns optimal strategies by trials and errors (Huang & Vasarhelyi, 2019). Additionally, RL has been applied in multi-robot systems, optimizing cooperative tasks such as warehouse automation in which a group of robots collaborate to have the orders fulfilled in an efficient way (Hallikainen et al., 2018).

7.2.5 Deep Learning Architectures for Robotic Systems

DL brought a remarkable advancement to the field of robotics by providing the capabilities of robots to process vast amounts of sensory data to make a much more intelligent decision-making process. One of the popular architectures of DL is convolutional neural networks (CNNs), particularly efficient in the case of robotic vision systems for image recognition and object tracking (Ivančić et al., 2019). Robots with CNNs can more successfully navigate complex environments, recognize and avoid obstacles, and interact more meaningfully with humans. Apart from CNNs and their derivatives, recurrent neural networks (RNNs) and their variants, namely, long short-term memory (LSTM) networks, are applied in the context of robot operation for tasks that entail information based on time. These could be speech recognition, motion prediction, and anything else (Kokina & Blanchette, 2019). DL architectures also help robots to more accurately grasp dynamic environments and respond appropriately to them, further enriching their versatility and capability in health, manufacturing, and logistics contexts (Leshob et al., 2018). Deep reinforcement learning (DRL) is a combination of DL and RL. DRL has bridged the horizon of robotic autonomy to date. DRL enables direct learning from a high-dimensional input: images. The achievement is quite worthy wherein complex tasks are performed in real-time environments based on DRL (Jiménez-Ramírez et al., 2020). Application of DRL was made to the control and navigation of robots, unlike traditional approaches of RL which often tend to fail due to the complexity of sensory data that needs to be operated on in applicable scenarios (Moffitt et al., 2018.)

7.3 ROBOTIC CONTROL SYSTEMS

Robotic control systems are essential for achieving automation with high precision and efficiency. Classical control systems include several approaches, such as proportional–integral–derivative (PID) control, state space control, and model predictive control. These approaches are used in most applications of robotic automation. The integration of AI with the classical control strategies enhances them to make robots perform much more complex and adaptive and independent operations. This section covers these classical control methods and their AI-driven enhancements and then moves on to intelligent control systems with applications in robotics, including fuzzy logic and neural networks.

7.3.1 Classical Control Approaches with AI Enhancements

Classical control methods play an important role in developing the control systems in robotics. In classical control, the system works on a deterministic model and makes use of feedback for proper control of the robotic system. However, due to increasing complexity of the robotics system and the demand to operate under uncertain environments, AI has become extremely crucial in improving such classical control techniques.

7.3.2 Proportional–Integral–Derivative Control

PID is one of the most basic control strategies employed in most industrial and robotic applications. PID controllers continuously adjust the output from the controller based on the error, defined as the difference between the desired set point and the measured process variable. It is widely appreciated that simplicity and effectiveness in linear processes characterize PID control strategies. However, AI has been exploited for optimizing PID controllers using optimizations of the parameter tuning itself and system adaptability. To that effect, optimization algorithms based on AI, which include ML models, are deployed to automate the adjusting of the PID parameters to optimize control performance in real time; these are more complex nonlinear cases (Hindel et al. 2020). This capability enables robotic systems to handle dynamic and less predictable environments, an important characteristic in several application fields, such as autonomous driving and precision manufacturing (Eikebrokk & Olsen, 2020).

7.3.3 State Space Control

State space control is the method, which uses state variables for the dynamics of a system and describes how the system's behavior evolves with the passage of time. This form of control method is very effective for multi-input/multi-output systems wherein one or more variables is/are to be controlled. The state space model presents a more general framework for control in systems and is especially beneficial in robotics applications. This is because AI has greatly contributed to state space control through enabling systems to better model nonlinearities and uncertainties in the real environment (Leno et al., 2020). There is always more room for better forecasting of state variables so that control performance is adequately improved in complex, unstructured environments (Asatiani & Penttinen, 2016). For instance, RL is the subarea of ML that has been combined with state space control that allows robots to learn optimal control policies in a trial-and-error manner (Jiménez-Ramírez et al., 2020).

7.3.4 Model Predictive Control (MPC)

MPC is advanced control using the dynamic model for predictions of future behavior and computes necessary control actions. It has widely and dominantly been used in high-performance and precision industries such as aerospace and manufacturing. It is different from PID control based on error reactions because MPC is an optimum scheme of future control actions, based on predicted states of the system in the future. AI-driven MPC systems are now very popular in that it allows one to integrate ML models into MPC systems to enhance predictive accuracy and performance (Patel et al., 2019). AI-enhanced MPC can easily adapt to changing environments by continuously improving the model based on new data accumulation, which enables robots to operate autonomously in dynamic and unpredictable environments (Güner et al., 2020). For example, neural networks are used in MPC systems to predict the system behavior of complex nonlinear processes where traditional modeling methods cannot be applied successfully (Gotthardt et al., 2019).

7.3.5 Fuzzy Logic Control

Fuzzy logic control mimics human reasoning by dealing with imprecision and uncertainty. Such a form of control will be highly suitable for environments in which data are uncertain or incomplete. Most of these areas have used typical classical control systems that rely on crystal clear inputs and outputs. However, fuzzy logic controllers can handle cases in which the system behavior is ambiguous. For example, the application of fuzzy logic in RPA has been dependent on improved decision-making in uncertain scenarios (Chacon-Montero et al., 2019). For instance, a robot guided by a fuzzy logic controller could navigate through variable environments or present mixed workloads in industrial applications. Fuzzy logic control is also used in more complex tasks that are found in robotic systems where specific models are unavailable, such as healthcare robotics or autonomous navigation (Santos et al., 2019). Linguistic variables and rules mean that these controllers can interpret qualitative data. This versatility has led to their application in scenarios like obstacle avoidance, adaptive control, and decision-making in uncertain or dynamic environments (Hwang et al., 2020).

7.3.6 Neural Networks

Neural networks form one type of the most feasible ML models in which the system acts just like the human brain. Neural networks are made up of nodes (neurons) connected to each other and can easily learn complex patterns from data. Neural networks have been applied more widely in robotic control systems in order to enhance adaptability and make robots more intelligent. Neural networks within the

RPA simulate complex, nonlinear systems, enabling robots to learn and adapt to new tasks based on data instead of human experts in real time (Leno et al., 2020). For example, training neural networks for understanding patterns in sensor data will make the robots learn how to move through their surroundings or manipulate objects independently (Asatiani & Penttinen, 2016). Neural networks are also essential in RL, whereby robots learn to perform certain tasks depending on feedback from their environment. The approach has been successfully applied in areas such as robotic grasping, path planning, and human–robot interaction (HRI) (Leopold et al., 2018). Equipped with neural networks, robots may learn from the interactions of their environment to increase their subsequent performances to make for more autonomous and intelligent systems.

7.4 MACHINE LEARNING TECHNIQUES IN ROBOTICS AND AUTOMATION

The process of robotics and automation is drastically transforming, and ML techniques are becoming an integral part of this change to upgrade the functionalities of robotic systems. Learning machines will enable robots to learn in their environment, complete tasks better, and base decisions on data without explicit programming. The vital ML techniques that may be applied to robotic systems includes decision trees and random forests, support vector machines (SVMs), neural networks, and RL algorithms.

7.4.1 Decision Trees and Random Forests in Robot Applications

Decision trees are a type of supervised learning algorithm mainly used for both classification and regression problems. Applying decision trees to robotic applications was described as pertaining to the decision-making process wherein a robot could choose an appropriate action depending on definite conditions or rules. One of the reasons why the decision tree is important is its interpretability; therefore, it is easily understood by human operators regarding the decision-making process of robots. Random forests are ensemble methods that create multiple decision trees and combine the results of these forests to yield an even more accurate and less noisy prediction. Random forests have appeared to be very promising in robotics applications and can be very useful to avoid the overfitting that decision trees typically suffer from. Averaging the results of multiple trees improves the robustness and accuracy of predictions in robotic systems. For instance, in the context of autonomous navigation missions, random forests have been applied to the real-time classification of environments as either obstacles or nonobstacles in order for the robot to take safer and more efficient navigational decisions (Syed et al., 2020). Moreover, random forests have been deployed in industrial robotics for predictive maintenance. Random forests can predict a probable failure through the analysis of sensor data coming from sensors mounted on robotic arms or components and thus schedule preventive maintenance, hence minimizing downtime (Schmitz et al., 2019). The wide application of random forests in robot systems is due to the high ability of random forests to handle vast volumes of data with minimal error probability in making predictions.

7.4.2 Support Vector Machines in Classification of Robot

Another widely used supervised learning algorithm in the area of robotics for classification tasks is the support vector machines, which can, first and foremost, be noted especially for their strong statistical predictability and ability to handle large feature spaces. SVMs have been implemented for difficult robotic tasks, such as object recognition, gesture recognition, and robotic vision systems. The SVM algorithm works by finding a hyperplane that best separates data points into different classes. The SVM has been applied effectively to various applications in classifying objects for robots, and as such, autonomous robots can identify and classify various types of objects in space. This aspect is very vital in warehouse automation, in which the robot needs to distinguish products for the proper organization or retrieval of items (Radke et al., 2020). It has also been applied in humanoid robots for the recognition of human gestures and expressions in HRI, making such interaction much more intuitive (Lewicki et al., 2019). One of the advantages of SVMs in robots is that they can work with linear and nonlinear datasets using the kernel trick, thus transforming the input data into a higher-dimensional space where separation might be easier by a hyperplane (Jalali and Wohlin, 2012). This characteristic qualifies SVMs for highly suitable robotic systems, especially when working in dynamic and unstructured environments with data not necessarily fitting into the simple linear pattern.

7.4.3 Neural Networks and Deep Learning Architectures

Neural networks and deep architectures have transformed the world of robotics. These designs resemble the neural structure of the human brain, making it possible for a robot to learn complex patterns in large datasets. Techniques like CNNs and RNNs have advanced DL, making robots' visual perception and decision-making remarkable. For example, CNNs have been applied a lot in the application of robotic vision for object detection, image segmentation, and recognizing different objects. In the case of the autonomous car, CNNs are applied for detecting pedestrians, traffic signs, and other automobiles so that the robot can drive safely in the city (Geyer-Klingeberg et al., 2018). DL architectures have been also applied in industrial robots for quality check where such architectures may be used for inspecting the product's defects with high sensitivity and speed (Wroblewska et al., 2018). However, in robotics applications, RNNs are there to do tasks that entail sequential data processing. They are appropriate for speech recognition and natural language understanding. On the field, within the scope of service robots, RNNs enable humans to interact with the robot in more natural ways since it can understand and respond to voice commands (Willcocks & Lacity, 2016). Overall, such neural network-based approaches have increased the flexibility and autonomy of robots in different fields of application. This is also one main disadvantage of neural networks in robotics, requiring large quantities of prelabeled data for training. Recent advancement in unsupervised and semi-supervised learning has, however, helped overcome parts of those challenges, making it possible for robots to learn from smaller datasets, and even without explicit supervision (Syed et al., 2020). Additionally, neural networks have been integrated with other approaches such as RL to create more intelligent and adaptive robotic mechanisms.

7.4.4 Reinforcement Learning Algorithms

RL is that kind of ML by which an agent, in this case, a robot, learns to take actions in an environment with the aim of maximizing cumulative reward. This has proved especially influential in robotics where robots are compelled to act in uncertain and dynamic environments. RL allows them to learn through experience based on feedback that their actions receive in return. One of the applications of RL to robotics is in robotic manipulation. By using RL, robots can learn how to pick up objects of different sizes and shapes through trial and error at the same time as improving dexterity (Wanner et al., 2020). RL has also been employed in locomotion by robots, which may learn on the way and navigate complex terrains with the use of interacting dynamics with their environment (Osmundsen et al., 2019). One of the crucial innovations in this domain is DRL, which is the integration of reinforcement learning with DL. DRL has been applied in various robotic tasks, such as control of robots and autonomous driving, and training humanoid robots (Riedl & Beetz, 2019). For instance, using the robotic arm, DRL enables a robot to learn the optimal control strategies to perform manufacturing tasks like assembly and packaging (Suri et al., 2018). However, even despite the successes related to the application of reinforcement learning in robotics, some challenges still remain, mainly regarding sample efficiency; in most practical applications, a large number of interactions are needed in order for the robots to learn appropriately. Recent work has been mainly done on improving the efficiency of the RL algorithms and making those more applicable to real-world systems of robotics (Ian et al. 2016).

7.5 ARTIFICIAL INTELLIGENCE IN AUTOMATION CONTROL SYSTEMS

Industrial and robotic automation has revolutionized control systems with the introduction of AI. Today, with AI integration, what was once represented using static rules and predefined models in automation control systems now undergoes adaptability, learning, and predictive capabilities. In such AI-based systems, it is possible to manage dynamicities in complex environments, simplify processes, and make efficient decisions while considering the particular scenario under which the traditional mechanisms of control would collapse as a result of dynamic or uncertain variables. The two essential areas in AI applied to automation control systems are AI-based adaptive and predictive control and AI application in hybrid control.

7.5.1 AI-Based Adaptive and Predictive Control

Adaptive control systems have been of fundamental necessity in dynamically or unpredictably changing environments for ages. Conventional systems have these parameters adapt based on variations in system behavior to maintain maximum performance. AI, however, adds a new twist such that these adaptive systems become even more intelligent and robust. Systems of adaptive control, powered by AI, operate using ML algorithms that continue to learn from their environment, as well as historical data, to make predictions on future states and operate accordingly. The other importance of AI in adaptive control is its ability to enhance performance in real time. For example, under industrial robotics, AI algorithm can be able to analyze sensor data to provide a prognosis of future failures on machinery allowing for preemptive maintenance (Hamamoto et al., 2018). Predictive control systems are designed in such a manner that they have an expectation of how the system will behave given historical data patterns and inputs in real time. In the context, the prediction of behavior in the system could be done using algorithms like neural networks and decision trees. These algorithms predict system behavior and make essential inputs at real time to optimize its efficiency (Sarker et al., 2020). DL techniques have also been applied significantly in predictive control, especially in manufacturing systems and industrial automation (Kim et al., 2018). Such AI-based predictive models can be trained on big volumes of sensor data in such settings and can therefore predict possible issues, including machine breakdowns or suboptimal process performance, and thus, the operations are adjusted to avoid lost time (Islam et al., 2020). It has hence provided value in the improvement of manufacturing operation efficiency by ensuring that machinery performs to the best. Additionally, reinforcement learning is one of the categories of AI that may be applied in adaptive control systems. It makes systems learn optimal policies through trials and errors. The use of algorithms based on reinforcement learning may be applicable when a robot has to learn how to navigate through an environment that is complex or time-varying so that it adapts to unforeseen variations and consequently learns effective paths or actions based on previous experiences (Kaelbling et al., 1996). The ability is especially useful in dynamic environments like the autonomous car, self-driving drones, or similar areas where conditions are constantly changing. Predictive capabilities of AI go far beyond industrial applications into health and energy management, where predictive control is necessary to optimize resource utilization. For instance, in energy-related systems, AI is capable of forecasting consumption patterns to adjust the supply according to the demand without wasting it, thus benefiting both the economies and the environment (González-Briones et al., 2018). In health care, AI-based predictive control has been applied to optimize schedules of treatment for patients and predict outbreaks of diseases, thus improving effectiveness in health delivery services (Islam et al., 2020).

7.5.2 AI in Hybrid Control Systems

Hybrid control systems combine classical control techniques with AI-based techniques to come up with stronger and more flexible solutions. In these systems, model-based control and AI algorithms are fused together so that there can be the benefits of the respective approaches. Traditional model-based controls give stability and the reliability necessary for well-understood processes, while AI provides the ability to learn and adapt for uncertainty and nonlinearity in complex systems. Hybrid control systems are found to have pervasive application in robots. Conventional control techniques, such as the PID, take effect only on very simple linear systems. However, their performance deteriorates in the presence of nonlinear dynamics and some uncertainty in the environment. Hybrid AI control systems are constructed in a manner that they can be adaptable to time-varying situations and can learn new data and make them function with better performance in complex tasks, including robotic manipulation and navigation (Dupond, 2019). In self-driving cars, for example, AI algorithms may be hybriding with classical control theories to manage complex driving scenarios. While the AI component predicts the patterns and variable speed of the traffic flow as it responds to sensor feedbacks, the traditional control mechanisms will ensure that the vehicle stays within safe limits (Islam et al., 2020). This integration of AI and classical control allows autonomous systems to efficiently and effectively deal with dynamic and uncertain scenarios better than either approach can do in isolation. With AI-enabled hybrid control in manufacturing, the production process can be optimized by mirroring traditional automation reliability with the adaptability of AI. These systems will adjust on variations in demand in the production, equipment wear, or even material quality to remain efficient and cost-effective in the production process (Gonzalez-Briones et al., 2018). Hybrid systems can predict and mitigate the potential disturbances with type of equipment malfunctions or supply chain delays, in which continuous operation occurs with minimal human intervention through learning from the data in real time. Hybrid control systems have also been used for energy management. In smart grids, such hybrid control systems help in balancing the supply and demand of electricity. Currently, with AI-based algorithms, the energy usage and the distribution of energy between such sectors are predicted. In contrast, the conventional control systems ensure that the operation of energy grids stays within the limits that are defined (Sarker et al., 2021). This reduces the risk of blackouts or overloading of the grid while promoting better utilization of energy. Hybrid control systems use fuzzy logic to make them more robust in dealing with uncertainty. Fuzzy logic allows a system to make a decision based upon uncertain or incomplete data – a situation that is very prevalent in real-world environments (Zadeh, 1965). For instance, in a smart irrigation system, fuzzy logic would be applied in the distribution of water depending on the changes in soil moisture and weather and thus optimizes water consumption while preserving crop production (Krishnan et al., 2019). AI hybrid systems with fuzzy logic will allow for adaptation to an incredibly wide set of conditions, which contributes to more resilient and efficient solutions. Another sector in health care where hybrid control is being applied is robotic-assisted surgery. Hybrid control is integrated into the AI algorithms with regard to the systems for aiding surgical decision-making and precision in control, while traditional control will support in stability and safety for the mechanical operation of the robot (Lamy et al., 2019). The systems ensure that more capability in surgical robots is added to enhance surgical procedures to be delivered with precision and accuracy.

7.6 CHALLENGES AND FUTURE DIRECTIONS

AI-powered and hybrid control systems offer many opportunities for significant benefits, but they are also fraught with a number of challenges. Among the most challenging aspects is the need for large volumes of data to properly train AI algorithms. The quality of available data in many industrial and healthcare applications might be low, making it difficult to acquire the required data, and thus, AI predictions would also be inaccurate. In addition, there is the integration aspect of AI with the traditional control systems that requires careful design so that both components will work seamlessly together. Another concern is interpretability of the AI model. Many control systems need to know why a system chose a given course of action; this is particularly important for applications like medical treatment or self-driving vehicles. Black-box models of AI, such as DL, are not very interpretable and therefore raise issues about safety and reliability of the model. Currently, there are research efforts to develop AI models that can be more transparent and interpretable in control systems (Goodfellow et al., 2016). Further into the future, advanced hybrid systems for automation control systems are in the pipeline that will allow for self-management of complicated environments using minimal or even no human interaction at all. By combining AI with newer technologies such as the Internet of things and edge computing, the total proficiency of the control system will be optimized for processing and responding in real time (Piccialli et al., 2020). Its application to various industries will continue to expand because more efficient and capable AI algorithms can learn well from a small amount of data.

7.6.1 Computer Vision and Perception in Robotics

Perception and understanding of the environment are quite important for autonomous operation, navigation, and execution of a given task. One of the flagship technologies that power AI, computer vision, has become part of every perception process in robotics. AI in computer vision is the use of AI to process images for analysis; it includes identification of objects, tracking movement, and decision-making based on visual input. It is a research area that is further discussed as follows: AI image processing, real-time object detection, and sensor fusion techniques in robotic systems.

7.6.1.1 AI in Image Processing and Feature Extraction

Computer vision in robotics is founded on image processing. It enhances the ability of AI to interpret and manipulate image data in getting useful features. While traditional techniques in image processing may be highly efficient in applications of a particular nature, they indeed falter when dealing with complex and unstructured environments that robots often deal with. The advent of AI and especially DL has revolutionized this field with more accurate and adaptive feature extraction techniques. DL algorithms, typically referred to as CNNs, are applied in robotics for the processing of images. CNNs would usually learn to automatically extract important features from images automatically – for instance, edges, textures, and shapes – without the manual feature engineering that tradition requires (LeCun et al., 2015). In the case of robotic applications, it benefits much in environments that have highly diverse visual inputs. One of the real applications of CNNs is in industrial automation systems for detecting defects in manufactured products, with quality control enhanced by the analysis of high-resolution images in real time (Gu et al., 2020). Generative adversarial network (GAN) – deep learning application in processing – is one of the most critical applications of AI when it comes to image processing. It helps in image synthesis and data augmentation. GANs can generate fake pictures from a given dataset. This is useful when real-world data are hard to collect or come with high costs, which is the case, for instance, for some research (Kim et al., 2018). In robotics, GANs are implemented to mimic lighting conditions and weather, increasing the robustness of visual perception systems for autonomous vehicles and drones in difficult environments. Indeed, a crucial role for medical robotics is found in AI-based image processing. This is because the algorithms used by AI help to facilitate robotic-assisted surgery by analyzing medical images, such as X-rays or MRIs, to guide robots through surgical procedures more accurately. These feature extraction algorithms include tissue boundaries or tumor locations and allow for more precise, less invasive interventions (Lamy et al., 2019). Moreover, AI-driven image processing enhances the robotic vision system, making it functional for better performance in medical environments as used in diagnostics and treatment planning. This includes hybridizing traditional feature extraction methods like edge detection, histogram analysis, and texture recognition, with AI. It is, for instance, that classical methods like the Sobel or Canny edge detector result in the achievement of more accurate or even context-aware feature extraction when used in conjunction with AI (Piccialli et al., 2020). This hybrid approach makes the robots more sensitive to their environment's context and consequently allows the decisions being made by them to be more informed.

7.6.1.2 Real-Time Object Detection Using AI Algorithms

In order to detect objects in real time, it's necessary for the recognition to be immediate so that an autonomous robot can determine what's ahead and what's its environment in real-time autonomous robots require real-time object detection capabilities. Advances made by AI algorithms propelled the field of object detection significantly to allow tracking and recognition of multiple objects in real time under challenging conditions such as under occlusions, changing light settings, or cluttered scenes. The most popular AI-powered object detection algorithms in robotics are you only look once (YOLO) and single-shot multibox detector (SSD) due to their ability to function at very high speeds and accuracy levels (Pan et al., 2018). Most objects have a detection and classification time below milliseconds using DL models, and that is why these applications are the perfect fit in real-time applications, such as autonomous vehicles, navigation systems for drones, and robotic production (Wang et al., 2018). For instance, the nature of application includes pedestrian and vehicle detections and road sign recognition through AI-based object detection mechanisms in an autonomous vehicle to drive through a complex urban environment. Real-time object detection becomes the defining characteristic of robotic arms in assembly processes because the utilization of such a device ensures that within an industrial environment, the part might be gripped and manipulated suitably for manufacturing purposes. Since AI algorithms can differentiate between various forms and dimensions of components, the robot may change its grip and position for the part in real time accordingly – such as to handle different manufacturing processes requiring precision and speed (Khosravani et al., 2019). Another important aspect of utilizing AI is the ability to process video streams in real time. Such robots would use real-time object detection algorithms to identify a specific danger, such as an intruder or an unattended package. AI algorithms scan continuously around the environment and mark those suspicious objects or activities for follow-up inspection (Harrou et al., 2019). This use of systems in public spaces, airports, and critical infrastructure increases security through autonomous monitoring. AI was added into the classic computer vision approaches using techniques based on the DL approach to object detection. Techniques involving the analysis of optical flow – that is, movement of an object from one frame to another – have now been supplemented by learning models that predict how objects will move or behave (Phan et al., 2019). Combining AI with classical vision techniques gives the best and most robust and reliable object detection especially in cases where objects are partially occluded or moving at very high speeds. One of the challenges of real-time object detection is ensuring that AI algorithms can work as efficiently as possible on the often-limiting resources available within many robotic systems. To address this, researchers have developed lightweight AI models that require less processing power yet achieve high levels of accuracy. For example, models like MobileNet and Tiny YOLO are designed to run efficiently on embedded systems used in drones and mobile robots, enabling real-time object detection in resource-constrained environments (Wang et al., 2018).

7.6.1.3 Sensor Fusion Techniques Powered by AI

Perception in robotics is, however, not limited to visual data. Most robots are fitted with multiple sensors collecting different types of information from the environment. Sensor fusion refers to the fusion process that brings together data from cameras, LiDAR, radar, and Inertial Measurement Units (IMUs) for a better understanding of the environment. AI has had a vital role in developing sensor fusion techniques in order to allow robots to more intelligently collect and interpret data coming from disparate sources. AI algorithms, particularly the ones which have a focus upon DL and probabilistic models, may process and combine data inputs from a large number of sensors in real time. For example, in driverless cars, AI synthesizes the data from cameras and LiDARs with radar to understand the surroundings, measure distances, and steer clear of traffic. The integration of the data provided by the sensors enables AI-based systems to compensate for the shortcomings of individual sensors that may include either inadequate lighting that compromises the capabilities of cameras or inaccuracies on LiDAR at greater distances (Wang et al., 2019). Kalman filter is among the major applied AI techniques in sensor fusion. The Kalman filter is a probabilistic model for merging data from varying sources to estimate the state of a system over time. In this regard, dynamic adaptation by AI-enhanced Kalman filters becomes conceivable depending on the quality and reliability of sensor data received, which is highly optimistic for more accurate and robust perception of robots (Srinivas et al., 2018). For example, in autonomous drones, data between GPS, IMU, and cameras are fused by Kalman filters, achieving stable flight and positioning. AI-based sensor fusion has also been applied in medical robotics where robots had to combine the data from multiple imaging modalities like ultrasound, MRI, and CT scans. By combining these data, AI facilitates more precise diagnosis and enhances the capability of the robot to perform complex surgical maneuvers (Lamy et al., 2019). In smart prosthetics, AI algorithms combine information coming from pressure sensors, accelerometers, and electromyography signals in giving more natural and adaptive control for the user (Krishnan et al., 2019). Sensor fusion is significant in robotic search and rescue applications because of the navigation in hazardous environments. AI algorithms will fuse data from a thermal camera, acoustic sensors, and LiDAR to recognize victims under low-visibility conditions, such as smoke-filled rooms or buildings that have collapsed. It ensures that robots help human responders better and reduce human exposure to dangerous situations (Khosravani et al., 2019).

7.6.2 Natural Language Processing and Human–Robot Interaction

NLP enabled new interactivity between humans and robots, to understand what a human says or gestures. That would become the foundation for developing HRI further, from a robot that may become an intuitive partner in health care or education and even in industrial automation. The three significant areas of HRI focused are speech recognition and natural language understanding, AI for gesture and emotion recognition, and AI-enhanced human–robot collaboration (HRC).

7.6.2.1 Speech Recognition and Natural Language Understanding

Speech recognition and natural language understanding are basics for the assumption that a robot can understand verbal commands and communicate with humans with ease. Innovations in natural language processing through AI-driven advancement have permitted robots to make real-time interpretation and response to complex commands in order to offer more natural and effective HRI. Applications of recurrent neural networks and more specifically of the long short-term memory (LSTM) networks to sequential data such as speech for capturing contextual meaning from sentences (Mikolov et al. 2010). These abilities to tackle dependencies in the speech also make LSTMs suitable for robots that aid humans in performing service tasks and perhaps taking verbal instructions across a wide range of contexts (Elliman & Pulido, 2002). In healthcare environments, robots utilize LSTM models to understand requests from patients and assist with daily activities aimed at making both care effectiveness and quality better (Bhargav et al., 2023). Similarly, NLP has served as the backbone in the conception of conversational agents that are built into social robots to support humans in performing tasks such as information provision, scheduling an appointment, or even operation of home appliances. The agents are only reliant on AI to infer the intent of their users, even if they are talking vaguely or vaguely, which has a tendency of making the interactions fluid and user-friendly as illustrated by Kang et al. (2020). For example, AI-based noise-cancellation techniques have subdued some of the challenges of noisy industrial or public places. DL models can be trained to filter noise to allow robots to understand speech in noisy environments (Poort et al., 2020). The ability to hear is very important for robots in the manufacturing environments where heavy machinery has obscured the verbal exchange, yet the instructions to be performed are required to be precise. In addition, advances in natural language understanding allow for robots not just to process the words spoken but also to comprehend the message and intent of a sentence. And whereas robots are merely performing commands, they can now engage in more advanced multi-turn dialogues in such a way that their ability depends more on AI models that really understand context more than earlier versions (Lamy et al., 2019). For instance, robots make use of NLP in the customer service context to answer more elaborate questions so that the customers get tailored answers to their needs, hence boosting the level of satisfaction among customers (Deng & Liu, 2018).

7.6.2.2 Artificial Intelligence for Gesture and Emotion Recognition

Apart from understanding spoken communication, AI enables robots to interpret nonverbal communication like gestures and facial expressions, which are a fundamental part of human communication. The ability of robots to recognize gestures and emotions helps them better understand human intent and emotional states in order to provide responses that are not only empathetic but also contextually appropriate. Gesture recognition is seen as one of the most successful applications of AI-based computer vision algorithms, especially using convolutional neural networks (CNNs). These capabilities have been witnessed with great ease in identifying and interpreting human gestures in real time (Pan et al., 2018). Such robots can identify hand signal gestures, body movements, and other forms of gesture, thus serving as a means for assisting humans by mimicking communication mechanisms where verbal methods may not work. An example of gesture recognition is in industrial settings whereby workers are able to issue commands to a robot through hand movements. This has proved very efficient in boosting productivity while using fewer complex input devices (Muller et al., 2020). However, emotion recognition deals with AI models identifying aspects associated with facial expression, voice intonation, and other physiological signals that relate to the emotional state of a human being. DL models such as deep neural networks (DNNs) are trained on massive datasets to recognize emotions, be it happiness or anger or pain. These have allowed the robots to adapt their actions in view of these emotions (Ale et al., 2019). Emotionally intelligent robots are being used in healthcare and educational sectors to provide emotional support, detect signs of stress or discomfort, and provide companionship (Wang et al., 2019). Emotionally responsive robots are increasingly common in applications to customer service and social interactions, where they would respond based on the condition of a customer's emotion. For example, the system used by retail or hospitality industries uses AI to sense whether a customer is frustrated or satisfied, and such finding enables them to adjust the given response and action (Muller et al., 2020). It would bring an integration of gestures and emotion recognition in robots, thus revealing a more holistic understanding of human interaction, making the response given by the robot more natural and empathetic. Another integration is gesture and emotion recognition, which enables the robot to nonverbally communicate by enhancing its interaction with a speech or hearing-impaired person. For example, in learning environments, robots can detect gestures or emotions of the students, whereby they can alter their teaching tactics based on the needs of the different learners (Ramzan et al., 2020). Such applications are highly relevant within the context of inclusive education, where robots provide tailored support to students with unique learning skills.

7.6.2.3 Hybrid Human–Robot Collaboration Using Artificial Intelligence

HRI changes industries through collaboration between humans and robots working side-by-side while dividing up work and responsibilities. In contrast to traditional automation, where the robot acts single-handedly, HRC provides close cooperation between the robots and human workers, who can rely on their strengths. The possibility of AI allowing a robot to predict human action, adapt to environmental changes, and improve the efficiency of a general task is an important role that AI plays in enhancing the cooperation. Cobots are short for collaborative robots. They are designed to be used in concert with humans in manufacturing, health care, and logistics, among other industries. AI improves the ability of cobots in predicting human movement and activity, making them safer collaborators and more efficient (Raghav et al., 2024). For example, in the assembly line, for instance, AI-powered cobots can be set to alter their actions real time based on changes in the task and other human workers' actions during that particular job for smooth coordination and less possibility of accidents (Bhavithra & Saradha, 2019). Surgeries in the health sector are one of the notable examples of AI-amplified HRC. In surgeries, these robots work with surgeons to achieve high precision with the surgeon and at real-time data analysis. The algorithms used by AI assist the robot to synchronize with the surgeon's movements; the robot offers advice and feedbacks based on the kind of patient and the nature of the surgical process (Muller et al., 2020). This synchronization and combination of human expertise and robotic precision elevate the surgery outcome and reduce recovery time for patients. Collaborative robots integrating NLP, gesture recognition, and emotion detection mean that they can effectively communicate with workers. For example, within warehouses, AI robots use gesture recognition to interpret workers' signals and NLP to understand the verbal commands that enable them to assist in picking and placing or material movement for dropping at particular locations (Ramzan et al., 2020). These machines, with AI, can interpret the multimodal input which supports their co-working within fast-paced environments where real-time communication is required. Through AI, such robots can also learn from the interactions that they share with human partners and eventually become more perfect with time. Reinforcement learning algorithms permit adaptation of robots to new tasks, environments, or both because of continuous learning from experiences (Pearl, 1988). In a manufacturing setting, for example, the robots can learn to optimize their movements and actions toward obtaining better results based on instructions by human operators, thus gradually becoming efficient collaborators. The future of AI about improving HRI tends to construct more autonomous systems that can learn in real time to new tasks and changing environments. The AI models will be powering the next-generation collaborative robots by endowing them with the capacity to sight behavior done by humans, predict what might happen, and shift their operations in real time, thus becoming indispensable partners across industries from health care to logistics (Schwab,2016).

7.6.3 Mathematical Perspectives of AI on Robotics

AI changed the way of robots by introducing mathematical modeling of decision-making, optimization, and control with advanced algorithms for computation. Focusing on control theory, optimization, and dynamical decision-making, AI systems in robotics are centered around key issues which influence the improvement of functionality and accuracy of robotic systems.

7.6.3.1 AI on Mathematical Modeling and Control Theory

More fundamentally, mathematical models form the basis for predicting and managing dynamic systems in robotics. Besides the classic control theory methods, such as the use of PID controllers in robotic systems, newer strategies in control which involve learning-based approaches have come with AI. An example in this direction is reinforcement learning (RL), and here, the policy is optimized in dynamic environments. Reinforcement learning is particularly efficient in nonlinear systems where classical control methods might not even provide the expected flexibility (Sutton & Barto, 2018). Robust control techniques are also constructed based on the use of Markov decision processes (MDPs) to model the process of decision-making under stochastic outcome environments. The MDPs allow for the robot's actions to be optimized through encoding the nature of sequences in tasks and occurrences in decision points. Applications of MDPs combine RL algorithms to enhance real-time decisions in robotics (Bellman, 1957). Techniques from deep learning, such as CNNs, are applied to high-dimensional data in applications such as visual perception and object recognition in robotic systems (Li & Deng, 2020). In a similar way, Bayesian networks and probabilistic graphical models are mathematical tools implemented to deal with uncertainty in robotic systems, mainly when sensor fusion and decision-making under uncertainty occur. These probabilistic models enable robots to make appropriate decisions even with incomplete noisy data from sensors (John & Langley, 1995). This means that AI mathematical models allow robots to work independently, especially in complex, unstructured environments, such as disaster areas or other extraterrestrial surfaces (Han et al., 2011).

7.6.3.2 AI-Driven Optimization Techniques

Optimization is behind most of the current applications of AI in robotics. Genetic algorithms and evolutionary algorithms have widely been applied for optimizing robotic behaviors, which includes path planning, object manipulation, and energy-efficient motion (Reddy et al., 2018). By simulating biological evolution, the adaptive strategies make it highly suitable for complex multiobjective optimization problems. Swarm intelligence is inspired by the collective behavior of social organisms, including ants and bees, which has very largely been successful in applications such as multi-robot systems. Swarm algorithms, such as particle swarm optimization (PSO) and ant colony optimization (ACO), allow a group of robots to work together toward a common objective, such as searching an area or lifting heavy objects, to arrive at a goal (Yang et al., 2020). Another area where optimization using AI is applied is in the control of robotic arms. Techniques like model predictive control (MPC) make use of optimization at each time step to determine control actions that optimize future performance while satisfying constraints. That is a widely applied technique in the control of industrial robots and autonomous vehicles since the requirements are high precision and adaptability (Maedche & Staab, 2001). In the field of robotics, deep reinforcement learning (DRL) has recently become pretty popular as it enables agents to act in continuous spaces. In fact, robots that learn via DRL algorithms have been utilized in autonomous drone navigation and manipulation tasks (Aha et al., 1991). DRL thus captures the representational power of DL and the decision-making capability of reinforcement learning, thereby endowing robots with the ability to learn complex control policies in which it maximizes long-term rewards (Breiman, 2001).

7.7 ETHICAL, LEGAL, AND SOCIETAL ISSUES OF AI IN ROBOTICS

The fast pace of progress of AI technology in robotics goes hand-in-hand with the new series of ethical, legal, and societal issues. In this context, it is very important to consider the issue of robots' autonomy and their capacity to be capable of interaction and eventually decisions of their own.

7.7.1 Ethical Considerations

Ethical considerations are central in the deployment of AI-driven robots, most particularly those operating within human-centered environments. Among these, some of the issues cut across autonomy, accountability, and transparency. For instance, the utilization of AI in self-driving cars and healthcare robots has raised intense debate over the moral accountability in case of an accident or unintended harm (Srinivas et al., 2018). Probably, the most discussed ethical framework is Asimov's laws of robotics, where robots should not harm humans, must obey human orders, and preserve their own existence without violating the first two laws. However, while these laws give a conceptual notion, the present-day system of AI scenarios poses far more complex questions of ethics. For example, the existence of drone and military robotic weapons has brought about ethical questions over the apparent misuse of the technologies as well as the major problem associated with devolving decisions on lethal operations to machines (Talib et al., 2021). In addition, AI robotics should be developed with a fair bias mitigation approach. Most ML models trained on biased datasets are likely to worsen societal inequalities, which can result in discriminating behaviors by the AI-driven robotic systems. In this challenge, development of socially responsible AI technologies has been proposed through ethical AI frameworks, such as fairness, accountability, and transparency in ML (FAT-ML) (Corrales et al., 2020).

7.7.2 Legal Standards for AI in Robotics

At the same time as AI becomes increasingly integrated into robotics, there is a need for developing robust legal infrastructure that may keep pace with and control them. To date, there are very few general comprehensive standards on AI-driven robots. The General Data Protection Regulation (GDPR) in the European Union is the most significant legal framework that imposes several requirements on the regulated automated decision-making system, which includes robotics. The need to be in line with GDPR brings with it the dictate of transparency within AI system decision-making processes and users are to be allowed the contest of automated decisions (Breiman et al., 1984). Additionally, there are ongoing discussions about introducing **liability frameworks** for autonomous systems. A critical question is whether manufacturers, programmers, or end users should be held accountable when AI-driven robots cause harm. This issue becomes more complex when considering robots that learn from their environment and evolve their behaviors over time (Scott et al., 1992).

7.7.3 AI and Employment

A massive social impact of AI on robotics would be the effect on employment. Highly advanced tasks done by robots will sooner or later replace many jobs in a manufacturing, logistics, and even healthcare sector (Blumenstock, 2020). Mechanization not only can spur more productivity and prosperity but also raises the possibility of big job displacement. The studies also reveal that automation can easily affect lower-skilled workers because most of the tasks they perform can be given to AI-driven robots (Syed et al., 2021). On the contrary, AI and robotics are also opening up new employment avenues in an area of AI research, robot maintenance, and software engineering. Thus, although AI in robotics is displacing certain categories of employment, it is also likely to spawn employment in still more specialized areas (Sahar et al., 2019).

7.8 FUTURE TRENDS AND EMERGING TECHNOLOGIES

7.8.1 Explainable AI Robotics

As AI systems become more sophisticated, there is growing interest in making them more transparent and interpretable. Explainable AI (XAI) refers to the technique of developing AI models that can provide human-understandable explanations of their decisions. Of course, this has to involve tremendous importance in robotics. Robots controlled by AI have to take tough, real-time decisions based on systems changing at every moment. XAI is going to help build trust from humans to machines through clear explanations of robotic actions, which is extremely important in applications such as health care, law enforcement, and autonomous vehicles (Turk et al., 1991). Explainability is useful in debugging and improving AI models in robotics. When a robot does not behave as desired or cannot perform a required function, explainability will be valuable in tracing the root of the malfunction and building the AI model along the relation. The attention mechanisms and saliency maps are developed to reveal more of the decision-making process that occurs in AI models and thereby increase their transparency for human users (Deng & Liu, 2018).

7.8.2 Quantum Computing and AI Integration

Another promising development of AI in robotics is the inclusion of quantum computing. Quantum computers can handle large quantities of data almost at once and can solve complex optimization and ML problems that are currently infeasible with classical computers. The enhanced performance of AI algorithms in robotics can be used for real-time path planning, object recognition, and decision-making in uncertain environments (Keerthi et al., 2001). Breakthroughs would be expected in the area of quantum enhanced reinforcement learning. The parallel exploration of multiple solutions by quantum computers would increase the efficiency of reinforcement learning algorithms and speed up the convergence rates. This allows robots to learn optimal behaviors in highly dynamic, very complex environments; such applications might include space exploration, as well as underwater missions (Kohonen, 1990).

7.8.3 Advancements in AI-Driven Swarm Robotics

Swarm robotics is one area that draws motivation from the behavior of social insects such as ants and bees. For example, a swarm of simple robots can be used collectively to search an area or for construction of structures or any other activity. The use of AI algorithms from distributed ML and reinforcement learning has made robots collaborate much more effectively and efficiently (Krishan et al., 2020). Another area of application for AI-powered swarm robotics is disaster recovery, environmental monitoring, and agriculture. The example can be seen in the ability of swarms of drones fitted with AI algorithms to map vast areas, critically evaluate the severity of natural disasters, or provide instant crop health information (Das et al., 2001). Swarm robotics offers many advantages related to flexibility and scalability that can be most helpful in applications requiring robustness, redundancy, and adaptation (Corrales et al., 2020).

7.9 CONCLUSION

AI, at its core, has transformed robotics and automation, introducing a wave of exciting innovations in nearly all sectors where more sophisticated and autonomous operations are feasible. Implementation of machine, deep, or reinforcement learning abilities into robotic systems has given them the ability to achieve highly accurate results and adapt more flexibly in real time. Such technologies are applied across vast industries, ranging from manufacturing to health and logistics; generally, AI-driven robots tend to improve efficiency, lower operational costs, and streamline process systems. Bringing AI together with conventional control systems has led to more intelligent and responsive robotics that could navigate through unpredictable environments and perform tasks autonomously with minimal human intervention. Robots can now boast of more advanced sensory perception, natural language processing, and even autonomous path planning, which allows it to interact with its surroundings in more dynamic ways. The development of AI in robotics also has significantly important ethical, legal, and societal considerations. As with new technologies such as quantum computing and explainable AI, questions regarding employment or accountability and transparency are gaining their relevance now. Such signals point toward unbridled growth going forward and are an indication that AI will play a very large role, not only driving innovation but also shaping the future of robotics and automation across several sectors.

REFERENCES

Aguirre, S., & Rodriguez, A. (2017). Automation of a business process using robotic process automation (RPA): A case study. In *Workshop on Engineering Applications,* 65–71, Springer.

Aha, D. W., Kibler, D., & Albert, M. K. (1991). Instance-based learning algorithms. *Machine Learning*, *6*(1), 37–66.

Ale, L., Sheta, A., Li, L., Wang, Y., & Zhang, N. (2019). Deep learning-based plant disease detection for smart agriculture. *In 2019 IEEE Globecom Workshops (GC Wkshps)*, 1–6, IEEE.

Arora, K., Joshi, G.P., Ragab, M., Rawa, M., Milyani, A.H., Mansour, R.F., & Yang, E. (2022). Bilateral contract for load frequency and renewable energy sources using advanced controller. Computers, Materials & Continua, Tech Science Press, 73 (2), 3165–3180, 10.32604/cmc.2022.026966.

Arora, K., Kumar, A., Kamboj, V.K., Prashar, D., Jha, S., Shrestha, B., & Joshi, G.P. (2020). Optimization methodologies and testing on standard benchmark functions of load frequency control for interconnected multi area power system in smart grids. In *MDPI Journal of Mathematics*, Mathematics, 8(6), 980, https://doi.org/10.3390/math8060980

Asatiani, A., & Penttinen, E. (2016). Turning robotic process automation into commercial success – Case OpusCapita. *Journal of Information Technology Teaching Cases, 6*(2), 67–74.

Asquith, A., & Horsman, G. (2019). Let the robots do it! – Taking a look at Robotic Process Automation and its potential application in digital forensics. *Forensic Science International: Reports*, 1, 100007. https://doi.org/10.1016/j.fsir.2019.100007.

Auth, G., & Bensberg, F. (2019). Impact of Robotic Process Automation on Enterprise Architectures. In *INFORMATIK 2019: 50 Jahre Gesellschaft für Informatik – Informatik für Gesellschaft* (Workshop Beiträge), Gesellschaft für Informatik eV.

Bellman, R. (1957). A Markovian decision process. *Journal of Mathematical Mechanics, 2*, 679–684.

Bhatt, B., Sharma, H., Arora, K., Joshi, G.H., & Shrestha, B. Levy flight-based improved grey wolf optimization: A solution for various engineering problems. In *MDPI Journal of Mathematics*, Mathematics 2023, 11(7), 1745, https://doi.org/10.3390/math11071745

Bhavithra, J., & Saradha, A. (2019). Personalized web page recommendation using case-based clustering and weighted association rule mining. *Cluster Computing, 22*(3), 6991–7002.

Blumenstock, J. (2020). Machine learning can help get COVID-19 aid to those who need it most. *Nature*, *20*, 20.

Bosco, A., Augusto, A., Dumas, M., & La Rosa, M. (2019). Discovering automatable routines from user interaction logs. In *International Conference on Business Process Management*, 144–162, Springer.

Breiman, L. (2001). Random forests. *Machine Learning*, *45*(1), 5–32.

Breiman, L., Friedman, J., Stone, C. J., & Olshen, R. A. (1984). *Classification and regression trees*. CRC Press.

Bruno, J., Johnson, S., & Hesley, J. (2017). Robotic disruption and the new revenue cycle. *Healthcare Financial Management*, 71(9), 54–62.

Cewe, C., Koch, D., & Mertens, R. (2017). Minimal effort requirements engineering for robotic process automation with test driven development and screen recording. In *International Conference on Business Process Management*, 642–648, Springer.

Chacón-Montero, J., Jiménez-Ramírez, A., & Enŕíquez, J. G. (2019). Towards a method for automated testing in robotic process automation projects. In *14th International Workshop on Automation of Software Test*, 42–47, IEEE Press.

Cohen, M., Rozario, A. M., & Zhang, C. A. (2019). Exploring the use of robotic process automation (RPA) in substantive audit procedures. *The CPA Journal*, 89 (7), 49–53.

Corrales, D. C., Ledezma, A., & Corrales, J. C. (2020). A case-based reasoning system for recommendation of data cleaning algorithms in classification and regression tasks. *Applied Soft Computing*, *90*, 106180.

Das, A., Ng, W.-K., & Woon, Y.-K. (2001). Rapid association rule mining. In Proceedings of the Tenth International Conference on Information and Knowledge Management, 474–481, ACM.

Deng, L., & Liu, Y. (2018). *Deep learning in natural language processing*. Springer.

Dupond, S. (2019). A thorough review on the current advance of neural network structures. *Annual Review of Control, 14*, 200–230.

Eikebrokk, T. R., & Olsen, D. H. (2020). Robotic process automation and consequences for knowledge workers; a mixed-method study. In *Conference on e-Business, e-Services and e-Society*, 114–125, Springer.

Elliman, D., & Pulido, J. R. G. (2002). Visualizing ontology components through self-organizing maps. In *Proceedings Sixth International Conference on Information Visualization*, 434–438, IEEE.

Enriquez, J. G., Jiménez-Ramírez, A., Dominguez-Mayo, F. J., & Garcia-Garcia, J. A. (2020). Robotic process automation: A scientific and industrial systematic mapping study. *IEEE Access*, 8, 39113–39129.

Fernandez, D., & Aman, A. (2018). Impacts of robotic process automation on Global Accounting Services. *Asian Journal of Accounting and Governance*, 9, 123–132.

Flechsig, C., Lohmer, J., & Lasch, R. (2019). Realizing the full potential of robotic process automation through a combination with BPM. In *Logistics Management, Lecture Notes in Logistics,* 104–119, Springer.

Fung, H. P. (2014). Criteria, use cases and effects of Information Technology Process Automation (ITPA). *Advances in Robotics & Automation*, 3(3), 1–14.

Gao, J., van Zelst, S. J., Lu, X., & van der Aalst, W. M. P. (2019). Automated robotic process automation: A self-learning approach. *Lecture Notes in Computer Science*, 11877 LNCS, 95–112.

Gartner (2019). *Magic Quadrant for Robotic Process Automation*. Technical Report.

Geyer-Klingeberg, J., Nakladal, J., & Baldauf, F. (2018). Process mining and robotic process automation: A perfect match. In W. M. P. van der Aalst, F. Casati, R. Conforti, M. de Leoni, M. Dumas, A. Kumar, J. Mendling, S. Nepal, B. T. Pentland, & B. Weber (Eds.), *Proceedings of the Dissertation Award, Demonstration, and Industrial Track at BPM 2018,* pp. 124–131, CEUR-WS.org. researchr.org

González-Briones, A., Prieto, J., De La Prieta, F., Herrera-Viedma, E., & Corchado, J. M. (2018). Energy optimization using a case-based reasoning strategy. *Sensors, 18*(3), 865.

Goodfellow, I., Bengio, Y., & Courville, A. (2016). *Deep learning*. MIT Press.

Gotthardt, M., Koivulaakso, D., Paksoy, O., Saramo, C., Martikainen, M., & Lehner, O. M. (2019). Current state and challenges in the implementation of robotic process automation and artificial intelligence in accounting and auditing. *Journal of Finance & Risk Perspectives, 8*, 31–46.

Gu, B., Ge, R., Chen, Y., Luo, L., & Coatrieux, G. (2020). Automatic and robust object detection in X-ray baggage inspection using deep convolutional neural networks. *IEEE Transactions on Industrial Electronics*, 68 (10), 10248–10257. https://doi.org/10.1109/TIE.2020.3026285

Güner, E. O., Han, S., & Juell-Skielse, G. (2020). Robotic process automation as routine capability: A literature review. In *Proceedings of the 28th European Conference on Information Systems (ECIS)* (Paper 153). Association for Information Systems. https://aisel.aisnet.org/ecis2020_rp/153/.

Hallikainen, P., Bekkhus, R., & Pan, S. L. (2018). How OpusCapita used internal RPA capabilities to offer services to clients. *MIS Quarterly Executive*, 17(1), 41–52.

Hamamoto, A. H., Carvalho, L. F., Sampaio, L. D. H., Abrão, T., & Proença, M. L. Jr. (2018). Network anomaly detection system using genetic algorithm and fuzzy logic. *Expert Systems with Applications, 92*, 390–402.

Han, J., Pei, J., & Kamber, M. (2011). *Data mining: Concepts and techniques*. Elsevier.

Harrou, F., Zerrouki, N., Sun, Y., & Houacine, A. (2019). An integrated vision-based approach for efficient human fall detection in a home environment. *IEEE Access*, *7*, 114966–114974.

Huang, F., & Vasarhelyi, M. A. (2019). Applying robotic process automation (RPA) in auditing: A framework. *International Journal of Accounting Information Systems*, 35, 1–23.

Hwang, M. H., Na, U. K., Lee, S., Cho, B. J., Kim, Y., Lee, D. H., & Shin, J. K. (2020). MIORPA: Middleware system for open-source robotic process automation. *Journal of Computing Science and Engineering, 14*(1), 19–25.

Ian, H., Dhayalan, A., & Scott, A. (2016). The future of professional work: Will you be replaced or will you be sitting next to a robot? *Management Services Journal*, 60(2), 22–27.

Islam, M. Z., Islam, M. M., & Asraf, A. (2020). A combined deep CNN-LSTM network for the detection of novel coronavirus (COVID-19) using X-ray images. *Informatics in Medicine Unlocked, 20*, 100412.

Ivančić, L., Suša Vugec, D., & Bosilj Vukšić, V. (2019). Robotic process automation: Systematic literature review. In *Business Process Management: Blockchain and Central and Eastern Europe Forum*, 280–295, Springer.

Jalali, S., & Wohlin, C. (2012). Systematic literature studies: Database searches vs. backward snowballing. In *Proceedings of the 2012 ACM-IEEE International Symposium on Empirical Software Engineering and Measurement*, pp. 29–38. Association for Computing Machinery. https://doi.org/10.1145/2372251.2372257.

Jiménez-Ramírez, A., Chacón-Montero, J., Wojdynsky, T., & González Enríquez, J. (2020). Automated testing in robotic process automation projects. *Journal of Software: Evolution and Process*, (1), 1–11.

John, G. H., & Langley, P. (1995). Estimating continuous distributions in Bayesian classifiers. In *Proceedings of the Eleventh Conference on Uncertainty in Artificial Intelligence*, 338–345, Morgan Kaufmann.

Kaelbling, L. P., Littman, M. L., & Moore, A. W. (1996). Reinforcement learning: A survey. *Journal of Artificial Intelligence Research, 4*, 237–285.

Kang, X., Porter, C. S., & Bohemia, E. (2020). Using the fuzzy weighted association rule mining approach to develop a customer satisfaction product form. *Journal of Intelligent and Fuzzy Systems, 38*(4), 4343–4357.

Keerthi, S. S., Shevade, S. K., Bhattacharyya, C., & Murthy, K. R. K. (2001). Improvements to Platt's SMO algorithm for SVM classifier design. *Neural Computation, 13*(3), 637–649.

Khosravani, M. R., Nasiri, S., & Weinberg, K. (2019). Application of case-based reasoning in a fault detection system on production of drippers. *Applied Soft Computing, 75*, 227–232.

Kim, J.-Y., Seok-Jun, B., & Cho, S.-B. (2018). Zero-day malware detection using transferred generative adversarial networks based on deep autoencoders. *Information Sciences, 460*, 83–102.

Kin, G. W., Mustaffa, R., Johari, M. A. J., & Amruddin, A. Y. (2018). Cognitive automation robots (CAR). In *E-Proceeding of the 7th International Conference on Social Science Research*, 1–11, Atlantis Press.

Kohonen, T. (1990). The self-organizing map. *Proceedings of the IEEE, 78*(9), 1464–1480.

Kokina, J., & Blanchette, S. (2019). Early evidence of digital labor in accounting: Innovation with robotic process automation. *International Journal of Accounting Information Systems*, 35(1), 1–16.

Krishnan, H. K., et al. (2019). Smart irrigation using IoT with fuzzy logic-based decision-making. In *Proceedings of the 2019 IEEE 9th International Conference on Advanced Computing (IACC)*, 169–173, IEEE.

Lacity, M., Willcocks, L., & Craig, A. (2015a). Robotic process automation at Xchanging. *The Outsourcing Unit Working Research Paper Series*, 15(3), 1–26.

Lacity, M., Willcocks, L., & Craig, A. (2017). Service automation: Cognitive virtual agents at SEB Bank (Working Paper No. 17/01). The Outsourcing Unit. *London School of Economics and Political Science*, 1–29.

Lamy, J.-B., Sekar, B., Gueye, D., Sèbe, N., Bouaud, J., & Seroussi, B. (2019). Explainable artificial intelligence for breast cancer: A visual case-based reasoning approach. *Artificial Intelligence in Medicine*, *94*, 42–53.

LeCun, Y., Bengio, Y., & Hinton, G. (2015). Deep learning. *Nature*, *521*(7553), 436–444.

Leno, V., Dumas, M., Maggi, F. M., & La Rosa, M. (2018). Multi-perspective process model discovery for robotic process automation. In *CEUR Workshop Proceedings,* 2114, 37–45.

Leno, V., Polyvyanyy, A., Dumas, M., La Rosa, M., & Maggi, F. M. (2020). Robotic process mining: Vision and challenges. *Business and Information Systems Engineering*, *63*(3), 301–314.

Leopold, H., van Der Aa, H., & Reijers, H. A. (2018). Identifying candidate tasks for robotic process automation in textual process descriptions. In *Enterprise, Business-Process and Information Systems Modeling,* 67–81, Springer.

Leshob, A., Bourgouin, A., & Renard, L. (2018). Towards a process analysis approach to adopt robotic process automation. In *15th International Conference on e-Business Engineering*, 46–53, IEEE.

Lewicki, P., Tochowicz, J., & van Genuchten, J. (2019). Are robots taking our jobs? A RoboPlatform at a Bank. *IEEE Software*, 36(3), 101–104.

Li, D. (2020). A tutorial survey of architectures, algorithms, and applications for deep learning. *APSIPA Transactions on Signal and Information Processing*, *3*, 20.

Lohrmann, M., & Reichert, M. (2016). Effective application of process improvement patterns to business processes. *Software & Systems Modeling*, 15(2), 353–375.

Maedche, A., & Staab, S. (2001). Ontology learning for the semantic web. *IEEE Intelligent Systems*, *16*(2), 72–79.

Masood, A., & Hashmi, A. (2019). Cognitive robotics process automation: Automate this! In *Cognitive Computing Recipes*, 225–287. Apress.

Miers, D., Kerremans, M., Ray, S., & Tornbohm, C. (2019, July 8). *Magic Quadrant for Robotic Process Automation Software*. Gartner, Inc.

Mikolov, T., Karafiát, M., Burget, L., Černocký, J., & Khudanpur, S (2010). Recurrent neural network-based language model. In *Proceedings of the 11th Annual Conference of the International Speech Communication Association (INTERSPEECH 2010)*, pp. 1045–1048, International Speech Communication Association. https://doi.org/10.21437/Interspeech.2010-343.

Moffitt, K. C., Rozario, A. M., & Vasarhelyi, M. A. (2018). Robotic process automation for auditing. *Journal of Emerging Technologies in Accounting*, 15(2), 1–10.

Mohanty, S., & Vyas, P. (2018). Application of robotic process automation in finance: An investigation based on interviews with industry practitioners. *Journal of Financial Transformation*, 47(1), 83–91.

Muller, J., Brodeck, M., & Markowetz, F. (2020). Industrial Internet of Things and Artificial Intelligence Towards Industry 4.0. *SN Computer Science*, *1*(1), 61.

Osmundsen, K., Iden, J., & Bygstad, B. (2019). Organizing Robotic Process Automation: Balancing Loose and Tight Coupling. In *52nd Hawaii International Conference on System Sciences*, 6918–6926.

Pan, X., Yan, W., & Du, S. (2018). A novel visual recognition based on deep learning for navigation mark classification. In *2018 IEEE 3rd International Conference on Image, Vision and Computing (ICIVC)*, 107–112, IEEE.

Panchapagesan, S., Sahai, A., & Nelakuditi, S. (2019). RoboAccounting: Using robotic process automation (RPA) for modernizing taxation frameworks. In *IEEE International Conference on Cloud Engineering,* 14–23, IEEE.

Patel, M., Shukla, A., Porwal, R., & Kotecha, R. (2019). Customized automated email response bot using machine learning and robotic process automation. *SSRN Electronic Journal*. https://doi.org/10.2139/ssrn.3367743

Pearl, J. (1988). *Probabilistic reasoning in intelligent systems: Networks of plausible inference*. Morgan Kaufmann.

Penttinen, E., Kasslin, H., & Asatiani, A. (2018). How to choose between robotic process automation and back-end system automation? *Proceedings of the 26th European Conference on Information Systems(ECIS 2018)*, 1–14, Portsmouth, UK.

Phan, T., Duong, C. N., et al. (2019). Improving sentiment analysis on tweets by combining feature ensemble model and fuzzy sentiment. *Computers & Electrical Engineering*, *76*, 321–330.

Piccialli, F., et al. (2020). Machine learning and IoT for smart cities: A survey. *IEEE Internet of Things Journal*, *7*(5), 4155–4168.

Pilla, S. R., Sen, S., & Lim, A. (2018). A stochastic framework for robotic process automation (RPA) under uncertainty. In *Proceedings of the 2018 ACM Conference on Economics and Computation*, 165–182, Association for Computing Machinery (ACM).

Poort, R. J., et al. (2020). Automated diagnostic analytics workflow for the detection of production events-application to mature gas fields. *Computers & Geosciences*, 137, 104420.

Radke, A. M., Dang, M. T., & Tan, A. (2020). Using robotic process automation (RPA) to enhance item master data maintenance process. *LogForum*, 16(1), 129–140.

Ramzan, M., et al. (2020). Intelligent hotel recommendation based on customer ratings: A recommender system. *Computers, Materials & Continua, 62*(2), 651–666.

Reddy, G. M., et al. (2018). Diagnosis of heart disease using fuzzy classifier and optimization techniques derived by adaptive genetic algorithm. In *2018 IEEE International Conference on Systems, Man, and Cybernetics (SMC)*, 1435–1440, IEEE.

Riedl, Y., & Beetz, K. R. (2019). Robotic process automation: Developing a multi-criteria evaluation model for the selection of automatable business processes. In *Proceedings of the 25th Americas Conference on Information Systems (AMCIS 2019)*, pp. 1–14, Cancun, Mexico.

Sahar, S., et al. (2019). Smart parking system: Using machine learning for parking assistance. In *Proceedings of the 2019 International Conference on System Modelling & Advancement in Research Trends (SMART)*, 216–219, IEEE.

Santos, F., Pereira, R., & Braga Vasconcelos, J. (2019). Toward robotic process automation implementation: An end-to-end perspective. *Business Process Management Journal, 26*(2), 405–420.

Sarker, I. H., et al. (2020). Cyber anomaly detection using machine learning techniques: A performance evaluation perspective. *SN Computer Science, 1*(5), 1–9.

Sarker, I. H., et al. (2021). Data science and analytics towards intelligent decision support systems. *SN Computer Science, 2*(3), 160.

Schmitz, M., Dietze, C., & Czarnecki, C. (2019). Enabling digital transformation through robotic process automation at Deutsche Telekom. In *Digitalization Cases, Management for Professionals*, edited by N. Urbach and M. Röglinger, 15–33, Springer.

Schwab, K. (2016). *The fourth industrial revolution*. World Economic Forum.

Scott, S., et al. (1992). Decision trees for learning. In *Machine Learning Proceedings 1992,* 309–315, Morgan Kaufmann.

Srinivas, S., et al. (2018). A prescriptive analytics framework for optimizing outpatient appointment systems. *Journal of Biomedical Informatics*, *86*, 1–9.

Suri, V. K., Elia, M., Arora, P., & van Hillegersberg, J. (2018). Automation of knowledge-based shared services and centers of expertise. In Digital *s*ervices and *p*latforms, *c*onsiderations for sourcing, global sourcing, lecture notes in business information processing, *344*(1), pp. 1-16, Springer.

Sutton, R. S., & Barto, A. G. (2018). *Reinforcement learning: An introduction* (2nd ed.). MIT Press.

Syed, J., et al. (2021). Conceptual ontology and cyber intelligence alert system for vulnerability management in IoT networks. *IEEE Internet of Things Journal*, *20*, 20.

Syed, R., Suriadi, S., Adams, M., Bandara, W., Leemans, S. J. J., Ouyang, C., Hofstede, A. H. M. T., van de Weerd, I., Wynn, M. T., & Reijers, H. A. (2020). Robotic process automation: Contemporary themes and challenges. *Computers in Industry*, *115*, 103162. https://doi.org/10.1016/j.compind.2019.103162

Talib, M. S., et al. (2021). Ontology-based cyber security policy implementation: A Saudi Arabian case study. *SN Computer Science*, *2*(3), 160.

Turk, M. A., et al. (1991). Eigenfaces for recognition. *Journal of Cognitive Neuroscience*, *3*(1), 71–86.

Wang, Y., et al. (2018). Human activity recognition with distributed deep convolutional networks in wearable embedded devices. *IEEE Transactions on Industrial Electronics*, *65*(12), 9885–9894. https://doi.org/10.1109/TIE.2018.2803205

Wang, Y., et al. (2019). Real-time collision prediction based on deep learning techniques for intelligent transportation systems. *IEEE Transactions on Intelligent Transportation Systems*, *20*(6), 1875–1894.

Wanner, J., Hofmann, A., Fischer, M., Imgrund, F., Janiesch, C., & Geyer-Klingeberg, J. (2020). Process selection in RPA projects – Towards a quantifiable method of decision making. In *40th International Conference on Information Systems, ICIS*, December 15–18, 2019, 1–17, Association for Information Systems (AIS).

Willcocks, L., & Lacity, M. (2015). Robotic process automation: The next transformation lever for shared services. *The Outsourcing Unit Working Research Paper Series*, 15 (7), 1–35.

Willcocks, L., & Lacity, M. (2016). Robotic process automation at Telefónica O2. *MIS Quarterly Executive*, 15(1), 21–35.

Wróblewska, A., Stanislawek, T., Prus-Zajączkowski, B., & Garncarek, L. (2018). Robotic process automation of unstructured data with machine learning. *Annals of Computer Science and Information Systems*, 16, 9–16.

Yang, Z., et al. (2020). A review of unsupervised learning methods and artificial neural networks used for anomaly detection. *Soft Computing*, *24*(20), 16013–16029.

Zadeh, L. A. (1965). Fuzzy sets. *Information and Control, 8*(3), 338–353.

8 Detection of Cyberattacks in Cyber-Physical Microgrid Using Machine Learning Techniques

Ankur Kumar, Niraj Kumar Choudhary, and Nitin Singh

8.1 INTRODUCTION

The power system infrastructure of the 21st century has seen a substantial transformation as a result of advancements in digital technology, automation, and connectivity. Smart power systems integrate traditional energy infrastructure with cutting-edge communication and information technology, leading to a new era of remarkable efficiency, reliability, and sustainability. Modern cyber-physical power systems are characterized by the integration of digital surveillance and control techniques into every aspect of the energy supply chain. ICT-enabled devices, sensors, and software platforms are used across the whole energy system, including generation, transmission, distribution, and consumption stages. Their objective is to enhance operational efficiency, ability to withstand disruptions, and facilitate better decision-making. A cyber-physical microgrid (CPM) is an essential component of the modern intelligent power systems. The CPM employs extensive processing and communication protocols to execute decentralized operations on electrical components and attain optimized performance. The control unit of an intelligent microgrid coordinates various distributed energy resources (DERs) and regulated loads to provide affordable and dependable electricity with negligible environmental impact [1].

Figure 8.1 illustrates a microgrid structure with three control levels that function at different time scales to accomplish control goals. The CPM consists of two individual layers. The physical layer ensures an efficient flow of energy to ensure regional load requirements, while the cyber layer facilitates the exchange of data between the microgrid elements through a sparse communication link.

In a CPM, most of the micro-generating modules produce DC power; hence, it is necessary to have inverter circuits to provide AC power to the loads [2, 3]. The cyber layer comprises communication devices outfitted with routers, connections,

DOI: 10.1201/9781003563440-8

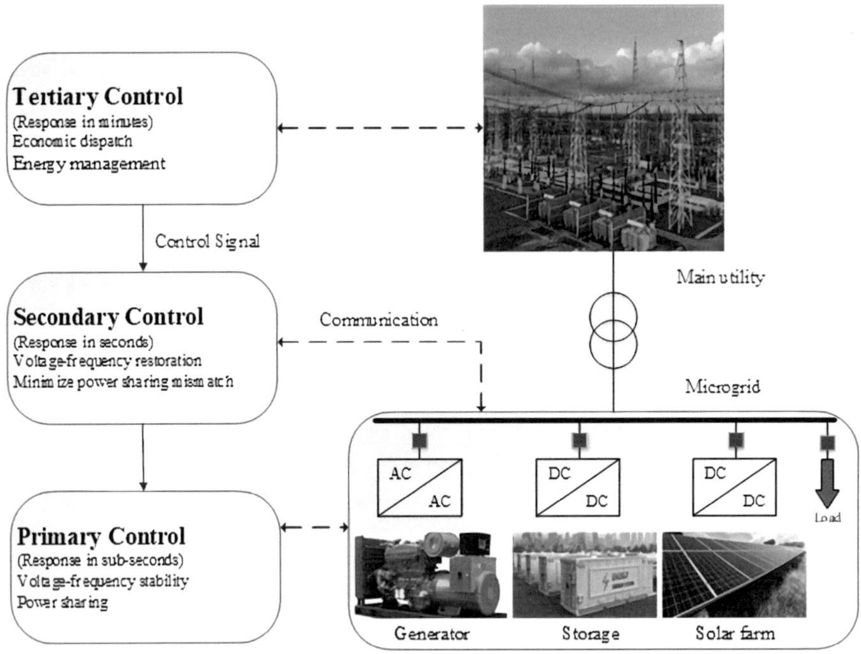

FIGURE 8.1 Hierarchical structure of microgrid.

regional control devices, and complex algorithms. Its purpose is to tackle the different issues encountered by microgrid operators and customers. A well-organized control setup has been widely employed to guarantee reliable and effective functioning of microgrids. When operating in the standalone mode, the principal controllers maintain the stability of the system by keeping the v/f values within a specific range [4].

The principal neighborhood controllers are synchronized using active and reactive power drooping techniques [5]. Therefore, the primary controller's effectiveness relies on the operation of these controllers, aiming to regulate the assigned proportions of reactive and active powers [6]. Centralized secondary control involves the transfer and analysis of real-time data, such as the voltage and current measurements from different buses, at a centralized node. As a result, it is more likely to experience single-point failure issues and requires a greater amount of bandwidth [7]. To address these problems, a decentralized auxiliary controllers is utilized. This controller is designed such that the generating unit carries its own regional controller and only requires information derived from its own source and its neighboring controller, rather than relying on information from the entire system. Therefore, a decentralized control mechanism is superior in terms of reliability and efficiency compared to centralized control systems. The secondary controllers operate on a larger time scale in relation to the primary controller. This facilitates the mechanisms of isolating and developing the primary and auxiliary levels of control [8].

It is essential to have a dependable protection system in place to provide a continuous and steady power supply, especially when fault occur. Traditional protective relays are insufficient in protecting the system in the event of inverter-based CPM owing to the presence of a weak fault current [9]. The most frequent faults are those closely associated with line and ground. An effective fault identification strategy, including algorithms for efficient fault classification and identification, is essential for a fault protection model. This method enhances the reliability of the entire protection procedure by categorizing faults and using an appropriate mitigation technique to minimize the time and expense required for restoration [10]. It is necessary to establish a fault identification technique that is both effective and utilizes an algorithm for accurately detecting and classifying faults. This strategy enhances the reliability of the overall protection process by categorizing the faults and executing a suitable mitigation method in minimizing the outage and system restoration period.

Cyberattacks occur when an adversary actively penetrates a communication network by either disseminating misleading information or disrupting its transmission channels among agents [11]. Both false data injection (FDI) and denial of services (DoS) attacks are the most common types of cyber threats in CPM. These attacks have the potential to interrupt the operation of the power grid and its robustness. To be more specific, distributed DoS attacks have an effect of rendering authorized consumers unable to access their products or services by leveraging the bandwidth constraints. Traditionally, network monitors are able to easily identify threats of this nature by examining the distribution rates of the threats [12].

Krishan Arora et al. [13] enumerate a discrete wavelet transform to detect faults in microgrids. Further, the attacks in the electrical grid are identified using a deep neural network [14]. It employs machine learning to assist microgrid administrators in detecting anomalies. A smart differential protection strategy was presented to detect cyberattacks in CPM. Using discrete Fourier transform, this approach reviews the anomalous voltage and current readings and finds distressed qualities at either terminal of the relevant feeder [15]. In order to arrive at a final conclusion, the decision tree is built using differential attributes that are computed from the appropriate attributes. From the above observation, following are the main contribution of this paper:

- To detect the cyberattack in cyber-physical layered microgrid using machine learning techniques
- To classify between a fault and cyberattack, when the system is under threat.
- To perform a comparative study of cyberattack detection using machine learning techniques

The remaining study is organized in four sections. Section 8.2 describes the potential cyber threats in distribution network. Section 8.3 presents an information about the system and the machine learning techniques. Section 8.4 covers the results and discussion of cyberattack detection in the microgrid and at last the study is concluded in Section 8.5.

8.2 POTENTIAL CYBER THREATS IN DISTRIBUTION NETWORK

A CPM is very vulnerable to serious cyber security risks. With the progress of automation and communication technologies in the distribution system, attackers now possess the capability to alter or interrupt the industrial control system, leading to disruptions. The incorporation of several DERs is also increasing, leading to the supply of excess electricity to the grid. The DERs are designed to carry out system monitoring and control, which increases the susceptibility to cyberattacks and the possibility of disrupting the power grid. Attackers use several methods to destabilize the distribution network with the aim of triggering a blackout or disturbing the system.

Malicious data injection refers to the deliberate inclusion of inaccurate information into a network connection. Typically, the attackers aim to introduce incorrect data into the cyber-physical layer's flow of information to diminish the controller's capability to make decisions. Intruders attempt to induce load shedding and wasteful tripping by executing fake data injections or transmitting false information to the microgrid carriers, therefore inflating the reported electrical consumption to the electrical grid operator. When there is a malicious risk, the system controller might get notified of a trailing power factor. In response, the controller acts to include negative kilovolt-ampere reactive (kVAR) in order to stabilize the power factor at unity in the system, hence causing the cyber-physical layer to operate at a leading power factor. Subsequently, this dominant power factor has the potential to elevate the voltage level inside the secondary distribution system, so posing a risk of damaging domestic appliances.

Electricity theft occurs when individuals deliberately underreport their power use to the operator in order to get financial advantages by decreasing their billing expenses. An intruder may ensure that the operator stays oblivious to load fluctuations through obstructing or delaying the network connection. The denial of service attack exploits vulnerabilities in the communication network by flooding it with a large number of requests, rendering the targeted components unable to function properly. As a result, the communication system or the host (server) is unable to operate efficiently. Thus, a proficient cyberattack or fault detection system is required. The details of the considered cyberattacks are highlighted in the following subsections.

8.3 DENIAL OF SERVICES

Despite lacking complete access to the electrical control system, the offenders of this cyberattack are capable of manipulating a limited number of devices with the intention of corrupting the system [8].

DoS attacks typically involve inundating the targeted system with an excessive volume of requests with the intention of overloading it and impeding the fulfilment of requests that are genuine [16]. In Figure 8.2, distributed DoS is employed as an attack to the system. The attackers use many altered distributed systems to send numerous malicious requests to the intended target, thereby impeding operators from responding to genuine questions.

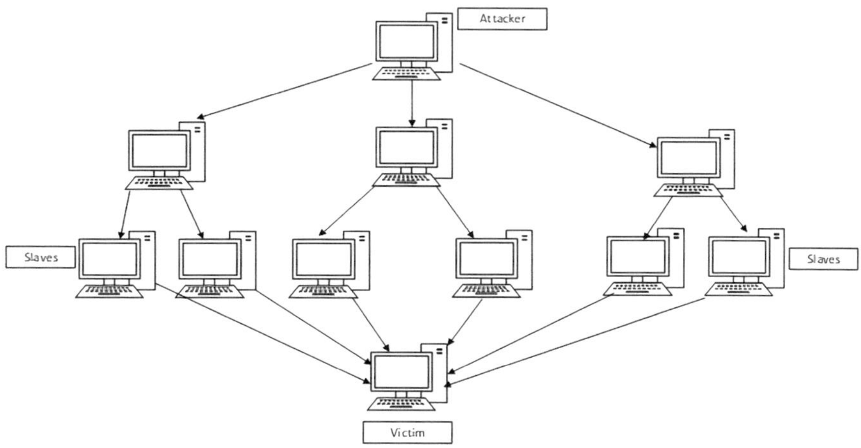

FIGURE 8.2 Denial of service.

8.4 FALSE DATA INJECTION

FDI refers to the unauthorized manipulation of sensors, such as voltage and current sensors, by attackers. Manipulating measurements using fake data might cause adversaries to mislead control systems and lead them to inaccurate decisions. The data used for state estimation may be compromised by an FDI attack, leading to inaccurate assessments of the system's status. This may lead to control systems performing erroneous actions, including inadequate generation adjustments or failure to identify faults. Disseminating inaccurate data within the electrical system might lead to a cascade of failures across the network. Identifying FDI attacks may be a challenge since they often attempt to resemble typical system operations.

8.5 TEST SYSTEM

This section presents an elaborate explanation of the intended system. This study focuses on the use of the CPL-based IEEE-15 bus and the modified IEEE-15 bus system. In both the systems, there are two layers, i.e., physical layer and cyber layer as shown in Figure 8.3. The physical layer is basically the benchmark or modified IEEE-15 distribution system which is connected to the cyber layer. The cyber layer facilitates communication between physical components over a limited network. An efficient network is required to link all the DERs with the central controller in order to provide the typical centralized secondary control.

In the current protective load (CPL), the system measures and utilizes physical state variables to control the flow of power. The cyber layer estimates the state parameters by recurrent computation and additional algebraic parameters. The measuring devices within the state estimation unit precisely assess measurements in real time and relay them to the processing section. Once the data has been processed, the computational unit delivers the estimated values to the controllers over a network

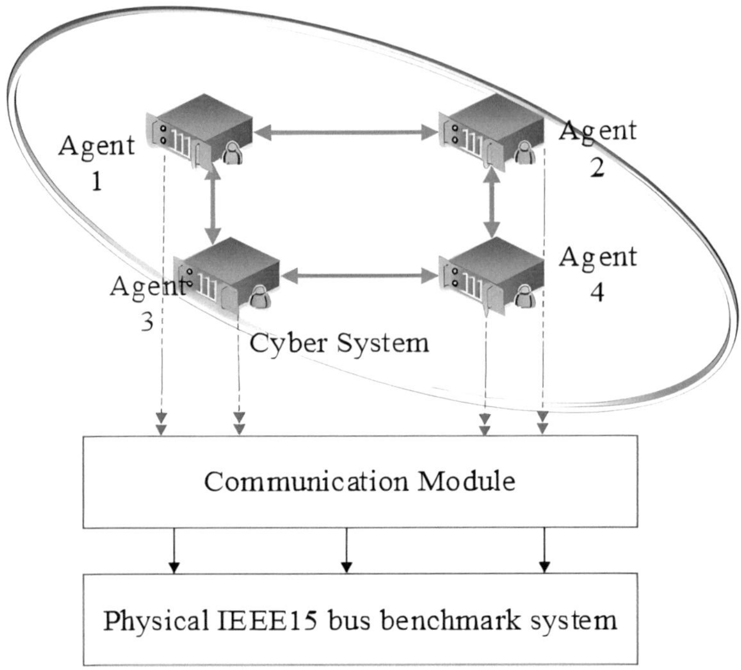

FIGURE 8.3 Cyber physical layered distribution system.

connection. Figure 8.3 depicts the components that contribute to the transmission of information from standard IEEE-15 bus radial distribution system.

8.6 ANOMALIES IN CYBER-PHYSICAL SYSTEM

In the IEEE-15 bus benchmark and modified IEEE-15 bus cyber physical system [17], three types of abnormalities are considered. First anomaly is due to insertion on false data from the bus node. A random attack is also considered by disconnecting the resources connect to the network. The stealthy data type attack is also considered by injecting the data at sensor nodes.

Physical faults are also considered to make a clear visibility between the fault and a cyberattack. These faults are separated majorly into two types: line to ground (L-G) and double line to ground fault (L-L-G). These faults are quite rare in real life. It is possible to find shunt faults by monitoring phase currents. The shunt faults indicated by the elevated current levels are categorized as symmetrical and asymmetrical faults. Asymmetrical faults, such as line-ground, line-line, and LL-G, are common examples. L-G, L-L-G, L-L-L, and L-L-L-G faults data, i.e., voltage and current values are considered to identify the attack or the state of fault in the distribution network (Figure 8.4).

In the proposed work, machine learning algorithms are employed to identify cyberattack in the CPL-based IEEE-15 and modified IEEE-15 bus systems. The

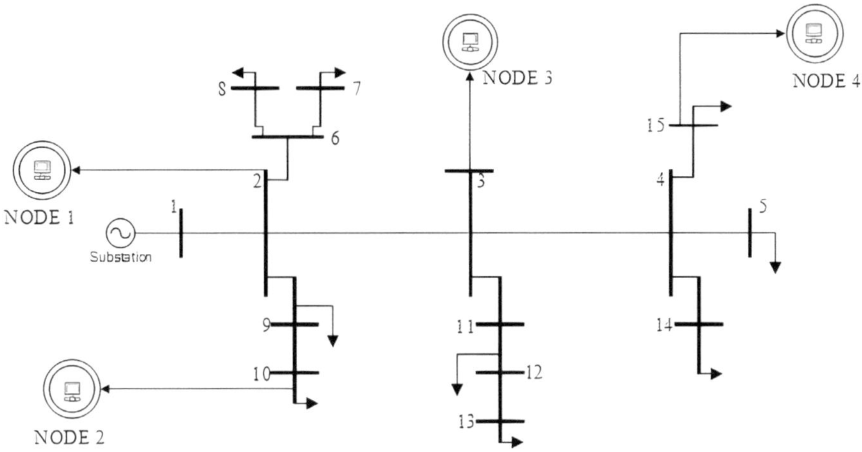

FIGURE 8.4 Single line diagram of IEEE-15 bus system with cyber physical layered node.

suggested solution combines the fine decision tree (DT), narrow neural network (NNN), and support vector machine (SVM) to accurately determine the cyberattacks.

Figure 8.5 shows an algorithmic to identify the cyberattack in the IEEE-15 bus benchmark and modified IEEE-15 bus networks. A databank of voltage and current for symmetrical and asymmetrical faults is generated for 60000 samples. The faulty and nonfaulty states of the distribution network are further categorized by featuring a special categorizing value. Using this databank, the machine learning classifiers, i.e. DT [18], NNN, and SVM [19] are trained. In the very next step, the anomalies of different stages are introduced like FDI and DoS at any node. The injected data is interpreted as the testing data which clarify the state of distribution network [20]. The false data of IEEE-15 bus is also generated by targeting the phase current values with the least and maximum multiplier of 0.4 and 6, respectively. The fabricated data is passed through the network trained by the machine learning classifiers. If any abnormalities are detected that correspond to the faulty data, the system is classified as being in a real-time fault state. Conversely, if no such anomalies are found, it is determined that the system is under a cyberattack. The following are the machine learning approaches used for cyberattack detection.

8.7 RESULTS AND DISCUSSION

The performance analysis of cyberattack and fault detection in intelligent distribution system is tested with three machine learning algorithms, i.e. DT, NNN, and SVM to identify whether the system is in faulty state or in cyberattack. Two type of attacks, i.e., DoS and FDI are considered in the proposed study. The three phase current values (I_a, I_b, and I_c) are categorized as the input. The condition of no fault, symmetrical, and unsymmetrical faults are simulated to collect the data samples for the training of machine learning algorithms. Distinct target values are assigned to distinguish between the system's presence and no faults state.

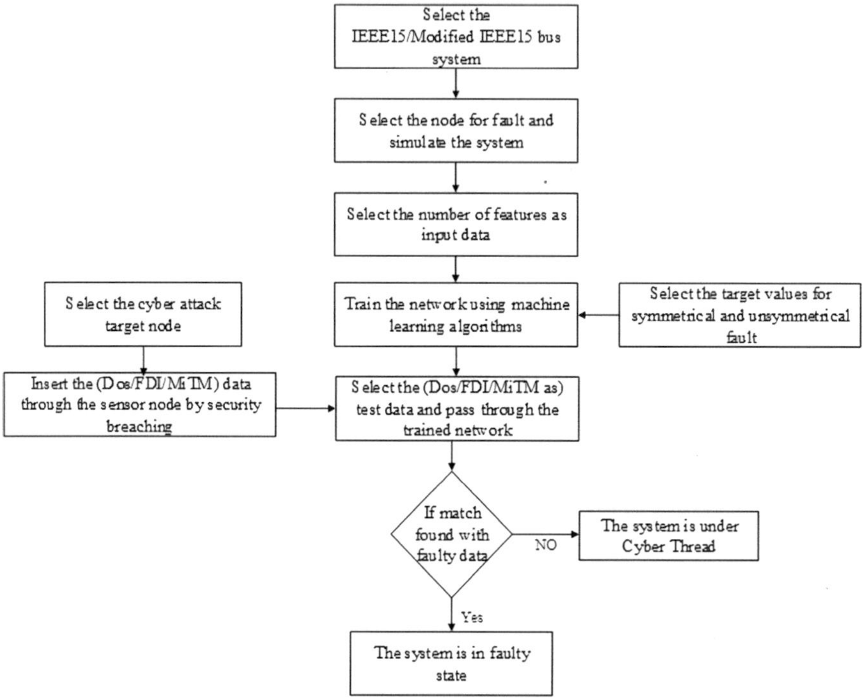

FIGURE 8.5 Algorithm for cyberattack detection.

The cyberattack datasets are selected by the anomalies injected in the smart distribution network. These anomalies disturbed the actual database and make a disruption at the sensor node. These disruptions behave like a fault in the network system, and the network collapse till the actual issue is not configured (Figure 8.6).

Positions 1, 2, and 3 correspond to L-G fault validation, whereas positions 4, 5, and 6 indicate L-L-G fault. Positions 7, 8, and 9 indicate a line-to-line fault, whereas the matrix 10 component indicates an L-L-L-G fault. The DT network has a validation error rate of 20.4% and a training time of 13.14 seconds. The DT achieves a validation accuracy of 79.56% and maintains a validation error rate of 20.6%. The NNN classifier has an overall validation accuracy of 89.1%, whereas the SVM classifier has an overall validation accuracy of 90.8%.

Figure 8.7 represents the validation matrix for NNN with an error rate of 10.9%, and Figure 8.8 shows the validation matrix of SVM with an error rate of 9.2%.

After the network validation, the test data is passed through the trained network. Test confusion matrix are observed for symmetrical and unsymmetrical faults. In Figure 8.9, the confusion matrix shows the prediction of fault classification over cyberattack using DT classifier.

The true positive rate (TPR) represents the classification rate of cyberattack with the fault data. The confusion matrix represents that DT classifies the cyberattack data

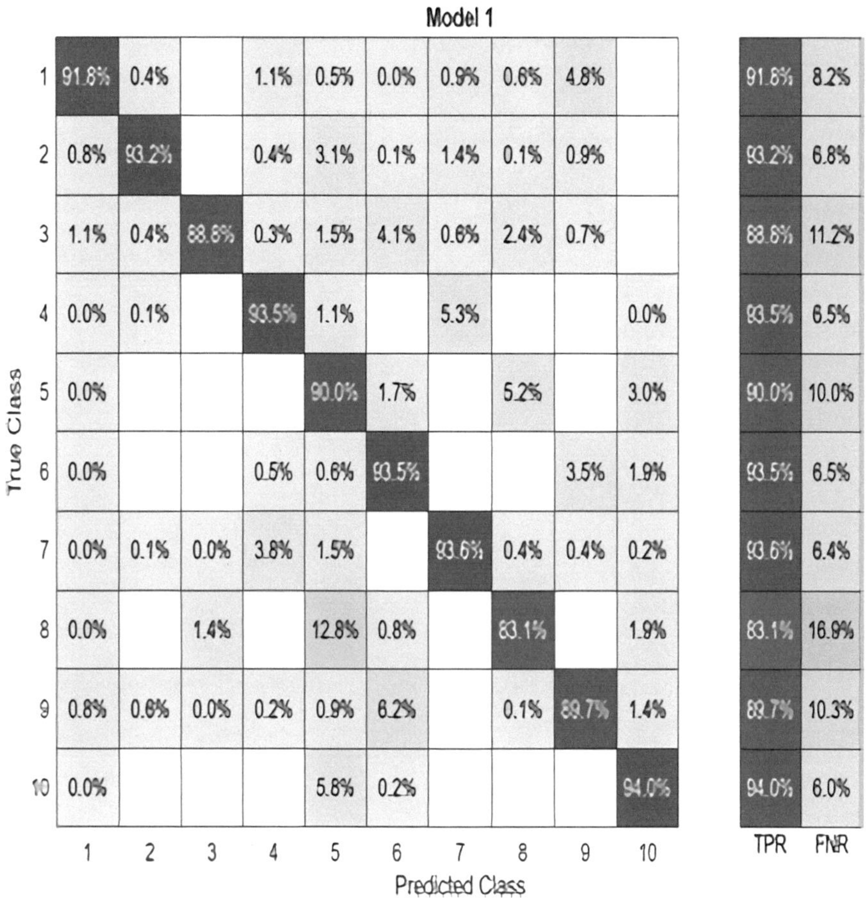

FIGURE 8.6 Validation of DT classifier.

as the fault data. The false negative rate (FNR) predict the better accuracy for the classification of cyberattack. Hence, the overall misclassification rate represents the accuracy of cyberattack detection. DT classifier shows the classification of cyber data as fault with a rate of 70.4%.

The same observation has been found for the NNN and SVM. Figures 8.10 and 8.11 represent the test confusion matrix obtained from NNN and SVM. In case of NNN, the findings of cyberattack in fault data is 32.8% which clearly measures that the cyberattack detection is performed well and the accuracy is 67.2%. The classification aspects from SVM are shown in Figure 8.11, where the least classification range cyberattack from fault data of 9.9% is observed. The better misclassification results shows the fine prediction of cyberattack in the active distribution network. Here, SVM is found as the better classifier of cyberattack with a success rate of 90.1%.

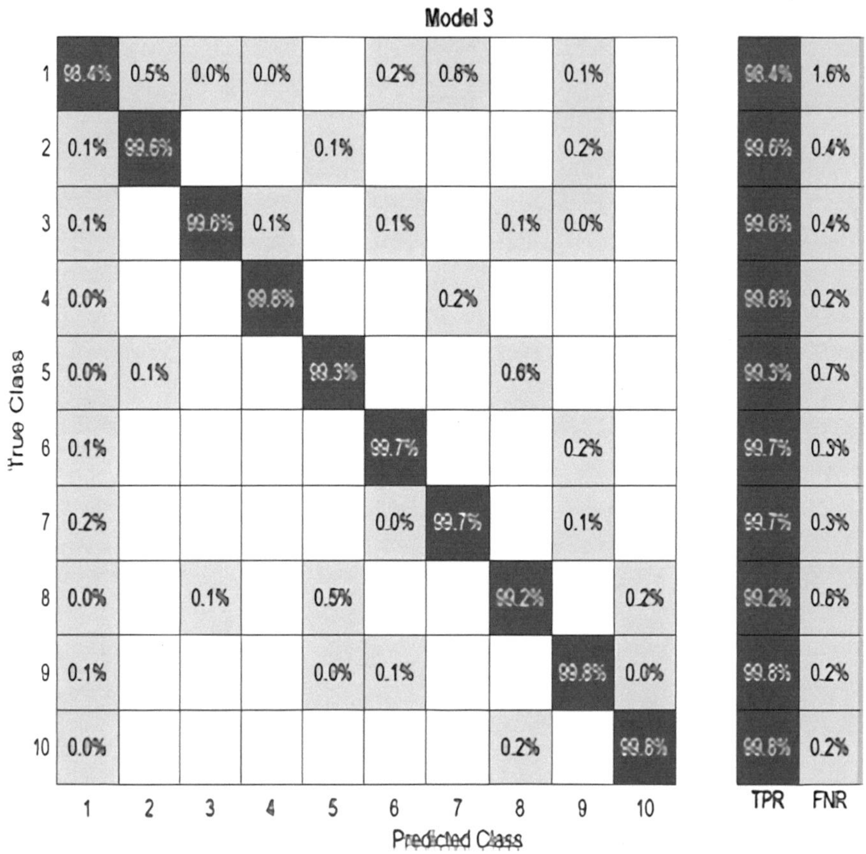

FIGURE 8.7 Validation matrix of NNN.

8.7.1 Case 1: Denial of Service

In the first scenario, the DoS cyberattack is initiated at Node2 in the CPL-based IEEE-15 bus system (Figure 8.2). In this scenario, the attack involves injecting data into the system when an uneven load is connected as a real-time request, causing the system to enter a breakdown or failure state. The same observation is assumed for DG-based modified IEEE-15 bus system. These observation shows a fine variation in the three-phase voltage and current. These data are considered as the testing data and passed through the trained network classifier.

Figures 8.12 and 8.13 shows the symmetrical faults data, where the DoS threat has been classified in the IEEE-15 bus radial distribution network. The DT classifier classified the cyberattack with least classification value of 58.5% for L-L-L fault and 19.5% for L-L-L-G fault conditions. In case of symmetrical faults, NNN classify the DoS attack with an accuracy of 62 and 54.4%, respectively. Similarly, the SVM classified the cyberattack the same faults with a maximum classification of 99.6% and 100%.

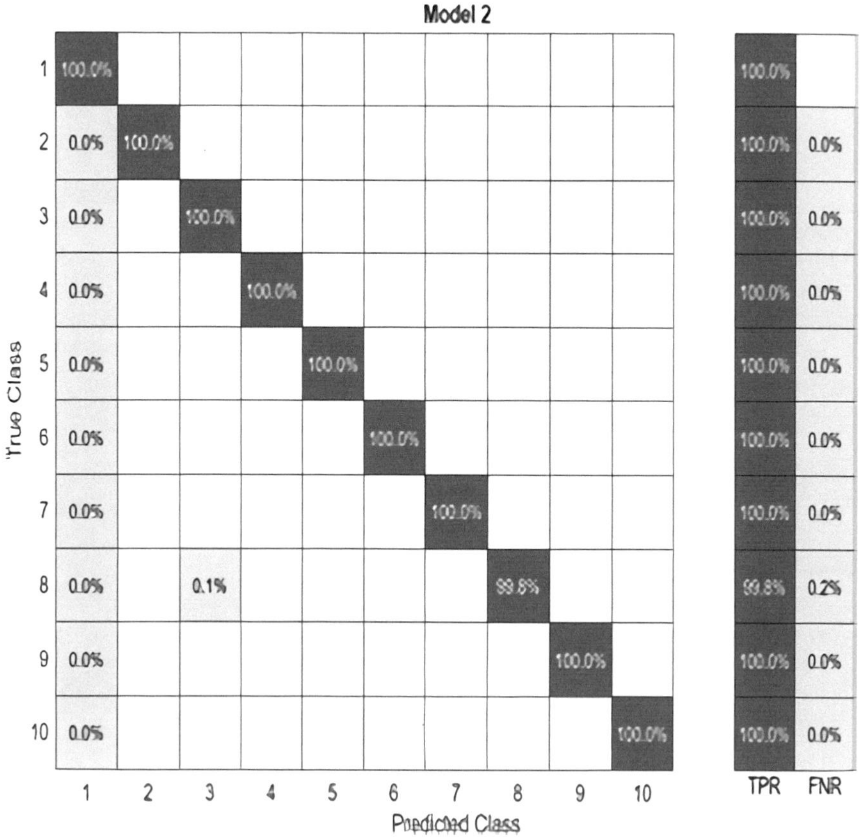

FIGURE 8.8 Validation matrix of SVM.

In case of unsymmetrical fault, DT classifies the cyberattack with an accuracy of 8.6%, 13.43%, and 18.8% for L-G, L-L, and L-L-G fault, respectively. The NNN predict the attack with an accuracy of 76.66%, 55.46% and 82.4%. SVM recognized the cyberattack with a precision of 100% classification under L-G, L-L, and L-L-G fault conditions.

Similarly, the DoS attack has been proposed for DG-based modified IEEE-15 bus system. Figures 8.14 and 8.15 show the classification accuracy. DT has least classification with 66.6, 31.46, and 17.2% for unsymmetrical types of faults. NNN has improved classification accuracy of L-G, L-L, and L-L-G faults with a rate of 66.60%, 77.20%, and 69.46%, respectively. In case of symmetrical faults, DT and NNN classify the cyberattack as a fault in the system. SVM has maximum accuracy of 80.8% and 81% in cyberattack classification as shown in Figure 8.16. In unsymmetrical fault data, the DoS is classified by SVM with an accuracy of 100%.

Figures 8.16, 8.17, and 8.18 show the ROC test curve for DT, NNN, and SVM classifier, respectively. The curve shows the misclassification of L-L-G and L-L-L-G fault with DoS attack in the IEEE-15 bus radial distribution system. Figures 8.19,

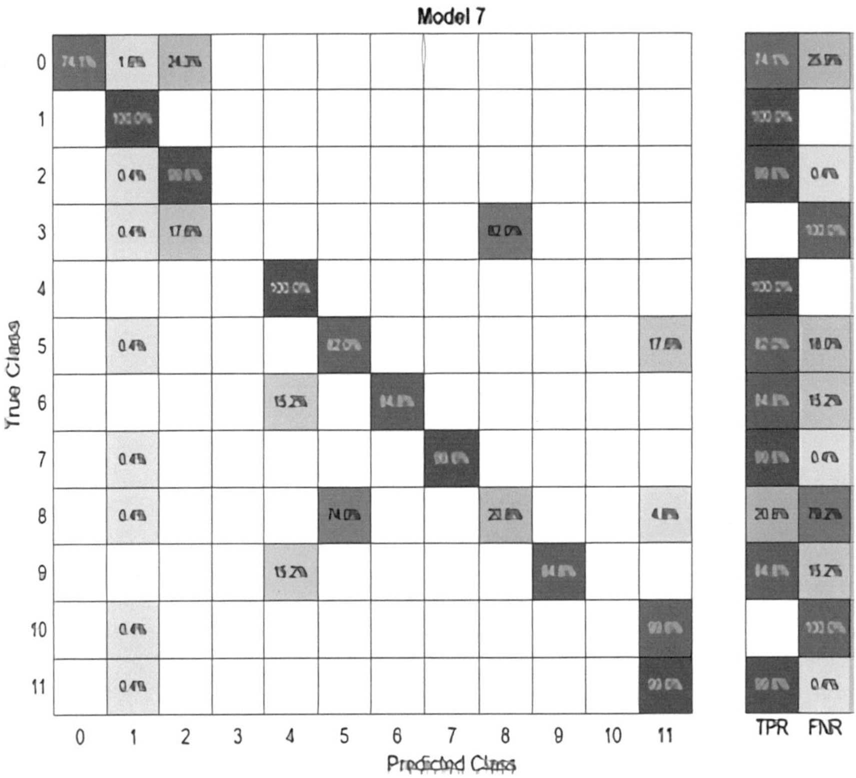

FIGURE 8.9 Test confusion matrix using DT classifier.

8.20, and 8.21 show the ROC test curve for DT, NNN, and SVM classifier in classification for DoS attack in the modified IEEE-15 bus system.

Table 8.1 represents the classification result of DoS attack in the cyber physical layered IEEE-15 bus system and modified IEEE-15 bus system. During the classification of DoS in the IEEE-15 bus system, the SVM has maximum validation of 90.76%, and it classifies the DoS attack with the highest accuracy of 90.07%. The DT classifier has the least accuracy of 29.56%.

In case of modified system, the DT, NNN, and SVM are quite efficient with a validation percentage of 75.76%, 86.52%, and 93.46%, respectively. During the classification of cyberattack, the DT has minimum accuracy in attack classification with a rate of 23.69%, and the SVM classifier recognized the cyberattack with highest accuracy of 92.96%. From the above results, it is clearly observed that in case of DoS attack, SVM performed as the best classifier to recognize the cyberattack in CPL distribution system.

8.7.2 Case 2: False Data Injection

The FDI involves the change in sensor node data by manipulating the measurement with a typically fabricated data. Under the observation of fault data and FDI threat,

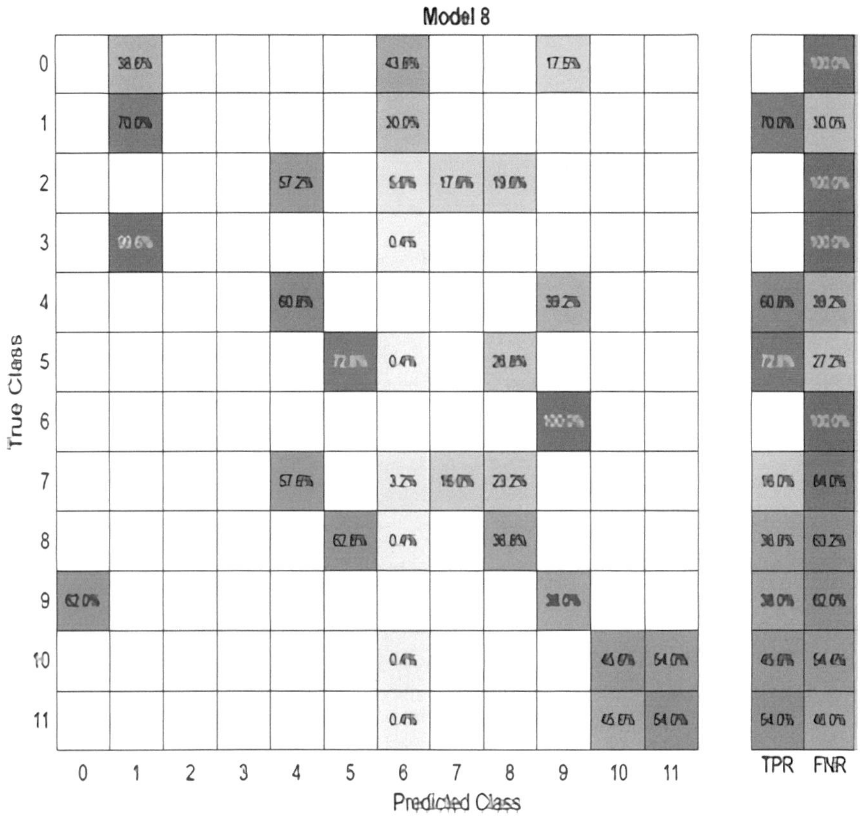

FIGURE 8.10 Test confusion matrix using NNN classifier.

the fabricated data is again passed through the trained network. This fabrication is initiated such that a minor change in any one or two phase fault current data can be changed.

Node 2 is again considered for the cyberattack classification in the IEEE-15 bus and modified IEEE-15 bus systems. Figures 8.22 and 8.23 show the misclassification of cyberattack under symmetrical and unsymmetrical fault states in the CPL-based IEEE-15 bus system. It is observed that the DT classifier has least accuracy to predict the cyberattack. The NNN classifier predicts the attack with a precision of 66.5%, whereas the SVM makes a clear prediction for classification of cyberattack with a maximum accuracy of 99.82%. Figures 8.24 and 8.25 show the classification accuracy in the modified IEEE-15 bus system.

Table 8.2 displays the classification details of FDI attacks for both the CPL IEEE-15 bus and the modified IEEE-15 bus systems. The DT classifier in the IEEE-15 bus system has a low level of response when it attempts classifying cyberattacks, with an accuracy rate of just 21.60%. The use of the NNN has resulted in an enhanced FDI classification accuracy of 66.20%. The SVM accurately classifies the FDI attack as a fault with a maximum classification accuracy of 98.82%.

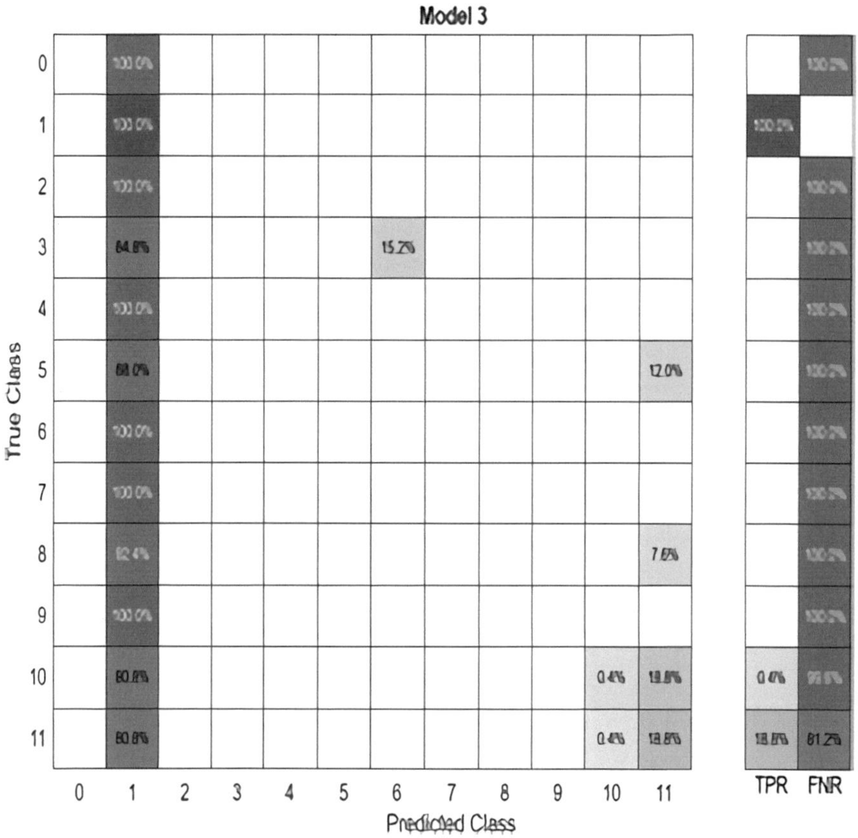

FIGURE 8.11 Test confusion matrix using SVM classifier.

In the modified IEEE-15 bus system, the SVM remain the best classifier to identify the FDI attack with a rate of 92.36%, whereas the NNN and DT classifier classifies the attack with an accuracy of 59.36% and 33.05%, respectively.

8.8 CONCLUSION

This chapter examines the performance of machine learning techniques to cyberattack monitoring, specifically focusing on detecting cyberattacks. Machine learning techniques are used to detect cyberattacks in CPL active distribution systems. As the intensity of attacks changes, the assessment indices often show an upward trend in response to the increasing severity of the attacks. The performance of the DT, NNN, and SVM algorithms has been evaluated using MATLAB® software. The study is conducted on the CPL IEEE-15 bus and modified IEEE-15 bus systems. The misclassification rate of a fault is an indicator of the accuracy of the trained network. The correlation of symmetrical and unsymmetrical faults is used to detect the cyberattack. The comparison of Tables 8.1 and 8.2 shows that the SVM is the best effective classifier for detecting cyberattacks.

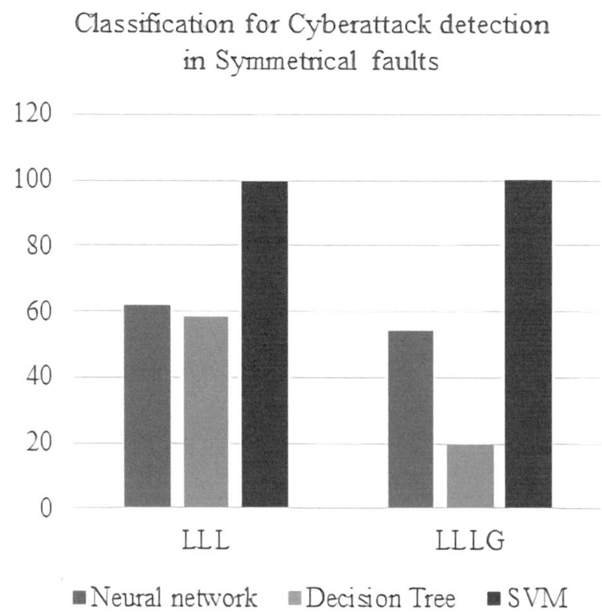

FIGURE 8.12 Detection of DoS under symmetrical fault in CPL-based IEEE15 bus radial distribution system.

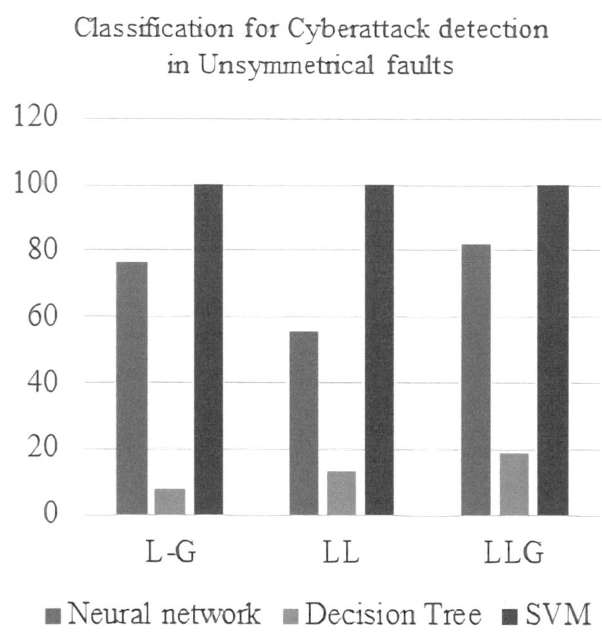

FIGURE 8.13 Detection of DoS under unsymmetrical fault in CPL-based IEEE15 bus distribution network.

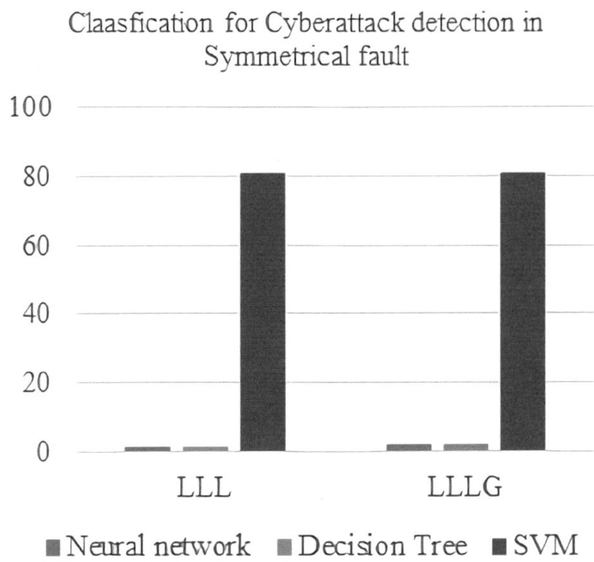

FIGURE 8.14 Classification of DoS in the modified CPL-based IEEE-15 bus system under symmetrical faults.

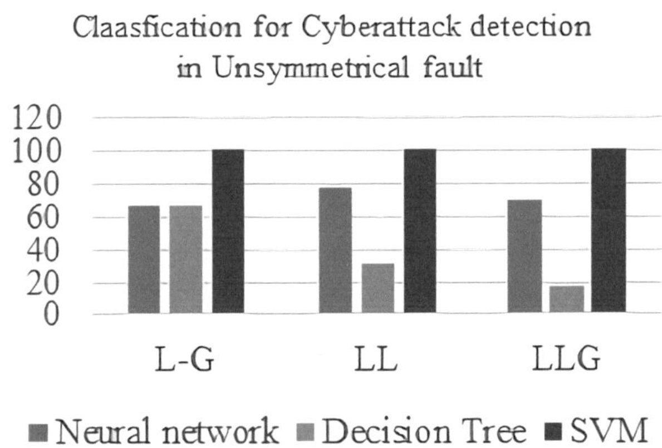

FIGURE 8.15 Classification of DoS in the modified IEEE-15 bus system under unsymmetrical faults.

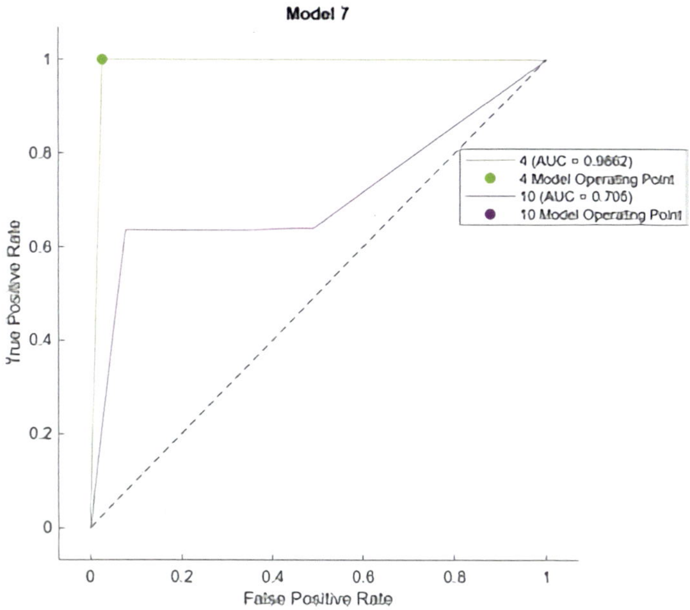

FIGURE 8.16 ROC test curve for the DT classifier for misclassification of L-L-G and L-L-L-G fault.

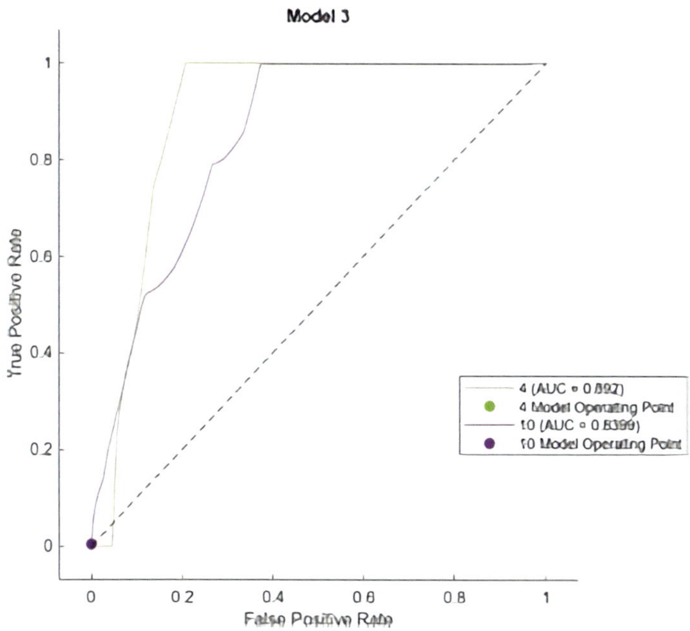

FIGURE 8.17 ROC test curve for NNN classifier for misclassification L-L-G and L-L-L-G fault.

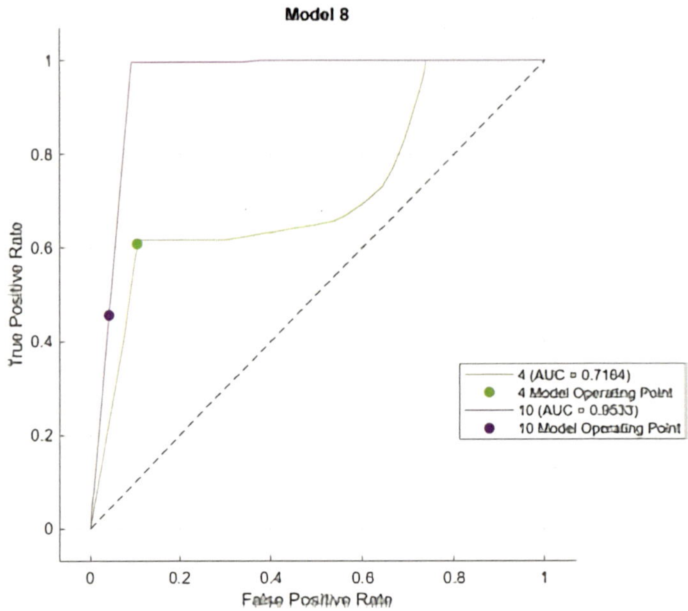

FIGURE 8.18 ROC test curve for SVM classifier for misclassification of L-L-G and L-L-L-G fault.

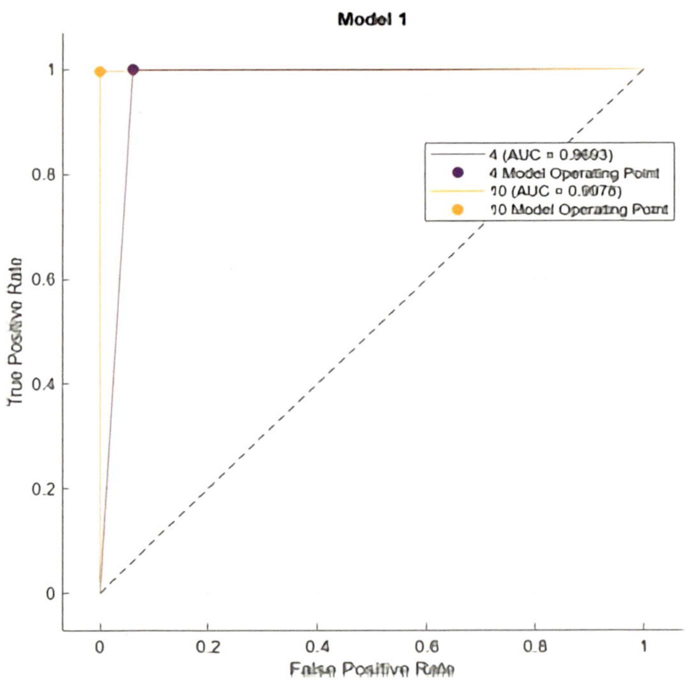

FIGURE 8.19 ROC test curve for the DT classifier in the modified IEEE-15 bus system.

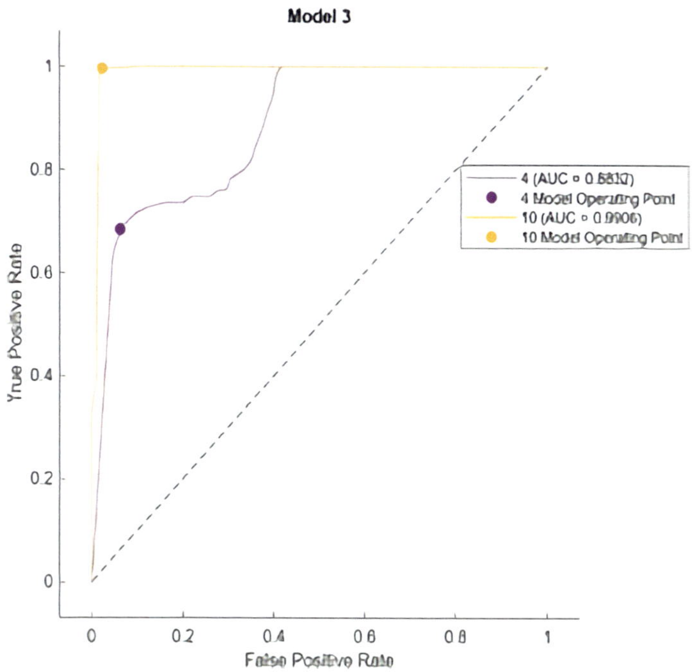

FIGURE 8.20 ROC test curve for the NNN classifier in the modified IEEE-15 bus system.

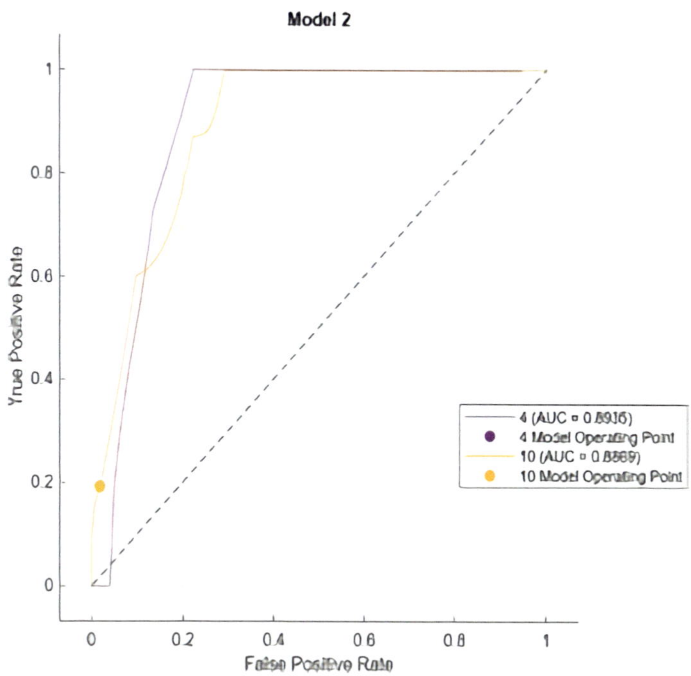

FIGURE 8.21 ROC test curve for the SVM classifier in the modified IEEE-15 bus system.

TABLE 8.1
Comparative analysis of DoS cyberattack using machine learning techniques

Sr. No.	Machine learning classifier	Features	Preset	CPL-based IEEE-15 bus system		CPL-based modified IEEE-15 bus system	
				Validation	Accuracy	Validation	Accuracy
1	Decision tree	Va, Vb, Vc,	Fine tree	79.56%	29.56%	74.76%	23.69%
2	Neural network	Ia, Ib, and Ic	Narrow neural network	89.15%	67.18%	86.52%	43.33%
3	SVM		Fine Gaussian SVM	90.76%	90.07%	93.46%	92.96%

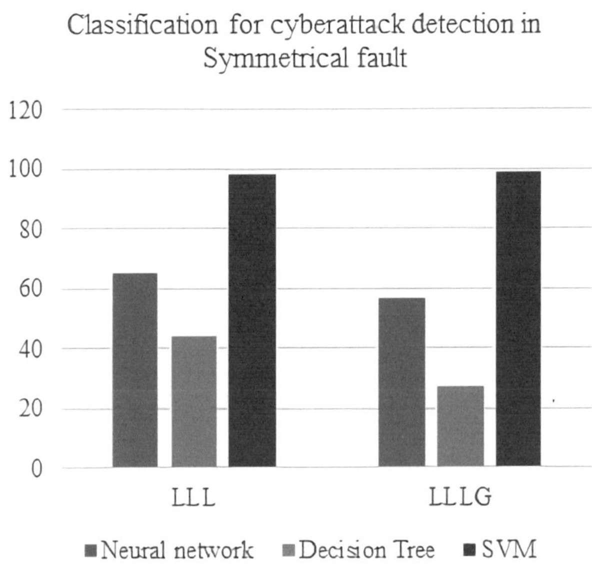

FIGURE 8.22 FDI classification in the IEEE-15 bus system under symmetrical fault.

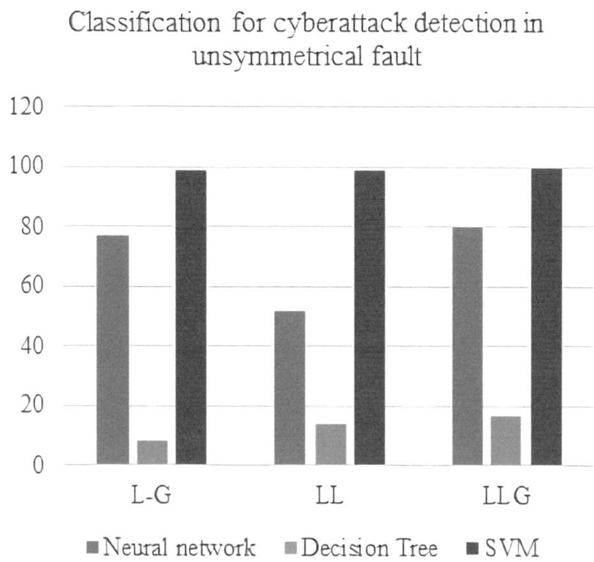

FIGURE 8.23 FDI classification in the IEEE-15 bus system under unsymmetrical fault.

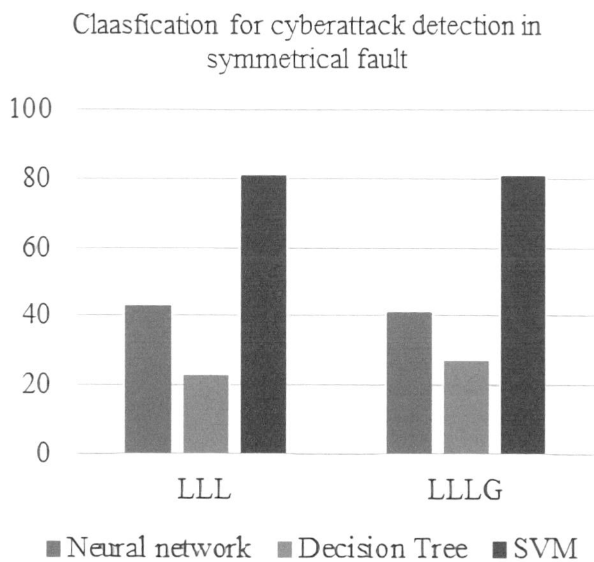

FIGURE 8.24 FDI classification in the modified IEEE-15 bus system under symmetrical fault.

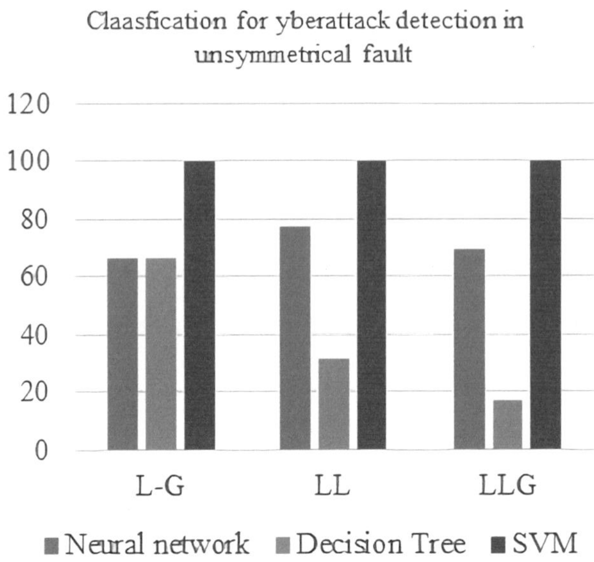

FIGURE 8.25 FDI classification in the modified IEEE-15 bus system under unsymmetrical fault.

TABLE 8.2
Comparative analysis of the FDI using machine learning techniques

				CPL-based IEEE-15 bus system		CPL-based modified IEEE-15 bus system	
Sr. No.	Machine learning classifier	Features	Preset	Network validation	Accuracy	Network validation	Accuracy
1	Decision tree	Va, Vb,Vc, Ia, Ib, and Ic	Fine tree	82.32%	21.96%	77.96%	33.05%
2	Neural network		Narrow neural network	91.15%	66.20%	93.48%	59.45%
3	SVM		Fine Gaussian SVM	92.66%	98.82%	95.62%	92.36%

REFERENCES

[1] R. Divya, S. Umamaheshwari, and S. Albert Alexander, "Machine learning based smart intrusion and fault identification (SIFI) in inverter based cyber-physical microgrids," *Expert Syst. Appl.*, vol. 238, no. 14, p. 122291, 2024.

[2] S. Ali, Z. Zheng, M. Aillerie, J.-P. Sawicki, M.-C. Péra, and D. Hissel, "A review of DC microgrid energy management systems dedicated to residential applications," *Energies*, vol. 14, no. 14, Art. no. 14, Jan. 2021, doi:10.3390/en14144308

[3] M. Jayachandran, C. R. Reddy, S. Padmanaban, and A. H. Milyani, "Operational planning steps in smart electric power delivery system," *Sci. Rep.*, vol. 11, no. 1, p. 17250, Aug. 2021, doi:10.1038/s41598-021-96769-8
[4] X. Wan, Y. Tian, J. Wu, and X. Ding, "Distributed event-triggered secondary recovery control for islanded microgrids," *Electronics*, vol. 10, no. 15, p. 1749, 2021.
[5] J. Pinto, A. Carvalho, and V. Morais, "Power sharing in island microgrids," *Front. Energy Res.*, vol. 8, Jan. 2021, doi:10.3389/fenrg.2020.609218
[6] G. Mathesh and R. Saravanakumar, "A novel digital control scheme for power management in a hybrid energy-source environment pertaining to electric vehicle applications," *Front. Energy Res.*, vol. 11, Feb. 2023, doi:10.3389/fenrg.2023.1130401
[7] S. Ullah, L. Khan, M. Jamil, M. Jafar, S. Mumtaz, and S. Ahmad, "A finite-time robust distributed cooperative secondary control protocol for droop-based islanded AC microgrids," *Energies*, vol. 14, no. 10, Art. no. 10, Jan. 2021, doi:10.3390/en14102936
[8] J. Hu and P. Bhowmick, "A consensus-based robust secondary voltage and frequency control scheme for islanded microgrids," *Int. J. Electric. Power Energy Syst.*, vol. 116, p. 105575, Mar. 2020.
[9] D. P. Mishra, S. R. Samantaray, and G. Joos, "A combined wavelet and data-mining based intelligent protection scheme for microgrid," *IEEE Trans. Smart Grid*, vol. 7, no. 5, pp. 2295–2304, Sep. 2016, doi:10.1109/TSG.2015.2487501
[10] B. Bhatt, H. Sharma, K. Arora, G. P. Joshi, B. Shrestha, "Levy flight-based improved grey wolf optimization: a solution for various engineering problems", *MDPI J. Math.*, vol. 11, no. 7, p. 1745, 2023, https://doi.org/10.3390/math11071745
[11] P. Risbud, N. Gatsis, & A. Taha, "Vulnerability analysis of smart grids to GPS spoofing". In *Proceedings of the IEEE International Conference on Smart Grid Communications (SmartGridComm)*, pp. 1–6, IEEE, 2018. https://doi.org/10.1109/SmartGridComm.2018.8587602
[12] Sureshbabu, S. Padmanabhan, G. Subramanian, A. A. Stonier, G. Peter, and V. Ganji, "Design and analysis of a photovoltaic-powered charging station for plug-in hybrid electric vehicles in college campus," *IET Electr. Syst. Transp.*, vol. 12, no. 4, pp. 358–368, 2022, doi:10.1049/els2.12060
[13] K. Arora, A. Kumar, V. K. Kamboj, D. Prashar, B. Shrestha, and G. P. Joshi, "Impact of renewable energy sources into multi area multi-source load frequency control of interrelated power system," *MDPI J. Math.*, vol. 9, no. 2, p. 186, 2021, https://doi.org/10.3390/math9020186
[14] Y. He, G. J. Mendis, and J. Wei, "Real-time detection of false data injection attacks in smart grid: a deep learning-based intelligent mechanism," *IEEE Trans. Smart Grid*, vol. 8, no. 5, pp. 2505–2516, Sep. 2017, doi:10.1109/TSG.2017.2703842
[15] S. Kar, S. R. Samantaray, and M. D. Zadeh, "Data-mining model based intelligent differential microgrid protection scheme," *IEEE Syst. J.*, vol. 11, no. 2, pp. 1161–1169, Jun. 2017, doi:10.1109/JSYST.2014.2380432
[16] T. Krause, R. Ernst, B. Klaer, I. Hacker, and M. Henze, "Cybersecurity in power grids: challenges and opportunities," *Sensors*, vol. 21, no. 18, Art. no. 18, Jan. 2021, doi:10.3390/s21186225
[17] A. Kumar, N. K. Choudhary, and N. Singh, "A comparative analysis for optimal placement and sizing of distributed generator in grid connected and islanded mode of microgrid," *Int. J. Power Energy Convers.*, vol. 12, no. 2, pp. 137–152, Jan. 2021, doi:10.1504/IJPEC.2021.114486
[18] K. Arora, G. P. Joshi, M. Ragab, M. Rawa, A. H. Milyani, R. F. Mansour, and E. Yang, "Bilateral contract for load frequency and renewable energy sources using

advanced controller", *Comput. Mater. Continua*, vol. 73, no. 2, pp. 3165–3180, 2022, doi:10.32604/cmc.2022.026966

[19] C. Cortes and V. Vapnik, "Support-vector networks," *Mach. Learn.*, vol. 20, no. 3, pp. 273–297, Sep. 1995, doi:10.1023/A:1022627411411

[20] "A training algorithm for optimal margin classifiers," Proceedings of the fifth annual workshop on computational learning theory, *ACM Conferences*, Accessed: Jun. 06, 2024. [Online]. Available: https://dl.acm.org/doi/10.1145/130385.130401

9 Double-Gate Junctionless Field Effect Transistor Circuit Based on Low Power Applications

Madhu Naik Nenavath and Sanjeet K. Sinha

9.1 INTRODUCTION

Double-gate junctionless field effect transistors (DG-JL FETs) represents significant advancement in semiconductor technology. Unlike traditional MOSFETs, DG-JL FETs do not have a p–n junction, which simplifies the fabrication process and enhances device performance. These transistors are particularly well-suited for low power very-large-scale integration (VLSI) applications due to their excellent control over short-channel effects and reduced leakage currents.

The evolution of transistor technology has seen a continuous push towards miniaturization and improved performance. From the early days of bipolar junction transistors to the advent of MOSFETs, each generation has brought about significant improvements. DG-JL FETs are the latest in this lineage, offering unique advantages that make them ideal for modern VLSI circuits.

9.1.1 Structure and Operation

The DG-JL FET consists of a thin semiconductor film sandwiched between two gate electrodes. The need for doping concentration gradients is eliminated in the absence of the junction between the drain and source regions, which are typically required in conventional MOSFETs. This junctionless design allows for better scalability and improved electrostatic control [1] (Figure 9.1).

The fabrication process of DG-JL FETs involves several key steps. First, a thin semiconductor layer is deposited on an insulating substrate. Next, the drain and source regions are defined using lithography and etching techniques. Finally, the gate electrodes are formed on either side of the semiconductor layer, completing the double-gate structure [3].

DOI: 10.1201/9781003563440-9

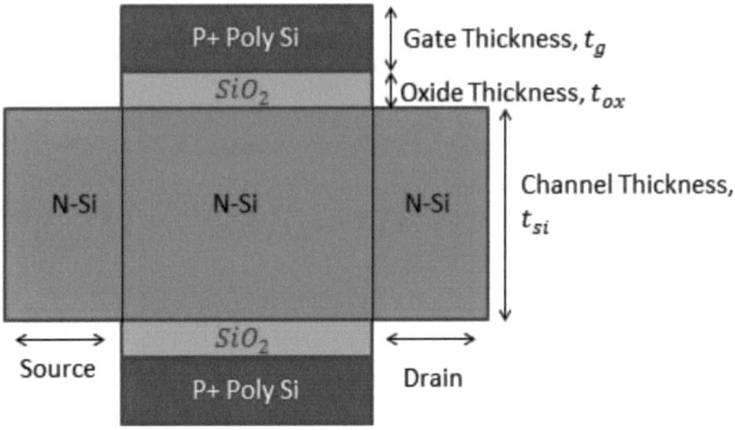

FIGURE 9.1 Structure of double-gate junctionless FET [2].

9.1.2 Advantages of DGJLFET

Double-gate junctionless field effect transistors (DG-JL FET's) offer several advantages that make them highly suitable for low power VLSI applications.

9.1.2.1 Low Power Consumption

- Reduced Leakage Currents: DG-JL FETs exhibit significantly lower leakage currents compared to traditional MOSFETs. The junctionless design eliminates the p–n junctions, which are prone to leakage, thereby reducing the off-state leakage current (I_{off}) [4]. This is crucial for low power applications where minimizing power dissipation is essential [5].
- Efficient Subthreshold Slope (SS): The SS of DG-JL FETs is steeper, meaning that the transition from the off-state to the on-state is more abrupt. This allows for lower threshold voltages, reducing the power required to switch the device on and off [5].
- High-k Dielectrics: The use of high-k dielectric materials in DG-JL FETs enhances gate control and reduces gate leakage currents. This further contributes to the overall reduction in power consumption [6].

9.1.2.2 High Performance

- Higher Drive Current: DG-JL FETs can achieve higher drive currents due to the improved gate control and reduced leakage. This results in faster switching speeds and better overall performance in digital circuits [7].
- Enhanced SS: The steep SS not only reduces power consumption but also improves the device's switching characteristics, leading to higher performance in both digital and analog applications.

9.1.2.3 Scalability

- Junctionless Design: The scalability of the fabrication of DG-JL FET can be increased in the absence of p–n junctions in DG-JL FETs. As technology nodes shrink, maintaining precise doping profiles becomes challenging in traditional MOSFETs. DG-JL FETs, with their uniform doping, are better suited for scaling down to smaller dimensions.

- Reduced Short-Channel Effects: The double gate structure helps in mitigating short-channel effects, which become more pronounced as devices are scaled down. This ensures that DG-JL FETs maintain their performance even at smaller technology nodes.
- Compatibility with Advanced Lithography: DG-JL FETs are compatible with advanced lithography techniques, like EUV an extreme ultraviolet lithography, that are very useful to scale down to sub-10nm technology nodes.

9.1.2.4 Thermal Stability

- Stable Operation at High Temperatures: DG-JL FETs exhibit better thermal stability compared to traditional MOSFETs. The junctionless design reduces the generation of thermally activated carriers, which can lead to leakage currents at high temperatures [5].
- Material Innovations: The use of high-k dielectrics and other advanced materials in DG-JL FETs enhances their thermal stability. These materials can withstand higher temperatures without degrading, ensuring reliable operation in various environmental conditions.
- Reduced Self-Heating: The double-gate structure helps in dissipating heat more effectively, reducing the self-heating effects that can degrade device performance. This is particularly important for high-performance applications where devices operate at high frequencies and generate significant heat.

DG-JL FETs offer a combination of low power consumption, high performance, scalability, and thermal stability, making them ideal for modern VLSI circuits. Their unique junctionless design and double gate structure provide significant advantages over traditional MOSFETs, addressing many of the challenges faced in scaling down semiconductor devices.

9.1.2.5 Applications

Double-gate junctionless field effect transistors (DG-JL FETs) have a wide range of applications in both real-life scenarios and VLSI circuits because of their unique benefits like less consumption of power, high performance, scalability, and thermal stability. Here are some detailed explanations of their applications:

9.2 DIGITAL CIRCUITS

- **Logic Gates**: DG-JL FETs are used in the design of basic logic gates such as AND, OR, NOT, NAND, and NOR gates. Their low power consumption and high switching speed make them ideal for these applications. The improved SS and reduced leakage currents ensure that the logic gates operate efficiently with minimal power dissipation [5].

- **Memory Cells:** DG-JL FETs are employed in the design of dynamic random access memory and static random access memory cells. The low leakage currents and high drive currents of DG-JL FETs enhance the performance and reliability of memory cells, making them suitable for high-density memory applications [5].

9.3 ANALOG CIRCUITS

- **Amplifiers:** In analog circuits, DG-JL FETs are used in the design of amplifiers. Their high trans-conductance and low noise characteristics improve the gain and signal integrity of amplifiers. This makes them suitable for applications in audio and radio frequency amplification.
- **Oscillators:** DG-JL FETs are also used in the design of oscillators, which are essential components in communication systems. The stability and low power consumption of DG-JL FETs ensure that oscillators operate reliably over a wide range of frequencies [6].

9.4 MIXED-SIGNAL CIRCUITS

- **Analog-to-Digital Converters (ADC)**: DG-JL FETs are used in the design of ADCs, which convert analog signals into digital signals. The high performance and low power consumption of DG-JL FETs improve the accuracy and efficiency of ADCs, making them suitable for applications in data acquisition and signal processing [6].
- **Digital-to-Analog Converters (DAC)**: Similarly, DG-JL FETs are used in DACs, which convert digital signals into analog signals. The precise control and low noise characteristics of DG-JL FETs enhance the performance of DACs in audio and video applications.

9.5 LOW POWER APPLICATIONS

- **Wearable Devices**: The low power consumption and high performance of DG-JL FETs make them suitable for wearable devices, such as fitness trackers and smart watches. These devices require efficient power management to extend battery life, and DG-JL FETs help achieve this goal [6].
- **Internet of Things (IOT)**: DG-JL FETs are used in IOT devices, which require low power consumption and high performance for efficient data processing and communication. The scalability and thermal stability of DG-JL FETs ensure that IOT devices operate reliably in various environments [6].

9.6 HIGH-PERFORMANCE COMPUTING

- **Processors**: DG-JL FETs are employed in the design of high-performance processors. Their high drive currents and low leakage currents improve the speed and efficiency of processors, making them suitable for applications in supercomputing and data centers [5].

- **Graphics Processing Units (GPUs):** DG-JL FETs are also used in GPUs, which require high performance and low power consumption for rendering graphics and processing large datasets. The scalability of DG-JL FETs ensures that GPUs can be designed with higher transistor densities, improving their performance [9].

DG-JL FETs offer significant advantages in various applications, from digital and analog circuits to biosensors and high-performance computing. Their unique characteristics, such as low power consumption, high performance, scalability, and thermal stability, make them ideal for modern VLSI circuits and real-life applications [9].

9.7 CASE STUDIES AND RESEARCH FINDINGS

Recent studies have shown that high-k dielectric materials in DG-JL FETs can significantly enhance their performance. For instance, a study by Kumar et al. demonstrated that DG-JL FETs with HfO2 as the gate dielectric exhibit superior performance in terms of threshold voltage stability and SS. Another study by Raja and Sekhar [10] compared the performance of DG-JL FETs with various bandgap materials, highlighting the potential for further optimization.

Simulation studies have also provided valuable insights into the behavior of DG-JL FETs. These studies have shown that DG-JL FETs can achieve lower power consumption and higher performance compared to traditional MOSFETs. Experimental results from recent research papers and conferences further validate these findings, demonstrating the practical viability of DG-JL FETs in VLSI for low power applications.

1. **Applications of Ultra Low Power for High-k Dielectric Double Gate Junctionless (DG-JL) MOSFET.** In this case study, the impact of high-k dielectrics on the performance of DG-JL FETs can be learned. An analytical model to evaluate the drain-induced barrier lowering (DIBL), threshold voltage, SS, and I_{ON}/I_{OFF} is developed in the paper. The results showed that using high-k dielectrics like HfO_2 significantly improved the device performance by reducing short-channel effects and off-state leakage currents. The study reported a 61.9% improvement in the figure of merit and a 34.29% improvement in the SS compared to devices with conventional gate oxides [6].
2. In this research, the development and compressive assessment of a junctionless multigate FinFET for low-power and high-frequency uses are discussed. A novel form of FinFET is introduced, featuring asymmetric gate and source/drain connections, specifically engineered for low-power and high-frequency applications. The research revealed that the new multigate FinFET displayed superior short-channel characteristics and higher current density. The device exhibited enhanced performance in terms of low-power consumption and high-frequency operation, making it well-suited for advanced VLSI circuits [11, 12].
3. An in-depth examination of junctionless Tri-Gate FinFET for high-frequency and low-power applications at a gate length of 5 nm was conducted. The focus of this

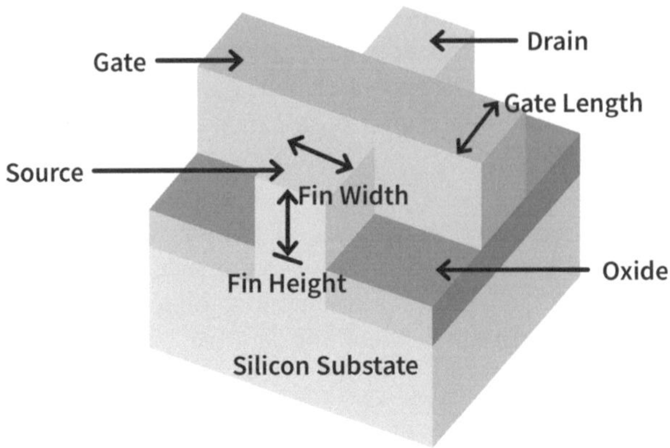

FIGURE 9.2 Basic structure of FinFET [8].

study was on the optimization of 5 nm gate length n-channel trigate junctionless SOI FinFET through the implementation of various spacer engineering techniques alongside hafnium-based high-k dielectrics [2]. It was discovered that the device displayed exceptional electrical characteristics, such as a low DIBL, sharp SS, and a high ON–OFF performance metric. The incorporation of high-k dielectrics into the gate stack notably improved the device's performance, rendering it suitable for high-frequency and low-power applications [7] (Figure 9.2).

9.8 CHALLENGES

Double-gate Junctionless field effect transistors (DG-JL FETs) offer many advantages, but they also face several challenges that need to be addressed for widespread adoption in VLSI circuits. Here are some of the key challenges:

9.9 FABRICATION COMPLEXITY

- Precision Requirements: The fabrication of DG-JL FETs requires precise control over the layer thickness of the semiconductor and the alignment of the double gates. Any deviations can significantly impact device performance.
- Advanced Lithography: The need for advanced lithography techniques increases the complexity and cost of manufacturing DG-JL FETs.

9.10 MATERIAL SELECTION

- High-k Dielectrics: While high-k dielectric materials can improve performance, finding the right material that balances performance and reliability is challenging [6].
- Semiconductor Materials: The choice of semiconductor material affects the device's electrical properties. Research is ongoing to identify the best materials for DG-JL FETs.

9.11 THERMAL MANAGEMENT

- Heat Dissipation: As devices scale down, managing heat dissipation becomes more critical. DG-JL FETs need efficient thermal management solutions to prevent overheating and ensure reliable operation.
- Thermal Stability: Ensuring that DG-JL FETs maintain their performance at high temperatures is essential for their use in various applications.

9.12 SHORT-CHANNEL EFFECTS

- Control Over Short-Channel Effects: While DG-JL FETs are designed to mitigate short-channel effects, achieving this consistently across different device sizes and operating conditions remains a challenge.
- Leakage Currents: Managing leakage currents, especially in ultra-scaled devices, is crucial for maintaining low power consumption [5].

9.13 INTEGRATION WITH EXISTING TECHNOLOGIES

- Compatibility: Integrating DG-JL FETs with existing CMOS technology requires careful consideration of compatibility issues, such as process variations and interconnects.
- Design Tools: Developing design tools and methodologies that can accurately model and simulate DG-JL FETs is necessary for their successful integration into VLSI circuits.

9.14 RELIABILITY AND VARIABILITY

- Device Reliability: Ensuring long-term reliability of DG-JL FETs under various operating conditions is essential for their adoption in commercial applications.
- Process Variability: Variations in the fabrication process can lead to inconsistencies in device performance. Addressing these variations is critical for achieving uniform performance across large-scale production.

9.15 ECONOMIC VIABILITY

- Cost of Production: The advanced fabrication techniques and materials required for DG-JL FETs can increase production costs. Finding ways to reduce these costs is important for their commercial viability.
- Scalability: Ensuring that DG-JL FETs can be produced at scale without compromising performance or reliability is a significant challenge.

9.16 RESEARCH AND DEVELOPMENT

- Ongoing Research: Continuous research is needed to address the challenges mentioned above and to explore new ways to enhance the performance and reliability of DG-JL FETs.

- Collaboration: Collaboration between academia, industry, and research institutions is crucial for advancing the development of DG-JL FETs.

Addressing these challenges requires a multidisciplinary approach, combining advances in materials science, device engineering, and circuit design. With ongoing research and development, many of these challenges can be overcome, paving the way for DG-JL FETs to play a significant role in future VLSI technology.

9.17 FUTURE SCOPES

9.17.1 Material Innovations

- High-k Dielectrics: Continued research into high-k dielectric materials can lead to better gate control and reduced leakage currents. Innovations in material science can enhance the performance and reliability of DG-JL FETs [6].
- New Semiconductor Materials: Exploring alternative semiconductor materials, like germanium, III–V compounds, and 2D materials like transition metal dichalcogenides and graphene, can offer improved electrical properties and scalability.

9.17.2 Advanced Device Architectures

- Multigate Structures: Developing multigate structures, such as FinFETs and nanowire FETs, can further improve electrostatic control and reduce short-channel effects. These architectures can enhance the performance of DG-JL FETs [12].
- Vertical Transistors: Research into vertical transistor architectures can provide better scalability and integration density. Vertical DG-JL FETs can offer improved performance and reduced footprint.

9.18 THERMAL MANAGEMENT SOLUTIONS

- Advanced Cooling Techniques: Developing advanced cooling techniques, such as microfluidic cooling and thermoelectric cooling, can help manage heat dissipation in DG-JL FETs. Efficient thermal management is crucial for high-performance applications.
- Thermal Interface Materials: Research into thermal interface materials (TIMs) can enhance the heat transfer between devices and the cooling solution, enhancing overall thermal performance [14].

9.19 INTEGRATION WITH EMERGING TECHNOLOGIES

- Quantum Computing: DG-JL FETs with high-k dielectrics have shown promise for operation at cryogenic temperatures, making them suitable for quantum computing applications. Integration with quantum computing systems can open new avenues for DG-JL FETs [6, 13].

- Neuromorphic Computing: DG-JL FETs can be used in neuromorphic computing systems, which mimic the human brain's neural architecture. Their low power consumption and high performance make them ideal for these applications [6].

9.20 DESIGN AND SIMULATION TOOLS

- Accurate Modeling: Developing accurate models and simulation tools for DG-JL FETs can help designers optimize circuits and systems. Improved modeling techniques can enhance the predictability and reliability of DG-JL FET-based designs [9].
- EDA Tools: Enhancing electronic design automation (EDA) tools to support DG-JL FETs can streamline the design process and facilitate their integration into VLSI circuits.

REFERENCES

[1] Vishnu Vardhan Reddy, I., & Tripathi, S. L. (2021). Enhanced performance double-gate junction-less tunnel field effect transistor for bio-sensing application. *Solid State Electronics Letters*, 3, 19–26.

[2] Sreenivasulu, V. B., & Narendar, V. (2021). A comprehensive analysis of junctionless tri-gate (TG) FinFET towards low-power and high-frequency applications at 5-nm gate length. *Silicon*, *14*(1), 2009–2021.

[3] Garg, A., Singh, B., & Singh, Y. (2020). A new trench double gate junctionless FET: A device for switching and analog/RF applications. *AEU- International Journal of Electronics and Communications*, 118, 154130.

[4] Sahu, C., & Singh, J. (2014). Device and circuit performance analysis of double gate junctionless transistors at Lg = 18 nm. *Journal of Engineering*, 2014(3), 105–110.

[5] Paul, A., Kshirsagar, C., Sachin Sapatnekar, S., Koester, S., & Chris Kim, H. Leakage modeling for devices with steep sub-threshold slope considering random threshold variations. *University of Minnesota*. 2014.

[6] Arora, K., Kumar, A., Kamboj, V. K., Prashar, D., Shrestha, B., & Joshi, G. P. (2021). Impact of renewable energy sources into multi area multi-source load frequency control of interrelated power system. *MDPI Journal of Mathematics*, 9(2), 186, https://doi.org/10.3390/math9020186

[7] Sen, D., Sengupta, S. J., Roy, S., & Chanda, M. (2021). Analysis of D.C parameters of short-channel heterostructure double gate junction-less MOSFET circuits considering quantum mechanical effects. *Silicon, 13*(1), 1165–1175.

[8] "What is a FinFET", retrieved from www.circuitbread.com/ee-faq on June 4, 2021.

[9] Chowdhury, D., De, B. P., Appasani, B., Singh, N. K., Kar, R., Mandal, D., Bizon, N., & Thounthong, P. (2023). A novel dielectric modulated gate-stack double-gate MOSFET-based sensor for detecting biomolecules. *Sensors*, *723*(6), 2953.

[10] Raja, T., & Sekhar, K. (2024). Comparative analysis and study the performance of narrow bandgap and wide bandgap material for the homogeneous structure of double gate junction-less tunnel field effect transistor (DG-H-JL-TFET). *AIP Conference Proceedings*, 2966(1), 030017.

[11] Bhattacharya, S., Tripathi, S. L., & Nayana, G. H. (2024). Design transmission gates using double-gate junction-less TFETs. *Silicon, 16*(1), 3359–3372.

[12] Kumar, E. S., & Kumar, P. S. (2021). Design and compressive analysis of junctionless multigate FinFET towards low power and high frequency applications. *Silicon, 14*(1), 7623–7631.

[13] Arora, K., Joshi, G. P., Ragab, M., Rawa, M., Milyani, A. H., Mansour, R. F., & Yang, E. (2022). Bilateral contract for load frequency and renewable energy sources using advanced controller. *Computers, Materials & Continua, 73*(2), 3165–3180, doi:10.32604/cmc.2022.026966

[14] Rahman Adnan, M. M. & Rahman, M. S. (2018). Impact of wave function penetration into gate oxide on gate capacitance and threshold voltage of n-channel double gate junction less transistor. In International Conference on Advancement in Electrical and Electronic Engineering, vol. 22, pp. 123–126.

10 Enhanced Grey Wolf Optimizer to Improve the Power Quality in Multilevel Inverter

Himanshu Sharma and Krishan Arora

10.1 INTRODUCTION

In the recent years, swarm intelligence and meta-heuristics algorithm have played a major role in the field of optimization. Swarm intelligence is the population-based method, in which it converges to get the optimized solution [1], whereas the meta-heuristic is the problem-specific, in which the output may change with the time.

The swarm intelligence is the technique that is derived by observing the behavior of animals like fishes and birds. In this technique, the group of animals such as birds are in search of prey. Therefore, each animal of that group will work together to get the prey, and they will co-ordinate each other in the entire task. This complete process of finding the prey intelligently is called swarm intelligence. The techniques such as ant colony, bee colony, PSO, and many more are examples that can be used in swarm intelligence. The swarm intelligence's primary objective is that the output is already defined and has to reach a certain destination or has to find the prey etc. but the main part is that it should reach the final result in minimum amount of time or in definite amount of time.

In the field of optimization, there are many techniques and algorithms, as shown in Figure 10.1 through which one can reach the optimized solution. One of them is nature-inspired algorithm which consists of meta-heuristic algorithm. It is the result oriented [2]. In this algorithm, the output must be optimized as well as the output cannot be predicted in this kind of problem. Or it can be defined as that the meta heuristic is the problem-specific. Meta-heuristic is not population-based but can apply the population-based technique into it to have an optimized solution.

10.2 OPTIMIZATION

The process of optimization is used to determine the best solution to a given issue under the given conditions. Optimization is one of the field where anyone can design their algorithm or method that help us to reach nearer to the desired solution. This field of optimization fascinates researchers to work on this, as there are lot of

DOI: 10.1201/9781003563440-10

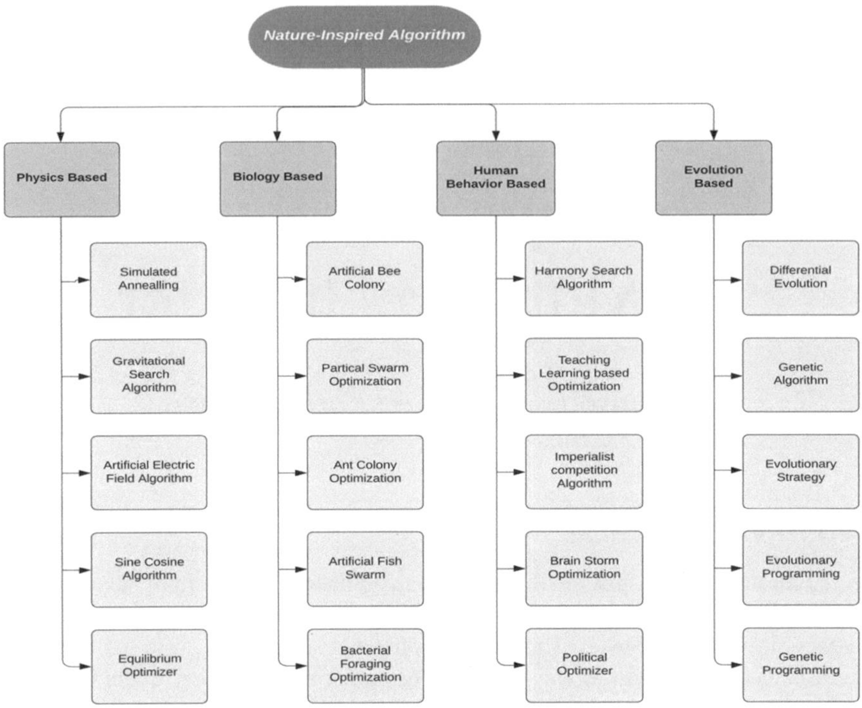

FIGURE 10.1 Flow chart of optimization techniques.

applications in day to day life and it has been used everywhere. Future scope in the field of optimization is increasing day by day. There are many algorithms that have been developed that are accurate as well as fast and reliable.

The reason behind the usage of this technique is that there is not a single algorithm that can be used in each and every problem, which is similar to the No Free Lunch (NFL) [3] which states that there is nothing free, or one can define that there is no technique that can be applied in every problem. So the researchers are creating the new algorithms based on the given problem so they get the best solution for that as well as on the basis of NFL; they used to create algorithms according to the different benchmark functions.

There are various techniques of optimization which are mentioned in Figure 10.1. Mainly most of the researchers are working on the nature-inspired algorithm i.e. Grey Wolves Optimization (GWO) [4], Particle Swarm Optimization (PSO) [2], Ant Colony Optimization (ACO) [5], Ant Lion Optimization (ALO) [6], and many more. Here are the few optimization techniques that have been used in this paper.

10.2.1 Grey Wolves Optimization

This method is part of a population-based technique, which mimics the hierarchical system and hunting technique of grey wolves [4]. Usually, these wolves are found in

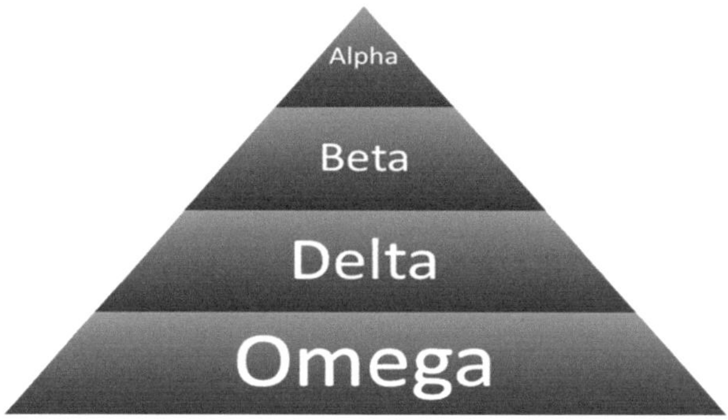

FIGURE 10.2 Hierarchy levels of GWO.

the groups for hunting, and they are divided into several groups such as alpha, beta, delta, and omega, as shown in Figure 10.2.

In this, the wolves of the alpha groups are the dominant, or it can be said that they are the leader of the group. They provide the best fitted solution in the GWO algorithm. Wolves from these groups are used to make the decisions such as hunting [7] and planning, and they set certain tasks for other wolves in their group. This wolves are considered as the strongest wolves of their group.

The next level of this group is beta. They provide the next best solution or best-fitted value after the alpha [7]. The role of this group is to help the alpha group in making decisions and command the other two groups below them.

The lowest group is omega. They play the role of victim. Their solutions does not affect the final solution in the GWO algorithm. As they are at the bottom of the group, they have to follow the order of their dominant wolves, and they are given the chance in the last of the task. For example after hunting, they are allowed to eat at the last after all other wolves complete it. The pseudo code for this algorithm is as follows:

```
I/P Problem Parameters
O/P Pbest
Main Loop
       Providing the search agent from scratch.
       Initialization of a, C, lb, ub and Max_iter.
       Estimation of search and grading agents'
         fitness values.
T = 0
While(t < Maximum number of iterations)
       For each search agent
              Update the position of the search agent.
       End for
       Updating a, A and C.
Calculate the fitness of search and grading agent and
  then update the position of Xa , Xb and XD.
End While
End
```

10.2.2 Ant Lion Optimization

The algorithm is inspired by nature. Ant lion algorithm follows the nature's ant lion hunting behavior. This approach is typically employed to resolve difficult problems involving various search spaces. Ant lion are from the family of Myrmeleontidae and net-winged insects. In this, to get a prey in the bottom of the pit, ant lion throw sands to the bottom of the hole [6]. The size of the trap differs with the hunger level of ant lion. It has also been observed that they are hungrier on the full moon [8].

The search space is obtained using a roulette wheel and ant lion selection randomly. Search space is also discovered by the random walk of the ants across ant lion. With the help of the random walk and roulette wheel, one can easily have local minima in the ALO. Pseudo code for this algorithm is given below:

```
Random initialization of the first ant and antlion
  population.
Evaluate the ant and antlions' fitness.
Locate the superior antlions and take them for
  the elite.
While (Current_iter < Max_iter)
  For (1: size of antlions)
    Choose an antlions using the Roulette wheel.
    Slide randomly walking ants in trap
    Build a random walk, then normalize it.
    update the ant's location
  End for
Determine each ant's fitness score.
When an antlions becomes more fit, swap it out with
  the appropriate ant.
If an antlions becomes fit than the elite, update
  the elite.
End while
End
```

Optimized shrinking method is to estimate the border of the random walk, which manipulate the response to increase the iteration [9]. During this movement, the ants are frequently reduced, which ensures the consistency of the Ant Lion algorithm. When the antlion moves throughout optimization to the location of the ants, that search space is saved so that all the ants appear in optimal solution.

10.2.3 Whale optimization

This algorithm is the part of swarm-based meta heuristic algorithm. This algorithm takes its ideas from the humpback whale's hunting strategy. In this, the individual or the group of whale will find the prey in their search space. As the prey is in their search space, the whale will go deep inside, or the whale will get to his best position from where whale can attack the prey easily [10]. The other member of that group will take their best position from where they can easily approach the main whale.

As the whales are in their best positions, they will now start creating the bubbles in a circle or in the form of spiral around the prey and create the net-type structure. Creating nets with bubbles is called bubble net feeding behavior of whales. Prey will get trapped in this bubble net and then the whale that was in the best position will attack towards the prey. The pseudo code for this algorithm is given below: -

```
Input Number of search agent, control parameter,
  Max_iter
Output gbest, pbest
Start
Generate the initial position of whales
t=0 for the iteration counter
Determine each whale's fitness
Decide which whale is the best depending on fitness
While (t < Max_iter)
    For each whale
           Calculate the control parameter.
If (rand < 0.5)
     if ( |A| < 1)
     Change the whale's location
     Else if(|A| > 1)
     Choose the random walk.
     Current whale's location updated
     End if
     Else if (Rand > 0.5)
     Current whale's location updated
     End if
     End for
  Calculate the whale's fitness
  According to the fitness increment of the most
    recent iteration, adjust the whale's value.
End while
Return gbest
End
```

The above discussion was based on humpback whale and how they attack prey naturally, but to implement it in an algorithm one has to process the creation of the bubble net feeding which is a shrinking encircling mechanism and an spiral updating position.

1. *Shrinking Encircling Mechanism:*
   In this, the values of the constant are reduced from 2 to 0 as the iterations increases. Now one can set the new position randomly between the prey and

the current position of the whale. So the new position is less than one so that the search agent can easily attack the prey [11].

2. *Spiral updating position:*
   In this, the spiral position method represents or imitates the whale's helix-shaped movement between its position and its prey. This helix movement of the whale will create the confusion of its position, which cannot be determined easily, and it will get easy to approach towards the prey.

10.2.4 Honey Badger Algorithm (HBA)

The Honey Badger Algorithm (HBA) follows the honey badger's hunting style that is located at Indian subcontinent, Asian southwest, and in the rainforest. Honey Badger usually smell or dig the hole for a food source and they either track the honey bird for it. Honey badgers are capable of digging around 50 holes in a radius of 40 km [12]. Honey Phase: The honey bird is pursued by the honey badger until they reach the beehive [13]. The first phase in the HBA process is initializing the solutions within the lower boundary (lb) and upper boundary (ub), and it is described as

$$X^i = lb_i + r_1 * (ub_1 - lb_1)$$

$$X^i = lb_i + r_1 * (ub_1 - lb_1) \tag{1}$$

The honey badger's solution is given here by X^i, r_1 is the random number between 0 and 1, a is the density factor that is used in HBA for balancing the exploration and exploitation.

Digging Phase: in this, movement is being updated as per the cardioid shape that is given by the expression:

$$x^{new} = x^{prey} + F * b * l * x^{prey} + F * ap * di * r_3 * \left| cos(2 * pi * r_4) * \left[1 - cos(2 * pi * r_5)\right] \right| \tag{2}$$

Here, x^{new} is new position of X^i, x^{prey} is the most well-founded outcome, F controls the search direction, $r_{3,}$ is the uniformly generated random number from 0 to 1 [14]. The pseudo code for honey badger is as follows:

```
Set parameter T,N,B,C
Compute the population at random initial
  coordinates.
Analyze each honey badger position x_i's fitness.
Save the optimal x^prey position, then assign fitness.
While (t ≤ T do)
    Update the decrement factor α.
    For I = 1 to N do
    Determine the intensity.
      If  r < 0.5 then
```

```
        Revisit the position x^new
      Else
        Revisit the position x^new
      End if
        Assign to fitness after evaluating the new
        position.
      If f ≤ f^1 then
        Set x^1 = x^new and f^1 = F
      End if
      if F ≤ f prey then
        Set x^prey = x^new and f^prey = F.
      End if
     End for
    End while
Return x^prey
```

10.2.5 Slime Mould Algorithm

It was created using an oscillation mode seen in natural slime mould. The three morph types of slime mould's foraging behavior are approaching food, wrapping food, and grabbing food [15]. Slime mould may get to high-quality food, surround it, and employ enzymes to break it down by modifying the cytoplasmic flow in the vein with the help of the biological oscillator's propagation wave. Also, it can locate additional foods of a higher quality through adaptive transformation. The position of Slime mould is given by

$$X(t+1)=\begin{cases} rand.(ub-lb)+lb, \\ X_b(t)+vb.\left(W.X_A(t)-X_B(t)\right) \\ vc.X(t) \end{cases} \tag{3}$$

Here, the positions of the slime mould at iterations $(t + 1)$ and t^{th} are represented by $X(t + 1)$ and $X(t)$. $X_A(t)$ and $X_g(t)$ are the two search agents that had been choosen randomly from the population. The slime mould is placed in the optimal location by $X_g(t)$. lb and ub are the limit of the search space. vc and vb are the two vectors that oscillates between [−1, 1]. The pseudo code for this algorithm is given below:

```
Initialize population (N) and Max_iter
Initialize the population of slime mould
While(t < Max_iter)
      Determine the slime mould's fitness
      Update X_b
      Update W
             For each in the search space
                    Update p, vb, vc
                    Update position
             End for
End while, Return X_b
```

10.2.6 Prairie Dog Algorithm

The Prairie Dog Algorithm is a swarm intelligence optimization algorithm inspired by the behavior of prairie dogs, which are social rodents that live in large underground burrow systems [17]. This algorithm was proposed by Yan et al. in 2014 as a way to solve complex optimization problems.

The algorithm works by simulating the behavior of prairie dogs, which are known for their social organization and communication skills. In the algorithm, each individual or "prairie dog" represents a candidate solution to the optimization problem. These individuals interact with each other to share information and collectively search for the best solution.

The algorithm begins by randomly generating an initial population of prairie dogs. These individuals then move around the search space based on a set of rules that simulate the behavior of prairie dogs. For example, they may move towards areas where other individuals have found good solutions or towards areas where the environment is more favorable for finding a good solution. As the prairie dogs move around, they communicate with each other using a set of communication rules that are based on prairie dog behavior. For example, they may share information about the quality of the solutions they have found or about areas of the search space that they have explored.

Over time, the population of prairie dogs converges towards the best solution to the optimization problem. This is achieved through a combination of exploration (searching new areas of the search space) and exploitation. The Prairie Dog Algorithm has been shown to be effective at solving a variety of optimization problems, including numerical optimization, engineering design, and feature selection. It is particularly well-suited to problems that require a balance of exploration and exploitation to find the best solution.

```
Initialize the population of prairie dogs with
  random candidate solutions.
Evaluate the fitness of each prairie dog.
Set the iteration counter to 0.
Repeat until the stopping criteria are met:
   For each prairie dog i in the population:
    Select a subset of the population to communi-
    cate with.
    Update the position of the prairie dog using a
      movement rule based on the communication with
      other prairie dogs.
    Evaluate the fitness of the new position.
    If the new position has better fitness than the
      current    position, update the position of
      the prairie dog.
   Sort the population by fitness.
   If the best solution found so far is better than
   the previous best solution, update the best
   solution.
```

```
    Increment the iteration counter.
  Return the best solution found.
```

10.2.7 Levy Flight-Based Improved Grey Wolves Optimization

Levy flight-based improved grey wolf optimization (IGWO) is an optimization technique that is depended upon the concept of Levy flights. The IGWO is a method of meta-heuristic optimization that draws inspiration from the natural behavior of grey wolves. The levy flight-based IGWO algorithm combines the investigating and utilizing properties of the IGWO algorithm with the random walk behavior of Levy flights to enhance optimization performance and the algorithm's capacity for searching globally [5]. This hybrid algorithm is used to address challenging optimization issues in a variety of disciplines, including engineering, finance, and data analysis. Initialize the population of wolves (solutions) and their corresponding fitness values. The pseudo code is provided as follows:

```
Set the grey wolf population (solutions)
Evaluate the accuracy of every response.
Arrange solutions with decreasing fitness order.
Make the best response the dominant wolf α.
Make the 2nd best response as the β wolf
Make the 3rd best response as the δ wolf
Evaluate the new position of the wolves and update
  it with the initial position.
For each iteration:
         Generate new solutions for each grey wolf
           using Levy flight
         Evaluate the accuracy of every new solution
  According to the new answer, adjust each grey
    wolf's location.
  The α wolf location has been updated based on
    a combination of its own so-lution and the
    solutions of the β and δ wolves
Based on a combination of the α wolf's own answer,
  the β wolf's solution,  and the δ wolf's solution,
  the α wolf's location should be updated.
Evaluate the fitness of the updated solutions
Arrange the grey wolves, according to their most
  recent fitness
Update α, β, and δ wolves based on the new sorting
Return the best solution (the α wolf) as the result
```

10.3 INVERTER

A power electrical equipment called an inverter is used to convert DC voltage to AC voltage. The voltage source inverter (VSI) and the current source inverter (CSI)

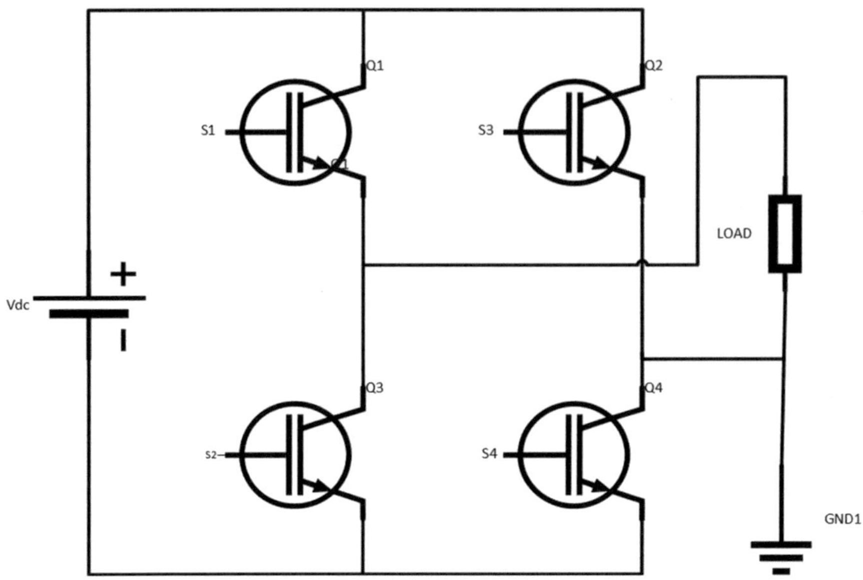

FIGURE 10.3 Circuit diagram of CHB.

are the two types of inverters that are available depending on the output topology as shown in Figure 10.3. At various firing angles, the inverter uses switches to convert DC to AC. The switches can be MOSFET, IGBT, SCR, FET, and BJT. Inverter are mainly of three types [17]:

- Diode clamped MLI
- Flying capacitor MLI
- Cascaded H bridge MLI

10.3.1 Diode-Clamped MLI

The diode-clamped inverter's main principle is to use diodes to restrict or lessen the voltage stress on the power switching devices, hence the name diode clamped. Each capacitor and switch is exposed to a voltage of Vdc [18]. A k level inverter requires $(k - 1)$ voltage sources, $2(k - 1)$ switching components and $(k - 1)(k - 2)$ diode.

10.3.2 Flying Capacitor MLI

The flying capacitor MLI or the capacitor clamped MLI is a development of diode clamped MLI and replaces the clamping diodes with clamping capacitors. This invention proved to be quite impactful as it provided switching redundancies; in other words, to achieve a voltage level, there were many switching combinations and out of these, the one which caused the least stress on the switching devices could be selected. Introduced by Meynard and Foch in the year 1992 [18], this represents an improvement over the Diode Clamped MLI and offers a control for both active and reactive power.

10.3.3 Cascaded H Bridge MLI

The design topology of cascaded H bridge multilevel inverter (CHB-MLI) is responsible for the most dependable, best in class MLI. It has been widely used in inverters ranging from medium to high voltage requirements. Simple design, better performance, reduced THD, and ease of implementation are some of its merits that helped it out the other designs. It can be referred to as a modern innovative design for the inverters of the future.

Speaking of the design of a CHB-MLI, the key attribute is the H-bridge inverter. The phrase "H Bridge" actually originated from the typical circuit a visual depiction of such a circuit that resembles the letter 'H'. Four switching devices are used to construct an H bridge (solid-state or mechanical). The switches can be either MOSFET, IGBT, BJT, and GTO in the given descending order of their switching frequencies. Also a diode is connected in antiparallel with the Switching Device to form a Power Switch.

When switches S1 and S4 are closed (and S2 and S3 are open), as shown in the following Figure 10.3, there will be a positive voltage applied to the entire load [19]. By opening the S1 and S4 switches and closing the S2 and S3 switches, negative voltage is enabled.

Switches S1 and S2 should never be closed at the same time, though, as this could result in a short circuit at the source of the input voltage. The same caution should be utilized while using switches S3 and S4. This situation is known as a "shoot-through".

A CHB-MLI may synthesize the necessary voltage from various DC sources by interconnecting H-bridges in series. Asymmetrical cascaded H-bridge multilevel inverters are known for DC sources with different levels of DC voltage and symmetrical cascaded H-bridge multilevel inverters for sources with the same level of DC voltage.

Apart from all other MLI's, CHB-MLI is the best and widely adopted, and the reasons are the need for renewable power generation and its growth over the year, solar power being more affordable, ease of implementation, various number of switching strategies and choice to be made based on the application requirement and reduced THD.

10.4 HARMONICS IN ELECTRICAL POWER

The basic definition of Harmonic is that the fundamental frequency's multiples are known as harmonics. Either 50Hz or 60Hz in case of electrical power systems, and it might be either both voltage harmonics and current harmonics are the outcome of nonlinear electric and electronic loads in an electric power system [20]. However, it is predominantly the electronic loads that are responsible for harmonics compared to electrical loads. Harmonic frequencies or harmonic content the electrical grid (power system) are frequent, major roots of power quality issues. Harmonics in power systems shorten the lifespan of electrical equipment, reduce efficiency, increase equipment and conductor heating, cause electronic equipment to malfunction, cause variable speed drives to misfire, cause power instability, and cause motor torque pulsations. Harmonics reduction is seen as being highly desired for minimizing all of these connected and related power quality issues.

Next speaking of THD, often known as total harmonic distortion, is a count or measure of the presence of harmonic distortion in a signal either a current or voltage signal. The ratio between the fundamental frequency and the combined power of all harmonic components is often used as a synonym for THD. Distortion factor is a related concept.

Lower THD means less undesired losses, peak currents, heating, equipment failure and damage, pollutants, and core loss in motors in power systems. Also possible is lasting motor damage. In other cases like audio systems and radio communication systems, lowered THD means distortion free signal, better and accurate signal reproduction, reduced interference with other sensitive electronic equipment. The need to describe the requirement of eliminating harmonics is to do with the fact that the inverters are made of power electronic switches or valves, who's switching ON and OFF results in harmonics and these generated harmonics in return are harmful to the inverter unit and the loads connected to the inverter.

10.4.1 Current Harmonics

Observe a typical alternating current (AC) power system. The current oscillates at the system frequency, which is a sinusoidal given frequency; 50 to 60 hertz is the usual. For an AC load that is linearly connected to this system, the load pulls an identical-frequency sinusoidal current i.e., the System Frequency as the voltage. However, the current and voltage may not always be in-phase with each other.

Nonlinear loads are mostly to blame for current harmonics or electronic loads in specific. For the load is nonlinear, a simple example being rectifier (AC to DC converter) is a part of the system, a load current which is not constant.

The variations in the AC current waveform resemble the variations of the voltage waveform. Depending on the type of load and how it communicates with other system elements connected to it, the AC current waveform can occasionally become highly complex. Despite of how complex or how irregular and nonperiodic the current waveform becomes, it can still be divided into a spectrum of frequencies using Fourier series analysis, and It can be broken down into a group of simple sinusoids that occur at integral multiples of the power supply's fundamental frequency.

Common electronic devices including mobile chargers, adapters, and office machines like fluorescent lighting, battery chargers, computers, and variable-speed drives with a power electronics foundation are other instances of nonlinear loads.

Harmonics are viewed as positive integral multiples of the fundamental frequency when referring to power systems. For instance, the third multiple of the fundamental frequency is the third order harmonic. These harmonics can also be produced by electrical loads that are not linear. Transistors, overloaded or nonlinear loads include, but are not limited to, saturated electrical motors and nonideal transformers. Disruptions in the fundamental harmonic, which are brought on by nonlinear loads, result in the production of various harmonics. However, the third order harmonic is the main emphasis because of its unique properties when seen in the context of power systems. Eliminating the Third Harmonic and its Multiples, commonly referred to as "Triplen Harmonics," is therefore of highest importance in the context of power systems

10.4.2 Voltage Harmonics

Current harmonics lead to voltage harmonics. Because of the non-linearity of the source impedance, the voltage that the voltage source produced is warped by the current harmonics. Only minor Current harmonics result in voltage harmonics. When the impedance of the voltage source is low [21]. In comparison to current harmonics, voltage harmonics are indeed less significant. Consequently, the fundamental frequency of voltage can typically be used to approximate the voltage waveform. While using this approximation, current harmonics have little to no effect on the real power provided to the load [22,23]. By tracing the voltage waves at the fundamental frequency and superimposing it, A current harmonic with no phase shift is shown (this helps to more easily observe the following phenomenon). The results show that for every period of voltage, there is the same amount of space above the horizontal axis and below the current harmonic wave as there is below the axis and above the harmonic wave. In other words, the average actual power contribution of the present harmonics is zero [24,25]. However, current harmonics do contribute to the actual power provided to the load if bigger voltage harmonics are taken into consideration. This gift is known as "Harmonic Power."

10.5 RESULTS AND DISCUSSION

As the harmonics content in the inverter is high due to the less number of switching, which results in the distortion of the sine wave and get the high THD values. To resolve this issue, the 31-level inverter is created which creates the waveform near to the sine wave. In this, total seven optimization techniques have different firing angles so that the better output waveform has low THD value, as shown in Figure 10.4, and in this, 49 frequency is taken to calculate the THD value. In this paper, various techniques are used such as Grey wolves' optimization, Ant lion Optimization, Whale Optimization, Honey Badger and Slime mould algorithms. Out of this seven optimization techniques, LF-IGWO has less THD. Table 10.1 depicts the comparison of the different firing angles that has been obtained from the above algorithm.

With the help of Figure 10.4, the convergence curve for all the seven algorithm has been compared and it has been found that the ALO has the overall good convergence curve but when the ALO and HBA compared then it has been found that HBA converges must faster than ALO and goes to its minimum at the half of the iteration and then remains nearly constant in the rest of the iteration count. Whale optimization (WOA), PDO, slime mould algorithm (SMA), and GWO are nearly the same, but they do not converge as better than HBA and ALO. However, the LF-IGWO is considered then it's found that the convergence is slow as compared to the HBA and ALO but the value is almost the same as the HBA at the end of the iteration.

From Table 10.2, it is clearly depicted that the LF-IGWO has the lowest value of the THD as compared to the other six algorithm. By the help of the output waveform,

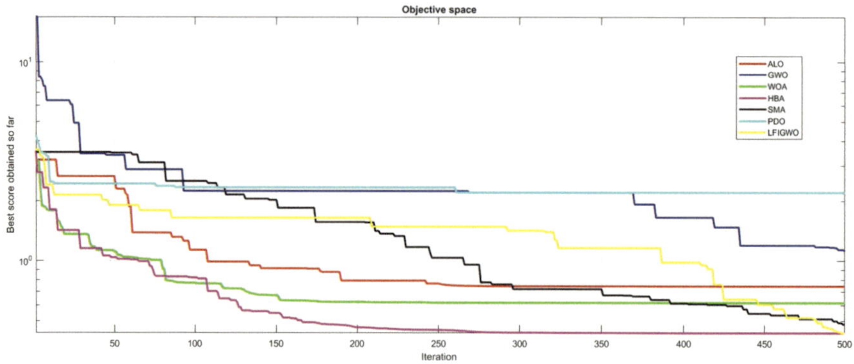

FIGURE 10.4 Convergence curve of optimization techniques.

TABLE 10.1
Firing angle of different optimization technique

	Algorithms						
Angles	SMA	ALO	GWO	WOA	HBA	PDO	LF-IGWO
A1	0.2696	0.5667	0.2074	0.2102	0.2393	0.91936	0.036788
A2	0.0630	0.4174	0	0.2519	0.7003	0.23802	0.84503
A3	0.2166	0.9582	0.6342	0.7513	0.1286	0.79347	0.24646
A4	0.0011	0.0004	0.0864	1.0535	0.0692	0.33216	0.94454
A5	0.5449	0.7888	0.5053	0.4831	1.5501	0	0.17788
A6	0.4250	0.9197	0.0428	0.6696	0.3163	0	0.69653
A7	0.1831	0.6744	0.0575	0.5663	0.0009	0.3611	0.47166
A8	0.3831	0.3456	0.1335	0.0280	0.5906	1.1891	0.29501
A9	0.7235	0.1783	0.74	0.3528	0.3329	0.21383	0.4211
A10	0.4887	1.27	0.3837	0.9311	0.1324	0.084389	0.77193
A11	0.4151	0.1059	0.3183	0.0902	0.8288	0.58353	1.2306
A12	0.8383	0.8516	0.5527	0.3417	1.5693	0.45054	0.56625
A13	0.1271	1.331	0.8574	0.4428	0.5594	1.136	0.11742
A14	0.3231	0.2872	0.2855	0.6468	0.4537	0.48994	0.63186
A15	0.0980	1.5355	0.4286	0.1343	0.3691	0.68733	0.36631

TABLE 10.2
THD percentage of different techniques

Algorithms	SMA	ALO	GWO	WOA	HBA	PDO	LF-IGWO
THD (%)	13.77	6.86	14.68	6.92	13.57	6.62	4.54

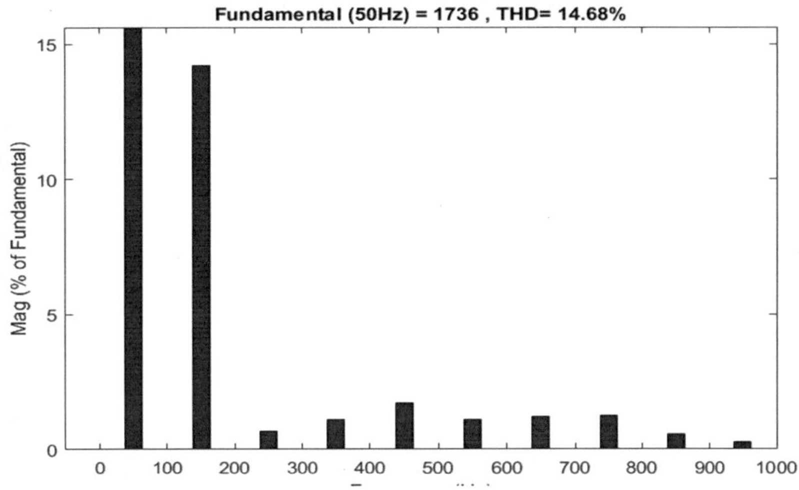

FIGURE 10.5 THD spectrum of GWO.

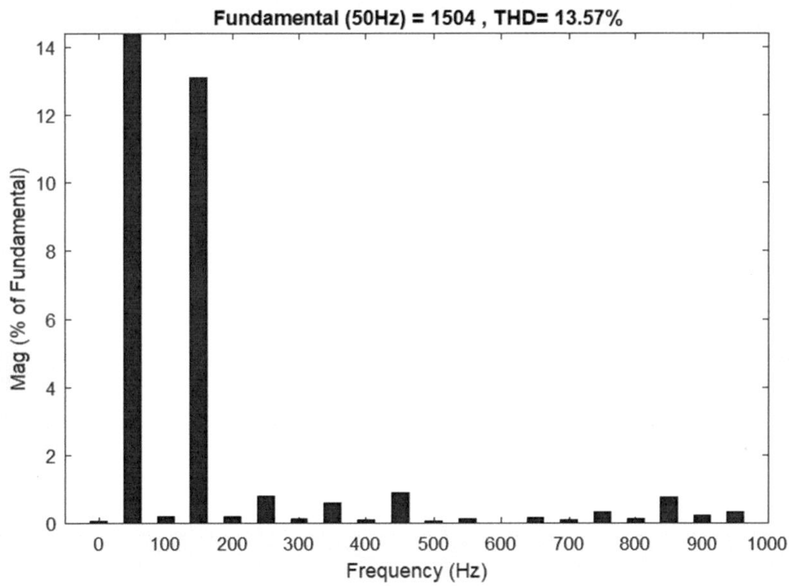

FIGURE 10.6 THD spectrum of HBA.

the FFT analysis has been conducted and the THD of each of the seven algorithm has been calculated. Figures 10.5–10.11 depict the FFT analysis and Figures 10.12–10.18 depicts the waveform that has been obtained from the different optimization techniques.

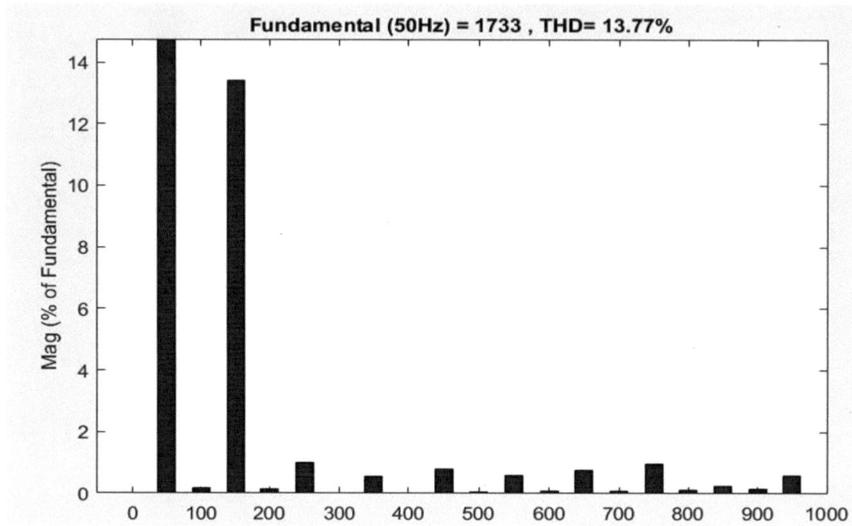

FIGURE 10.7 THD spectrum of SMA.

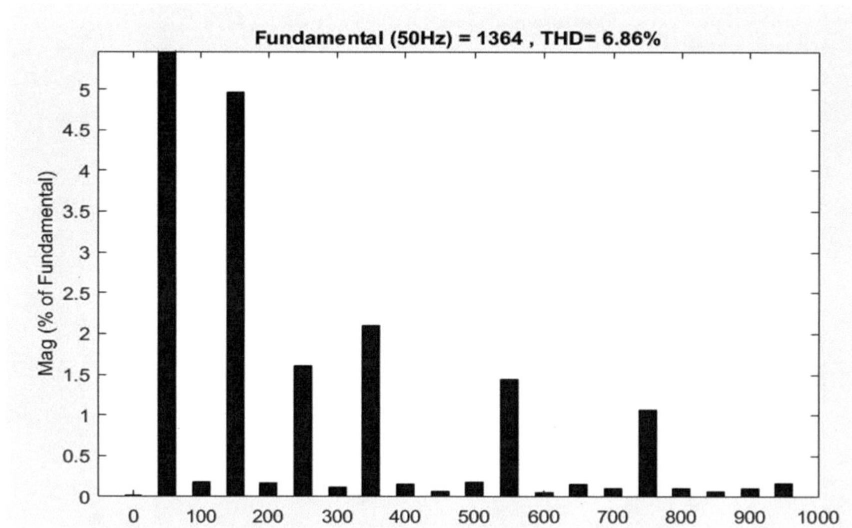

FIGURE 10.8 THD spectrum of ALO.

10.6 CONCLUSION

From the above decision, one can say that from the nature-inspired algorithm levy flight-based improved grey wolf (LF-IGWO) provides the best solution to the problem of the THD minimization of Inverters. And as per the IEEE standard 512, the THD of the inverter should be achieved below 5%, and from the help of this algorithm,

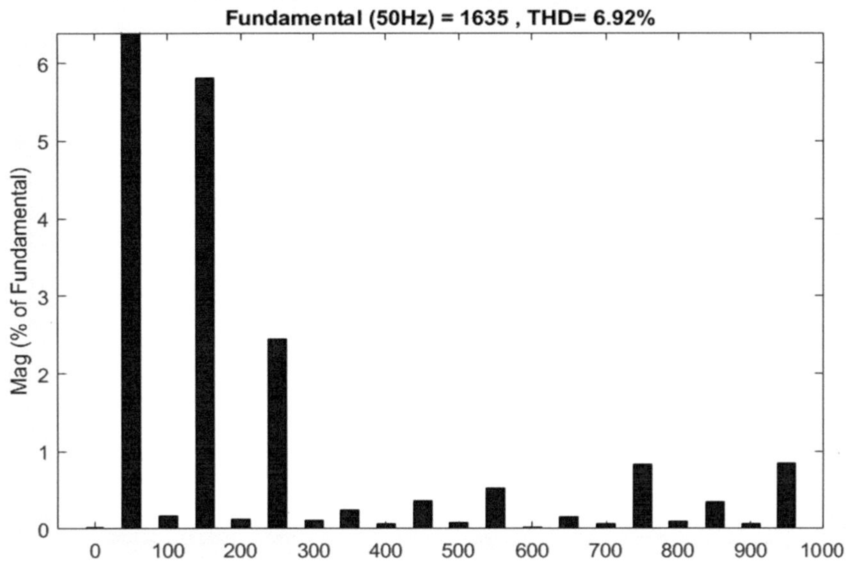

FIGURE 10.9 THD spectrum of WOA.

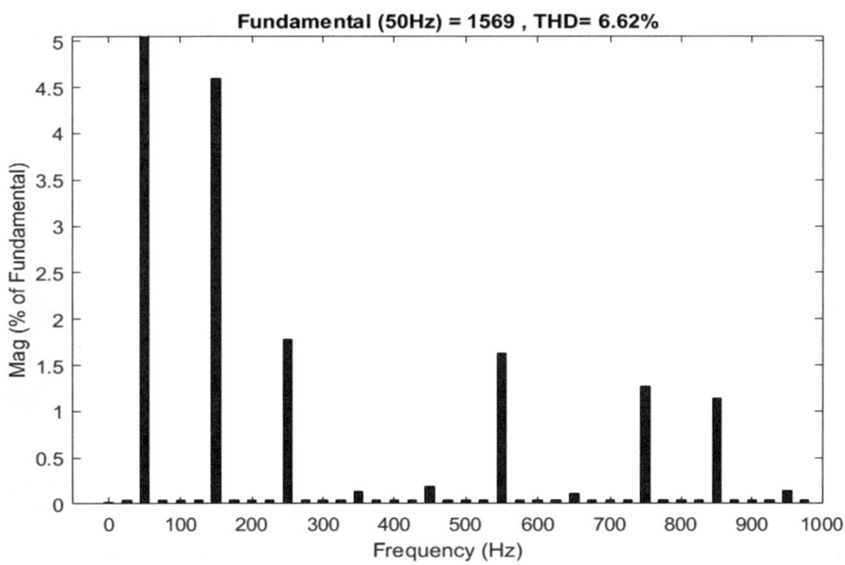

FIGURE 10.10 THD spectrum of PDO.

we have archived below the mentioned limit. As per the analysis with this all seven algorithms, one can conclude that the THD of LF-IGWO 4.54 PDO is 6.62 WOA is 6.92 SMA is 13.77, GWO is 14.68, HBA is 13.57, and ALO is 6.86. And from the

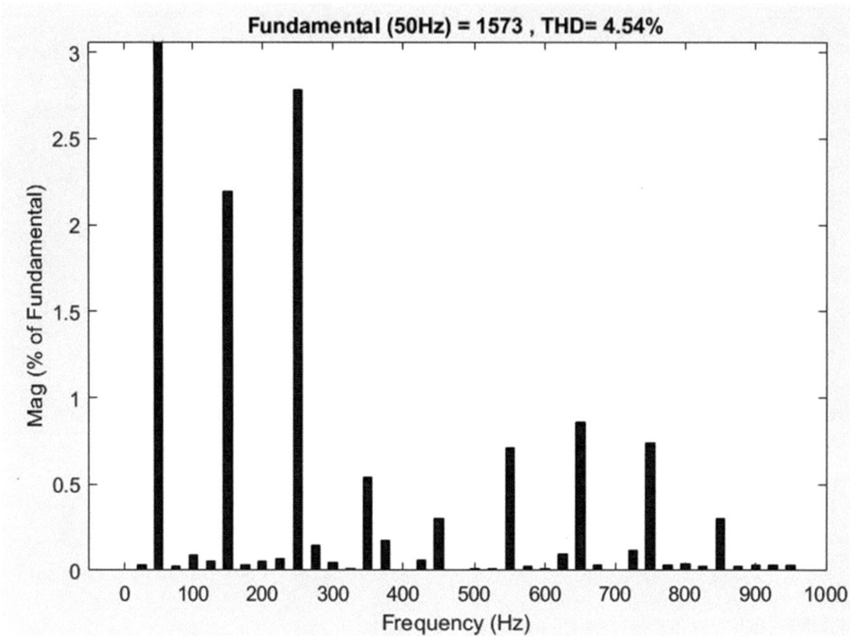

FIGURE 10.11 THD spectrum of LF-IGWO.

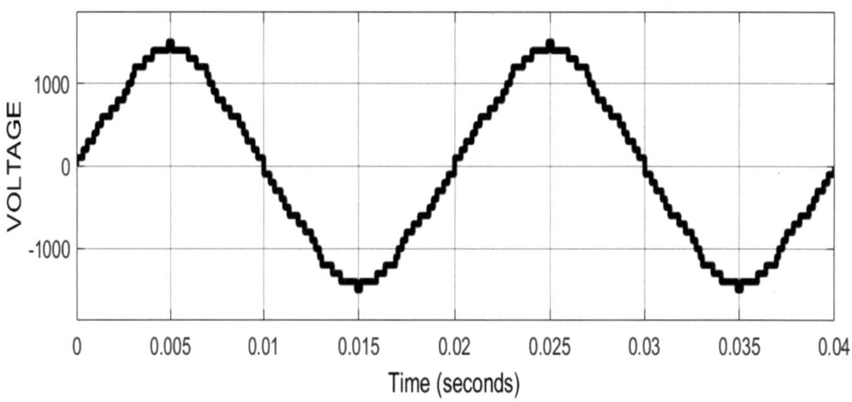

FIGURE 10.12 Output waveform of ALO.

Convergence curve, one can observe that, in comparison to other algorithms, the ALO and HBA, HBA have a high rate of convergence, whereas the GWO's rate of convergence is relatively low. As per the term NFL which is NFL that tells us that there are no such algorithms that can be applied to all the problems. Here, the effect of the NFL has come under notice that the SMA which is used to give the best results in the benchmark functions has failed to give us the better output.

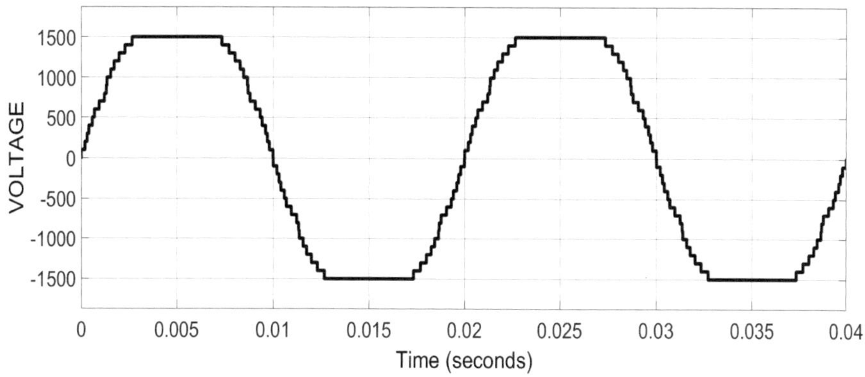

FIGURE 10.13 Output waveform of GWO.

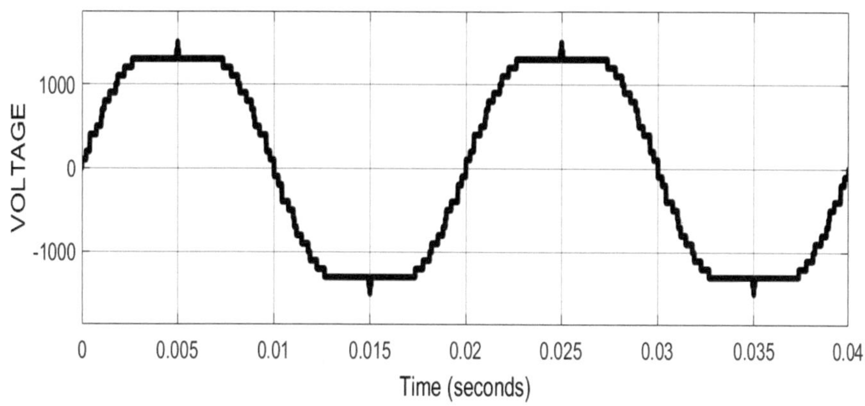

FIGURE 10.14 Output waveform of HBA.

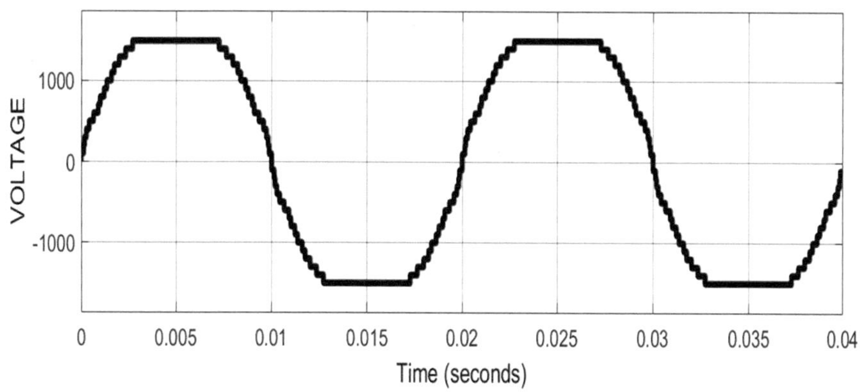

FIGURE 10.15 Output waveform of SMA.

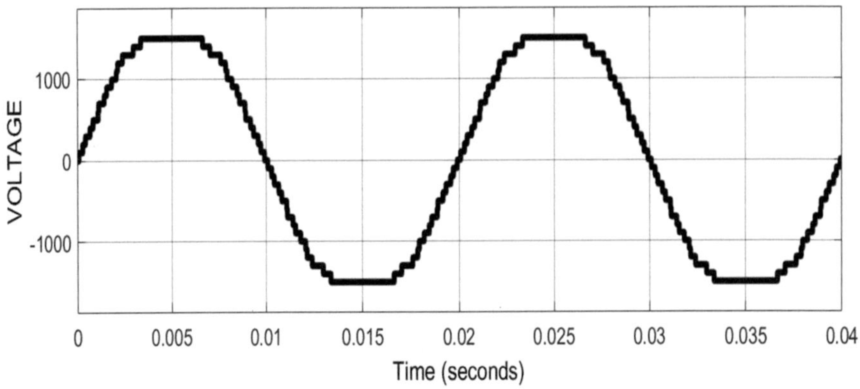

FIGURE 10.16 Output waveform of WOA.

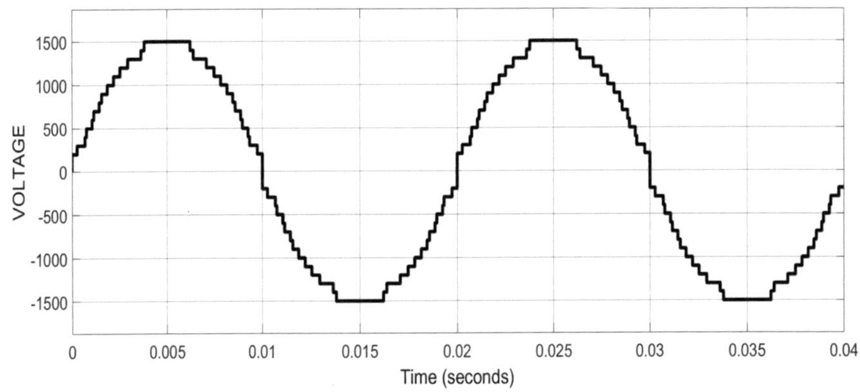

FIGURE 10.17 Output waveform of PDO.

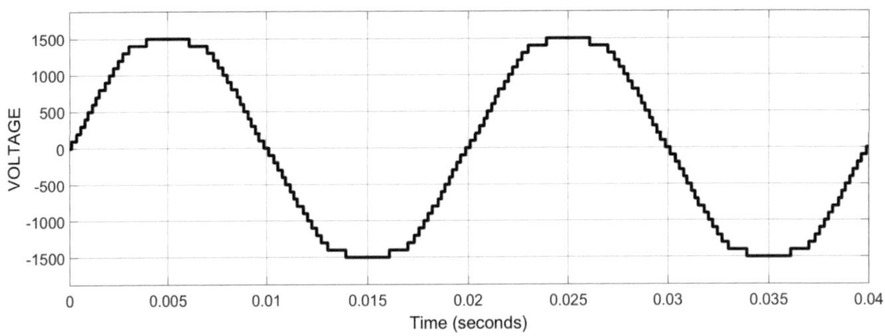

FIGURE 10.18 Output waveform of LF-IGWO.

REFERENCES

[1] K. Kaur and Y. Kumar, "Swarm Intelligence and its applications towards Various Computing: A Systematic Review," In 2020 International Conference on Intelligent Engineering and Management (ICIEM), London, UK, 2020, pp. 57–62, doi:10.1109/ICIEM48762.2020.9160177

[2] Mirjalili, Seyedali. "Particle Swarm Optimisation." *Studies in Computational Intelligence*, vol. 780, Springer Verlag, 2019, pp. 15–31. https://doi.org/10.1007/978-3-319-93025-1_2.

[3] Wolpert, D. H. and Macready, W. G. (1997). "No free lunch theorems for optimization," *IEEE Transactions on Evolutionary Computation*, 1(1), 67–82. doi:10.1109/4235.585893

[4] Mirjalili, Seyedali, Seyed Mohammad Mirjalili, and Andrew Lewis. "Grey Wolf Optimizer." *Advances in Engineering Software*, vol. 69, 2014, pp. 46–61. https://doi.org/10.1016/j.advengsoft.2013.12.007

[5] Bhatt, B., Sharma, H., Arora, K., Joshi, G. P., and Shrestha, B. (2023). "Levy flight-based improved grey wolf optimization: a solution for various engineering problems," *Mathematics*, 11, 1745. https://doi.org/10.3390/math11071745

[6] Mirjalili, S. (2015). The ant lion optimizer. Denmark: Elsevier Ltd.

[7] Faris, H., Aljarah, I., Al-Betar, M.A. et al. (2018). Grey wolf optimizer: a review of recent variants and applications. *Neural Computing and Applications*, 30, 413–435. https://doi.org/10.1007/s00521-017-3272-5

[8] Amaireh, Anas Atef, Asem S. Al-Zoubi, and Nihad I. Dib. "A New Hybrid Optimization Technique Based on Antlion and Grasshopper Optimization Algorithms." *Evolutionary Intelligence*, vol. 15, no. 4, 2022, pp. 1383–1422. https://doi.org/10.1007/s12065-022-00483-w.

[9] Reddy, D. P., Reddy, V.C., and Manohar, G. (2017). Ant lion optimization algorithm for optimal sizing of renewable energy resources for loss reduction in distribution system. New York: Elsevier Ltd.

[10] Mirjalili, S. and Lewis, A. (2016). The whale optimization algorithm. Denmark: Elsevier Ltd.

[11] Kumar, V. and Kumar, D. (2020). "Binary whale optimization algorithm and its application to unit commitment problem", *Neural Computing and Applications,* Vol. 1, pp. 1–17.

[12] Hashima, F. A., Housseinb, E. H., Hussainc, K., Mabroukd, M. S. and Al-Atabany, W. (2021). Honey badger algorithm: new metaheuristic algorithm for solving optimization problems. Hong Kong: Elsevier B.V.

[13] Elseify, M. A., Kamel, S., Abdel-Mawgoud, H. and Elattar, E. E. (2022). "A novel approach based on honey badger algorithm for optimal allocation of multiple DG and capacitor in radial distribution networks considering power loss sensitivity," *Mathematics,* Vol. 1, pp. 1–19.

[14] Almodfer, R., Mudhsh, M., Alshathri, S. and Abualigah, L. (2022). "Improving parameter estimation of fuel cell using honey badger optimization algorithm," *Frontiers in Energy Research,* Vol. 5, pp. 231-243.

[15] Li, S., Chen, H., Wang, M., Heidari, A. A., and Mirjalili, S. (2020). Slime mould algorithm: a new method for stochastic optimization. Denmark: Elsevier B.V.

[16] Ezugwu, A., Agushaka, O., Abualigah, L., Mirjalili, S., and Gandomi, A. (2022). "Prairie dog optimization algorithm," *Neural Computing and Applications*, 34. pp. 1–23. doi:10.1007/s00521-022-07530-9

[17] Dorigo, M., Birattari, M. and Stutzle, T. (2006). "Ant colony optimization," *IEEE Computational Intelligence Magazine*, 1(4), 28–39. doi:10.1109/MCI.2006.329691

[18] Jacob, T. and Suresh, L. P. "A review paper on the elimination of harmonics in multilevel inverters using bioinspired algorithms," In 2016 International Conference on Circuit, Power and Computing Technologies (ICCPCT), Nagercoil, India, 2016, pp. 1–8, doi:10.1109/ICCPCT.2016.7530273

[19] Krishna, A. R. and Suresh, L. P. "A brief review on multilevel inverter topologies," In 2016 International Conference on Circuit, Power and Computing Technologies (ICCPCT), Nagercoil, India, 2016, pp. 1–6, doi:10.1109/ICCPCT.2016.7530373

[20] BarathKumar, T., Vijayadevi, A., Brinda Dev, A., and Sivakami, P. S. (2017). "Harmonic reduction in multilevel inverter using particle swarm optimization," *International Journal of Innovative Science, Engineering & Technology*, 4(11). pp. 1–14.

[21] Firdoush, S., Kriti, S., Raj, A., and Singh, S. K. (2016). "Reduction of harmonics in output voltage of inverter," *International Journal of Engineering Research & Technology CMRAES*, 4 (02). pp. 1–17

[22] Najafi, E. and Yatim, A. H. (2012). "Design and implementation of a new multilevel inverter topology," *IEEE Transactions on Industrial Electronics*, 59, 4148–4154. doi:10.1109/TIE.2011.2176691

[23] Arora, K., Joshi, G. P., Ragab, M., Rawa, M., Milyani, A. H., Mansour, R. F., and Yang, E. (2022). "Bilateral contract for load frequency and renewable energy sources using advanced controller," *Computers, Materials & Continua*, 73(2), 3165–3180, doi:10.32604/cmc.2022.026966

[24] Brest, J., Maučec, M. S. and Bošković, B. (2020). "Differential Evolution Algorithm for Single Objective Bound-Constrained Optimization: Algorithm j2020," In 2020 IEEE Congress on Evolutionary Computation (CEC), Glasgow, UK, pp. 1–8, doi:10.1109/CEC48606.2020.9185551

[25] Shi, Y., Gao, H., and Wu, D. (2014). "An improved differential evolution algorithm with novel mutation strategy," In 2014 IEEE Symposium on Differential Evolution (SDE), Orlando, FL, USA, pp. 1–8, doi:10.1109/SDE.2014.7031540

11 Sustainable Application of Blockchain in Food Supply Chains Criteria in Selecting a Sustainable Blockchain Technology Service Provider

Amit Kohli and Hansinie Lorenzuhewa

11.1 INTRODUCTION

Implications of advanced technologies on the food industry and its supply chain are worth the probe as food is the first and foremost basic need of human survival. Today, massive developments and innovations in information technology (IT) have transformed many industries into faster and smoother versions of themselves. Guo et al. (2022) claim that as a recent trend in IT, blockchain contributes to a substantial portion of the commercial revolution. Zutshi et al. (2021) reported that blockchain technology (BT) had drastically transformed many industries. In simple terms, a blockchain is a common ledger that stores the details of transactions made by peers in a network. Such a system is free of central authority or administrators, thus increasing information transparency and security (Laroiya et al., 2020). With its new attributes, BT has widened the horizons of IT. Food Supply Chain (FSC), a setting where there is an information flow among peers (farmers, processors, transporters, storekeepers, retailers etc.), seems a perfect ground to apply BT—a public, peer-to-peer ledger system, to alleviate the existing issues.

As mentioned above, FSC is a complicated place involving many parties. In congruence with Chandan et al. (2023), FSCs are not only highly dispersed, collective, and diverse but also they are dissimilar from one another in terms of output, course of tasks, and destination. Such complexity has created inefficiencies in FSCs. Saha et al. (2022) state that the FSC, as a multifaceted system, has faced some challenges, including traceability and food safety. The same issues may interfere with the fundamental human right to have quality food; hence, they must be addressed and resolved.

However, the merger between technology and real life is not smooth. The positive outcome is a far view, and the quest of changing the traditional face and ways of

DOI: 10.1201/9781003563440-11

industry must first battle the confusion due to the unknown and fear of leaving the comfort zones for many. As stated by Chandan et al. (2023), the least utilization of IT is in the agri-food industry. The BT has been known for almost a decade, and the application of the same in FSCs is low (Saha et al., 2022). As per Vu et al. (2021), although BT has been recognized by the food industry, the practice has been hindered due to the lack of knowledge, related studies, and success stories. Choi et al. (2020) also find that the acceptance rate of BT is low, even though it is appreciated for creating sustainable supply chains. Moreover, higher sophistication, privacy issues, and the associated cost are a few reasons behind the resistance to BT (Choi et al., 2020). On that note, the vast potential of BT has not been recognized by some social and commercial segments, and its adoption rate into FSC is evidently slow.

Additionally, BT is perceived by the world as double-faced and subjected to public scrutiny. Some praise technology as a tool of sustainability. The World Bank (2019) recognizes BT as the pillar of the industrial revolution, like the steam engine and the internet, which reformed the industrial landscape radically. Regardless, some reject BT pointing toward its high energy consumption. BT adoption may cause environmental destruction through excessive power consumption and emissions, demotivating the adopters in circular economies (Rajeb et al., 2022).

However, BT is a versatile and rich technology that can be customized based on user needs. Rieger et al. (2022) elaborated on some BT solutions which are energy efficient, and they identified BT to be sustainability friendly. Here they demanded fairer debate. Sustainability is a hot topic today, and nations are marching toward sustainability goals. The government of Canada, in the Federal Sustainable Development Strategy (FSDS), described the need for a sustainable supply chain. Moreover, the House of Commons report identified BT as a tool to enhance FSCs. So, due to federal pressures, FSC players will soon be responsible for sustainable supply chains, and the BT will be their solution.

However, a majority of FSC players remain skeptical about BT. Considering their livelihoods which are bound with FSCs, and their limited knowledge, the prospective adopters are too apprehensive to test the waters at this stage. It is regrettable that in a high-tech world, despite all the technological expertise gained, the food industry (which addresses the most crucial human need for food) could not use the privileges.

In view of this, the research aimed to fill the knowledge gap and equip the perspectives on BT in FSCs so that awareness of the same can flow wherever needed. The study aspired to examine the advantages of BT to motivate its application into FSCs. The research also focused on the challenges of applying BT to FSCs to prepare the FSC players for reality. Plus, irrespective of attitudes and challenges, FSC players will sooner or later have to digitalize the FSCs with BT due to the sustainability-related rules and regulations. In addition to the benefits and challenges, the research explored another area to promote BT in FSCs. Most of the FSC players do not possess the multidimensional knowledge and technical literacy to use the BT correctly to achieve sustainability perspectives. Hence, they must find appropriate solution providers to support them in the endeavor. This study suggested environmental, economic, and social criteria for choosing a sustainable BT provider. The research also validated the literature-based findings through industry experts for improved pertinence.

11.2 LITERATURE REVIEW

BT is one of the most hyped terms in the past decade and argued to be the single most significant technological revolution since the internet invention. BT can be used in food industry—a safety sensitive sector ascertains food security.

11.2.1 What is Blockchain Technology?

Blockchain is referred to as the pillar of the industrial revolution, equalizing it with the inventions such as the steam engine and the internet, which brought about revolutionary industrial reforms in the past (The World Bank, 2019). As expressed by Kimani et al. (2020), Blockchain is among the most prominent technological novelties of the twenty-first century. With its applications mainly in the finance sector, insurance, transportation, logistics, and energy industries, the blockchain has gained increased attention from the government, trades, and researchers (Hamida et al., 2017). The Canadian government is exploring ways and means of encouraging BT application in the supply chain (House of Commons Canada, 2023).

The meaning of the term blockchain is implied in its name—a chain made from blocks. To begin with, Krishan Arora (2022) defined, "A blockchain is a chain of blocks that contains information." Such sequential blocks of information are stored in a database or a registry accessible to a network of peers. Liang et al. (2021) saw blockchain as a secure decentralized register that records sequential peer-to-peer transactions without third-party involvement. It is a sequence of data; the data are packed into blocks organized in chronological order (Gai et al., 2022). In accordance with Mohamed & Mohamed (2022), blockchain is a common ledger open to all the members of a dispersed network so that each member (node) has a replica of the whole database with updates related to transactions. Elaborating the concept more, Islam et al. (2020) reported that blockchain is made of blocks that include communication, proof of work, and references of prior blocks stored in a common database which can perform peer-to-peer transactions within the network. Considering these facts, it is evident that a blockchain is often interpreted as decentralized data storage shared by a network of peers for a specific purpose.

Blockchain operates in a process. Joshi et al. (2018) explained how blockchain functions in four steps. First, a certain network user or a node initiates a transaction within the network by recording and transmitting the information. That data related to the transaction is authenticated by another user within the network, and upon authentication, the data are converted to a block in the second step. Then, the transaction details are verified by all users by generating a proof of work algorithm or proof of stake algorithm to the block required verification. In the final step, the network's consensus algorithm stores the data in the added block to the chain. Moreover, all nodes within the network accept the new block, and based on the same, the chain continues (Joshi et al., 2018).

Contrary to centralized systems, blockchain is a decentralized system. Tang et al. (2020) declared that peers maintain the blockchain collectively without having to obey the centralized power. However, the firms may employ their liberty in allowing the degree of central authority upon the blockchain. Haleem et al. (2021) recognized three types of blockchains such as public, private, and hybrid. The public version is the origin of distributed ledger technology. It is fully open and decentralized. However, a private blockchain is a restricted platform. For instance, it may be a closed network, or there may be a controller with centralized power. A hybrid is a combination of both public and private blockchains. Organizations turn to hybrids seeking the best of both worlds. The hybrid model provides the firms with the discretion on data privacy as to what is public and what is private (Haleem et al., 2021). Based on their own requirements, businesses may customize BT.

11.2.2 The Current Applications of BT In FSC

Blockchain is an emerging technology, and the FSCs are still attuning. Conforming to Chandan et al. (2023), the agri-food sector has been reported for the lowest usage of IT. Agriculture is a primary industry that provides millions of direct occupations and many more through the supply chain, yet compared to other industries, it is left behind in terms of digitalization (Lim, 2022). Annosi et al. (2020) spotted the barriers agri-food firms face in their journey toward digitalization such as difficulty in accessing digital technologies, lack of supporting institutes, lack of incentives, high costs, lack of education, knowledge, and training, quality of internet usage, and so on (Annosi et al., 2020). These factors explained why advanced digital technologies like BT could not make the way toward FSCs. Mohammed et al. (2023) pointed out challenges related explicitly to BT adoption in FSCs. They are the issues of scalability, interoperability, higher costs, lack of experience, and regulatory limitations (Mohammed et al., 2023).

11.2.3 Implications on BT Towards Sustainability

Digitalization, especially BT, is involved in the sustainability debate. As defined by the United Nations (UN) Brundtland Commission in 1987, sustainability is satisfying today's needs without compromising future generations' ability to fulfill their needs (UN, n.d.). Jeronen (2013) emphasized that sustainability is a paradigm of envisioning the future in which economic, ecological, and social concerns are balanced in the expedition for a better life. Therefore, sustainability is an approach that considers the economy, the environment, and the community. From one angle, BT serves sustainability by communities by taking rough edges out of FSCs to ensure food safety and security. Polas et al. (2022) noted BT as a 'game changer' which can cater to intense ecological and economic sustainability challenges. To respond to the growing environmental issues, BT is transmuting sustainable inventions, maximizing sustainable economic practices and long-standing business models. As mentioned previously, many agree that BT can transform the FSCs to strengthen food safety and security. Hence, undoubtedly BT paves the way to sustainability through reliable and efficient FSCs.

On the other hand, BT is argued to have negative environmental effects. Rejeb et al. (2022) suggested that some applications of BT might lead to ecological degradation due to high consumption of electricity and emissions discouraging the adopters in a circular economy. These two contradicting ends left the prospective adopters confused and hesitant. However, as Haleem et al. (2021) mentioned above, there are various types of BT. Sarmah (2018) also explained different types of BT (private, public, semi-public, etc). Rieger et al. (2022) pointed out that various types of BT consume less energy. Rieger et al. (2022) also exposed convincing evidence to support the notion that organizations and governments can support sustainability with BT and not without. House of Commons Canada (2023) elaborated that the BT integrated supply chain could track carbon dioxide emissions and thereby enhance the carbon credit systems. Because of this, BT's positive contribution to sustainability is apparent.

11.2.4 Significance of Choosing the Correct Type Of The BT

Rieger et al. (2022) insisted on a fairer debate on BT and sustainability. As BT allows customization with its different variants mentioned above, prospective adopters can tailor the solutions based on their priorities and sustainability initiatives. In the exercise of adopting and operating BT, both financial and ecological factors should be addressed (Kramer et al., 2021).

11.3 HOW TO SELECT A BT SERVICE PROVIDER

With clouded perspectives and limited knowledge, it is highly challenging for FSC players to achieve sustainability through BT. Explicitly considering the sophistication of BT, the prospective adopters are less likely to choose such an appropriate BT solution by themselves. The most prominent challenge to adopting BT is the lack of understanding by the public (Mohammed et al., 2023). As observed by Annosi et al. (2020) above, there are many barriers and challenges in the BT journey. Accessing digital solution, lack of knowledge and training are among those (Annosi et al., 2020). Katsikouli et al. (2021) addressed the difficulty of choosing a technological solution to resolve the problems in FSCs. It is required to have expertise in natural, societal, political, scientific, ethical, economic, and legal dimensions to face the challenges in modern FSCs (Katsikouli et al., 2021). Based on the above-discussed factors, the adopters are already challenged by the higher technical expertise to apply BT to FSCs.

The study, however, stood by the sustainable adoption of BT; and tallying the sustainable intentions with BT adoption raised the bar of requisite expertise even more. In this case, a prospective adopter must possess the required proficiency for sustainable adoption of BT within the organization already, hire skilled employees, or outsource the service. Developing the solution in-house is unlikely to be successful for most of the adopters considering the multidimensional knowledge requirement discussed by Katsikouli et al. (2021) for BT adoption as well as the challenges of BT adoption discussed above (high cost etc.). Therefore, the study encouraged outsourcing a solution provider who can combine BT and sustainability within the FSC scope.

11.4 CRITERIA TO SELECT SUSTAINABLE BT SERVICE PROVIDER

This study advocated BT integration in FSCs based on a few reasons. Most prominently, as deliberated above, FSC is a setting where numerous parties share common information and work for the same goal of delivering the food from the origin to the consumer. In this case, the blockchain, as a shared database, is the ideal and the most modern solution in IT so far to facilitate the above-mentioned FSC operation. FSCs today report numerous issues, and considering the attributes of BT is the best appropriate remedy to address those. Secondly, BT is promising, as it has transformed many industries profoundly. On this note, FSC, bound with a critical element of survival (which is food), is undoubtedly worth such transformation and uplift.

However, BT application is less in FSCs as it is still an unknown territory for the majority; some are dubious about its environmental impact, thus are puzzled about how to achieve their sustainability intentions within BT. With its versatility, BT can also be tailored to comply with sustainability standards by choosing the correct variety. Adding all the goodness together, BT has a huge potential in FSC.

In the name of food security, adopting BT into FSCs must be encouraged by providing the real picture of BT. At the same time, the subject of balancing digitalization and sustainability requires rare expertise that most organizations do not possess in-house. Irrespective of all the barriers, the FSC players will have to adopt BT sooner or later in the face of immense pressures from sustainability-driven governments.

Unfortunately, these notions are not flowing to FSC players, and organizations do not own subject-related technical wisdom to view the technology's real capability. Vu et al. (2021) accepted that the food industry has opened the door for BT; however, they claimed that the application of the technology is obstructed due to relevant knowledge and literature on the topic. Mohammed et al. (2023) reported that the research on BT application in FSC is new and limited. Additionally, the studies exploring the sustainable adoption of BT are even lesser. Kramer et al. (2021) concluded that the studies discussing the effects of the different kinds of BT solutions on ecological and economic sustainability are less.

In this context, the author deems it essential to fill the research gaps by providing sustainability parameters over which they can evaluate and outsource the appropriate BT solution provider for sustainable adoption.

11.5 METHODOLOGY

The study was based on qualitative research. The case study approach was followed mainly in the research process in combination with semi-structured interviews.

11.5.1 Case Study Approach

The research boundaries were hard to express depth or widthwise, and the outcomes were argument-based clarifications. The study also searched the answers to the 'what' and 'how' types of questions. For instance, the research was an exploration and a presentation of arguments and justifications in explanatory form, creating various views. The research also investigated the success factors and challenges of applying BT in FSCs ('what' question) and how BT adoption worked in real-life FSCs ('how' questionl). On this account, using the qualitative case study approach was justified.

The next point was whether the case study approach was suitable for studying the supply chain. In the existing research on supply chains, the case study approach was widely used (Javaid et al., 2021). Kamath (2018) concluded that the case study approach was excellent in studying the application of intelligent systems or expert systems in logistics. Accordingly, the case study approach was chosen to find the inputs for the literature-based framework. Here the case means an 'adopter' who has already integrated BT in FSCs (ex. Walmart). However, when required data were limited or not to be found related to the chosen cases, the author looked at the literature generally.

11.5.2 Semistructured Discussions/Interviews

In the fourth stage of the research, semistructured discussions were employed to obtain feedback from industry experts. Discussions were conducted mostly through virtual platform—Google Meet, and one discussion was conducted in person. Questionaries were not required to prepare, as the task was obtaining mere feedback for a preprepared framework.

Semistructured interviews are helpful for researchers as they create a ground where the participants' insights can be developed and explored extensively to gain a deeper understanding of the subject (Mojtahed et al., 2014). Considering the nature of this research, open-ended, situational questions were ideal for drawing out detailed insights from the experts.

Harveys-Jordan and Long (2001) found semistructured interviews as a tool to collect qualitative data. Semistructured interviews utilize a series of open-ended queries related to the context chosen by the researcher, allowing different theories and sub-topics to form. Generally, the theories are recognized before the interview; however, the interview process should be flexible enough to let the themes developed throughout the discussion be investigated. (Harveys-Jordan & Long, 2001). Unlike an unstructured interview, where the direction is not considered at all, a semi-structured interview provides a direction for the conversation. The research subject is already established beforehand; therefore, even if there is a change, the researcher, through the questions, can redirect (Ruslin et al., 2022).

In the application of the above in this research, semistructured interviews were chosen against structured or unstructured interviews due to adaptability, flexibility, and the ability to go to greater lengths. The author first outlined the theories from the literature related to research variables (benefits and challenges of BT and criteria to select sustainable BT solution providers) and created a framework. The same was discussed in the interview; however, the interviewees were allowed to freely express their views, deep insights were obtained through it, and they were redirected to the subject through contextual questions when necessary.

11.6 DATA ANALYSIS

The study overall presented a qualitative content analysis. Content analysis is a popular technique in qualitative research (Hsieh & Shannon, 2005). In agreement with Momeni Rad (2013), generally, content analysis investigates concepts, conditions, and connections attempting to imply and uncover concealed trends in interviews, observations, and written documents. Government of Canada (2022) claimed that the aim of conducting a content analysis is creating a framework to explain the theory in a conceptual form.

This research gathered data from literature and interviews, explored the findings to create themes and theories, and created a framework to demonstrate those theories.

11.7 RESEARCH PROCESS

1. Outline criteria from the literature to choose a sustainable BT service provider.
2. Get the criteria evaluated by subject matter experts.
3. Create a framework of expert reviewed criteria to choose a sustainable BT service provider.

11.8 DISCUSSION

Table 11.1 depicted the criteria for selecting a sustainable solution provider for BT application in FSCs (Tundys, 2016; Choi et al., 2020; Liang et al., 2021; Tang et al., 2020).

11.8.1 Economic Criteria

Most of the parameters given in Table 11.1 were included in the economic segment. It basically covered the criteria related to finances, characteristics of the technology, and the solution provider's abilities and involvement. However, unlike in conventional evaluations, here, the cost covered the environmental cost of the adoption process. For instance, BT requires specific infrastructure and energy to operate.

11.8.2 Environmental Criteria

Secondly, the environmental segment inquired about the service provider's sustainability initiatives and compliance with environmental regulations. A sustainable service provider who is in line with all the environmental laws can be trusted to apply the same sustainable standards in client projects.

11.8.3 Social Criteria

The social segment also incorporated many valuable criteria. FSC consists of many intermediaries, and the BT adoption is a process of collective agreement of all of them. Hence, it is vital to check the interactions of the service provider with society and stakeholders. The social parameter also examined the virtue of the service provider demanding corruption and discrimination-free conduct.

11.9 EVALUATION OF EXPERT FEEDBACK

The interviews with experts led to new sustainable solution provider selection criteria related to BT integration in FSCs. However, there were no major contradictions compared to the literature.

In the case of criteria, experts advised renaming some, and new ones were suggested as well. The economic category criteria 'preparedness' was noticed as repetitive and difficult to relate; therefore, it was removed. 'Proactive' was added

TABLE 11.1
Sustainable Solution Provider Selection Criteria—BT Adoption in FSCs

Aspect	Criteria	Interpretations
Economic	1. Overall Cost (including environmental cost)	The entire financial cost incurred to the adopter due to BT adoption and total costs of environmental impact (due to infrastructure installation etc.) and energy consumption
	2. Price	The financial value of the specific solution as indicated by the solution provider
	3. Product quality	Error/bug-free and fast
	4. Technological capabilities	Compatibility with other technologies and systems
	5. Reliability	Minimal breakdowns
	6. Experience in the field	Solution provider's experiences with BT installations
	7. After-sales support	The services and support from the solution provider after installing the solution
	8. Budget	The adopter's financial capacity
	9. Flexibility	Ability to change with the requirements arising from FSCs
	10. System maintenance costs	The costs involving monthly electricity and internet bills, system updates, and renewal of certifications
	11. Contract duration	The time period the solution provider is legally bound for the services to the adopter
	12. Solution delivery time	The overall time the solution provider takes to deliver an operable system
	13. Preparedness	Solution provider's ability to anticipate issues and be prepared for the same
	14. Customization	The ability to tailor the solution
	15. Financial stability	Financial stability of the service provider
	16. Location	The physical distance between the service provider and the adopter
	17. Punctuality	Ability to tackle issues immediately and on-time services
	18. The scope	The exact FSCs covered by the solution (ex. The FSC of coconut oil from Thailand)
	19. Response time	The average response time of the solution provider during a breakdown
	20. Breakdown maintenance process	The steps followed by the solution provider during a system breakdown
	21. Fulfillment of the contract obligations	Solution provider's conduct within the contract terms and past legal issues involving other clients

(*continued*)

TABLE 11.1 (Continued)
Sustainable Solution Provider Selection Criteria—BT Adoption in FSCs

Aspect	Criteria	Interpretations
	22. Service recovery	The overall effort to correct a mistake and the supportive attitude during a system breakdown
	23. Professionalism	Ethical and professional behavior.
Environmental	1. ISO 1400	Compliance with ISO 1400 – environmental laws
	2. Sustainable image	How the solution provider has been perceived by society for the sustainable conduct
	3. Environmental initiatives	The ambitions and the conduct of the solution provider involve the environment
	4. Sustainable research and development	Research and development work by the solution provider related to the environment
	5. Controlled resource consumption (water, energy, raw materials. The use of renewable energy)	Ways the solution provider preserves natural resources and minimizes the environmental impacts
	6. Energy-saving measures	The techniques used by the solution provider to minimize energy consumption
	7. Disclosure of information (relating to environmental aspects)	The transparency of the solution provider related to environmental aspects
Social	1. Rights of the stakeholders	The solution provider's ability to safeguard the rights of the stakeholders
	2. Occupational Health and Safety	The solution provider's concern about the health and safety of his own employees and the stakeholders
	3. Initiatives for equality	Discrimination – free conduct and treating all parties equally
	4. Social engagement	The solution provider's service to the society
	5. Stand against corruption	Solution Provider's stance against fraud

to the economic category as it is a crucial approach in avoiding numerous mishaps during BT adoption. Experience in the field (economic category) was changed to 'experience in specific use cases' to match the industry jargon. 'flexibility' (economic category) was removed as it is covered by other criteria such as 'technological capabilities.' One expert suggested to explore B2B value propositions under economic criteria and recommended adding the same. He also explained the importance of looking at the clientele of the solution provider and the respective industries of those

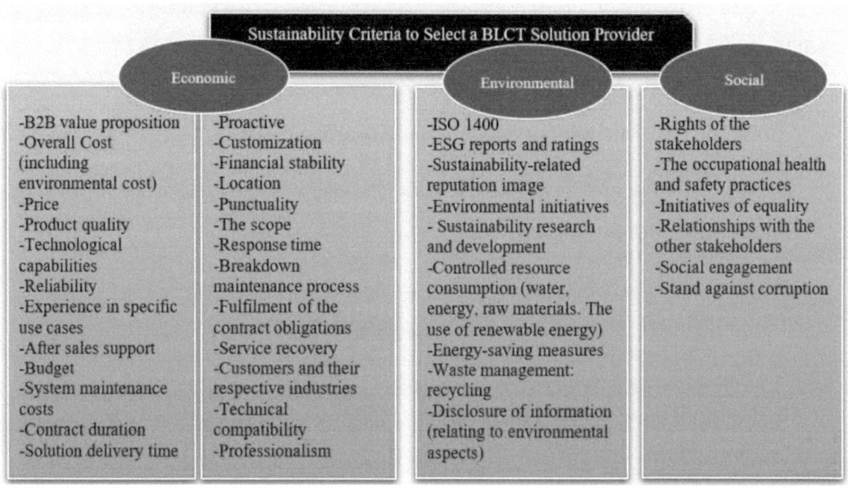

FIGURE 11.1 Sustainability criteria in selecting BT service provider for FSCs—an expert-reviewed framework. Source: Developed by the authors.

clients. A part of the same is covered in 'experience in specific use cases' suggested by the experts; however, the 'customers and their respective industries' were added separately for clarity. According to some experts, it is also vital to check the compatibility of the services provider's technology with the client and other stakeholders. For the environmental category of the criteria, the experts emphasized recycling practices and using renewable energy such as solar power. Recycling was added as a waste management practice, and the usage of solar energy was already included under 'controlled resource consumption.' Experts highlighted the need of inquiring ESG related reports and ratings, thus added. With the social category, both experts were satisfied; however, they advised checking the service provider's relationships with other stakeholders.

11.9.1 Expert-Reviewed BT Adoption Framework for FSCs

The literature-based framework was developed per the industry feedback received in the previous stage, and an expert-reviewed framework was created showcasing BT solution provider selection criteria related to sustainable BT adoption in FSCs. The same has been provided in Figure 11.1.

11.10 CONCLUSIONS

The study empowered the sustainable application of BT. As BT adoption into FSCs within sustainability involves rare technical expertise, the paper guided adopters towards outsourcing expertise by suggesting sustainability criteria to make an informed choice. The author first referred to the literature in search of the above and created a literature-based framework. The same was further improved based on industry experts' feedback to enhance the validity. The findings of the study are vital as it adds to the literature and the existing knowledge about BT in FSC.

The main implications of the research can be summarized as follows:

- Based on the need for food safety and security, BT adoption in FSCs must be promoted irrespective of the extensive efforts and other challenges on the way.
- BT benefits the FSCs in the long run, multiplying customer trust and sales. Therefore, organizations must view the initial efforts and costs as investments for the future.
- BT has been misjudged for high energy consumption; however, by knowing how to use BT appropriately, firms can incorporate BT in their sustainability journey successfully.
- As adaptors are not technically literate to balance the adoption of a high-end technology such as blockchain with sustainability, the experts should be hired for the same based on the given criteria.
- Sooner or later, FSC players will have to digitalize the FSCs due to government pressures on sustainability.

ABBREVIATIONS

B2B	Business to business
BC	British Columbia
BT	Blockchain technology
ERP	Enterprise resource planning
ESG	Environmental, Social and Governance
FSC	Food supply chain
FSDS	Federal Sustainable Development Strategy
IT	Information Technology
RO	Research objectives
RFID	Radio frequency identification detector
SDG	Sustainable development goals
TBL	Triple bottom line
UK	United Kingdom
UN	United Nations

ACKNOWLEDGMENTS

We would offer our sincere gratitude to all the industry experts contributed in producing this book chapter.

REFERENCES

Annosi, M. C., Brunetta, F., Capo, F., & Heideveld, L. (2020). Digitalization in the agri-food industry: the relationship between technology and sustainable development. *Management Decision*, *58*(8), 1737–1757. https://edepot.wur.nl/529133

Arora, K., Joshi, G. P., Ragab, M., Rawa, M., Milyani, A. H., Mansour, R. F., & Yang, E. (2022). Bilateral contract for load frequency and renewable energy sources using advanced controller. *Computers, Materials & Continua*, *73*(2), 3165–3180. doi:10.32604/cmc.2022.026966

Chandan, A., Rosano, M., & Potdar, V. (2023). Achieving UN SDGs in food supply chain using blockchain technology. *Sustainability*, *15*(3), 2109. https://doi.org/10.3390/su15032109

Choi, D., Chung, C. Y., Seyha, T., & Young, J. C. (2020). Factors affecting organizations' resistance to the adoption of blockchain technology in supply networks. *Sustainability*, *12*(21), 8882. https://doi.org/10.3390/su12218882

Gai, W., Gu, Y., & Qin, J. (2022). Financial automation audit method based on blockchain technology. *Computational Intelligence and Neuroscience*, *2022*, 1–12. https://doi.org/10.1155/2022/9941585

Government of Canada. (2022). *Achieving a Sustainable Future. Federal sustainable development strategy 2022 to 2026*. www.fsds-sfdd.ca/en

Guo, X., Zhang, G., & Zhang, Y. (2022). A comprehensive review of blockchain Technology-Enabled smart manufacturing: A framework, challenges and future research directions. *Sensors*, *23*(1), 155. https://doi.org/10.3390/s23010155

Haleem, A., Javaid, M., Singh, R. P., Suman, R., & Rab, S. (2021). Blockchain technology applications in healthcare: An overview. *Directory of Open Access Journals.* Retrieved May 5, 2023, from https://doaj.org/

Hamida, E. B., Brousmiche, K. L., Levard, H., & Thea, E. (2017). Blockchain for enterprise: Overview, opportunities and challenges. In *he Thirteenth International Conference on Wireless and Mobile Communications*. https://hal.science/hal-01591859/

Harveys-Jordan, S., & Long, S. (2001). The process and the pitfalls of semi-structured interviews. *Community Practitioner*, *74*(6), 219–221.

House of Commons Canada. (2023, June). Blockchain technology: Cryptocurrencies and beyond. Report of the Standing Committee on Industry and Technology (INDU), 44th Parliament, 1st Session. www.ourcommons.ca/Content/Committee/441/INDU/Reports/RP12522346/indurp15/indurp15-e.pdf

Hsieh, H. F., & Shannon, S. E. (2005). Three approaches to qualitative content analysis. *Qualitative Health Research*, *15*(9), 1277–1288. https://doi.org/10.1177/1049732305276687

Islam, I., Munim, K. M., Oishwee, S. J., Islam, A. K. M. N., & Islam, M. N. (2020). A critical review of concepts, benefits, and pitfalls of blockchain technology using concept map. *Directory of Open Access Journals.* Retrieved May 5, 2023, from https://doaj.org/

Javaid, M., Haleem, A., Singh, R. P., Khan, S., & Suman, R. (2021). Blockchain technology applications for industry 4.0: A literature-based review. *Technological Forecasting and Social Change*, *2*(4), 100027. https://doi.org/10.1016/j.bcra.2021.100027

Jeronen, E. (2013). Sustainability and sustainable development. In *Springer eBooks* (pp. 2370–2378). https://doi.org/10.1007/978-3-642-28036-8_662

Joshi, A., Han, M., & Wang, Y. (2018). A survey on security and privacy issues of blockchain technology. *Academic Search Ultimate.* Retrieved May 5, 2023, from www.ebsco.com/resources/academic-search-ultimate

Kamath, R. (2018). Food traceability on blockchain: Walmart's pork and mango pilots with IBM. *Journal of the British Blockchain Association*, *1*(1), 1–12. https://doi.org/10.31585/jbba-1-1-(10)2018

Katsikouli, P., Wilde, A. S., Dragoni, N., & Jensen, H. H. (2021). On the benefits and challenges of blockchains for managing food supply chains. *Journal of the Science of Food and Agriculture*, *101*(6), 2175–2181. https://backend.orbit.dtu.dk/ws/portalfiles/portal/222317355/jsfa.10883.pdf

Kimani, D., Adams, K., Attah-Boakye, R., Ullah, S., Frecknall-Hughes, J., & Kim, J. K. (2020). Blockchain, business and the fourth industrial revolution: Whence, whither, wherefore and how? *Technological Forecasting and Social Change*, *161*, 120254. https://doi.org/10.1016/j.techfore.2020.120254

Kramer, M. P., Bitsch, L., & Hanf, J. H. (2021). Blockchain and its impacts on agri-food supply chain network management. *Sustainability*, *13*(4), 2168. https://doi.org/10.3390/su13042168

Laroiya, C., Saxena, D., & Komalavalli, C. (2020). Applications of blockchain technology. In *Elsevier eBooks* (pp. 213–243). https://doi.org/10.1016/b978-0-12-819816-2.00009-5

Liang, T.P., Kohli, R., Huang, H.C. & Li, ZL (2021). What drives the adoption of the blockchain technology? A fit-viability perspective. Business Source Ultimate. Retrieved May 5, 2023, from www.ebsco.com/products/research-databases/business-source-ultimate

Lim, X. (2022. March 22). How technology can help address challenges in agriculture. World Economic Forum. www.weforum.org/agenda/2022/03/how-technology-can-help-address-challenges-in-agriculture/

Mohammed, A., Potdar, V., Quaddus, M., & Hui, W. (2023). Blockchain adoption in food supply chains: A systematic literature review on enablers, benefits, and barriers. *IEEE Access*, *11*, 14236–14255. https://doi.org/10.1109/access.2023.3236666

Mohamed, H. & Mohamed H. (2022). Instant payment framework based on blockchain technology and open API architecture to improve the e-payment system of the local banks in Egypt. *Business Source Ultimate.* Retrieved May 5, 2023, from www.ebsco.com/products/research-databases/business-source-ultimate

Mojtahed, R., Nunes, M. B., Martins, J. T., & Peng, A. Y. (2014). Equipping the constructivist researcher: The combined use of Semi-Structured interviews and Decision-Making maps. *The Electronic Journal of Business Research Methods*, *12*(2), 87. https://academic-publishing.org/index.php/ejbrm/article/view/1318

Momeni Rad, A. (2013). Qualitative content analysis in research tradition: Nature, stages and validity of the results. *Quarterly of Educational Measurement*, *4*(14), 187–222. https://jem.atu.ac.ir/article_92_en.html

Polas, M. R. H., Kabir, A. I., Sohel-Uz-Zaman, A. S. M., Karim, R., & Tabash, M. I. (2022). Blockchain technology as a game changer for green innovation: Green entrepreneurship as a roadmap to green economic sustainability in Peru. *Journal of Open Innovation*, *8*(2), 62. https://doi.org/10.3390/joitmc8020062

Rejeb, A., Rejeb, K., Keogh, J. G., & Zailani, S. (2022). Barriers to blockchain adoption in the circular Economy: A fuzzy Delphi and Best-Worst approach. *Sustainability*, *14*(6), 3611. https://doi.org/10.3390/su14063611

Rieger, A., Roth, T., Sedlmeir, J., & Fridgen, G. (2022). We need a broader debate on the sustainability of blockchain. *Joule*, *6*(6), 1137–1141. https://doi.org/10.1016/j.joule.2022.04.013

Ruslin, R., Mashuri, S., Muhammad, S., Alhabsyi, F., & Syam, H. (2022). Semi-structured interview: A methodological reflection on the development of a qualitative research instrument in educational studies. *IOSR Journal of Research & Method in Education*, *12*(1), 22–29. www.researchgate.net/publication/358906376_Semi-structured_Interview_A_Methodological_Reflection_on_the_Development_of_a_Qualitative_Research_Instrument_in_Educational_Studies_Ruslin

Saha, A. S., Raut, R. D., Yadav, V. S., & Majumdar, A. (2022). Blockchain changing the outlook of the sustainable food supply chain to achieve net zero? *Sustainability*, *14*(24), 16916. https://doi.org/10.3390/su142416916

Sarmah, S. S. (2018). Understanding blockchain technology. *Computer Science and Engineering*, *8*(2) 23–29. https://doi.org/10.5923/j.computer.20180802.02

Tang, Y., Xiong, J., Becerril-Arreola, R., & Iyer, L. S. (2020). Ethics of blockchain: A framework of technology, applications, impacts, and research directions. *Complementary Index.* Retrieved May 5, 2023, https://eds-p-ebscohost-com.ezproxy.myucwest.ca/eds/detail/detail?vid=2&sid=1175d9a7-e5e9-4506-ad98-0a60a7a40e94%40redis&bdata=JnNpdGU9ZWRzLWxpdmUmc2NvcGU9c2l0ZQ%3d%3d#AN=142131586&db=edb

Tundys, B. (2016). Sustainable supplier selection criteria in the context of developing of green supply chain. *2016 5th IEEE Int. Conf. On Advanced Logistics & Transport (ICALT)*, 147–153. www.researchgate.net/publication/304658416_Sustainable_supplier_selection_criteria_in_the_context_of_developing_of_green_supply_chain

Vu, N. H., Ghadge, A., & Bourlakis, M. (2021). Blockchain adoption in food supply chains: a review and implementation framework. *Production Planning & Control*, 1–18. https://doi.org/10.1080/09537287.2021.1939902

World Bank. (2019). *World Bank Annual Report 2019: Ending Poverty, Investing in Opportunity.* Washington, DC: World Bank. https://doi.org/10.1596/978-1-4648-1470-9

Zutshi, A., Grilo, A., & Nodehi, T. (2021). The value proposition of blockchain technologies and its impact on Digital Platforms. *Computers & Industrial Engineering*, *155*, 107187. https://doi.org/10.1016/j.cie.2021.107187

12 Application Scenario of AI-Enabled Architectures in Next-Generation Wireless Networks

Shakti Raj Chopra

12.1 INTRODUCTION

Artificial intelligence (AI) is becoming more and more important in creating next-generation architectures as wireless networks adapt to meet the increasing demands of connected devices. It is anticipated that the combination of AI and wireless communication technologies would result in more intelligent, efficient, and adaptable networks, which will help address the issues of increased data rates, extremely low latency, extensive connection, and better user experiences.

The capacity to handle a wide range of applications with various requirements, from high-bandwidth video streaming to low-latency IoT communications, is a defining feature of next-generation wireless networks, such as 5G and beyond. AI-enabled architectures are becoming a vital component to address these expectations [1].

AI approaches are utilized by AI-enabled designs to improve wireless network performance, efficiency, and flexibility. AI algorithms can make intelligent judgments in real-time, optimizing network operations and enhancing user experience, by analyzing large volumes of data and learning from past experiences [2].

The goal of next-generation wireless networks, especially 5G and beyond (6G), is to offer improved features like:

- Higher data speeds, reaching terabits per second.
- Extremely low latency (reaction times in milliseconds)
- Broad device connectivity—billions of Internet of Things devices
- Excellent dependability and energy economy

Nevertheless, as these networks get more intricate, dynamic and unexpected traffic patterns, spectrum allocation, and user demands become increasingly difficult to

 DOI: 10.1201/9781003563440-12

effectively manage for conventional static designs and rule-based techniques. This is the application of AI [3, 4].

12.2 AI-POWERED NETWORK ARCHITECTURE

AI is incorporated into several wireless network layers to improve decision-making, automate administrative duties, and maximize efficiency. The following are some of the main ways that AI-enabled designs are changing next-generation wireless networks:

12.2.1 Automation and Network Management

AI is utilized in predictive network maintenance, which minimizes downtime and enhances service reliability by using models to anticipate possible network faults or performance degradations before they happen. Self-organizing networks (SONs): AI makes it possible to dynamically optimize network topologies without the need for human intervention, including load balancing, handovers, and resource allocation.

12.2.2 Spectrum Handling

The following AI methods, such as machine learning (ML), are used to maximize the utilization of wireless spectrum:

Dynamic spectrum allocation: By using AI models to analyze traffic patterns, spectrum resources can be distributed more effectively, reducing interference and congestion.

Cognitive radio networks: AI makes cognitive spectrum sensing possible, enabling gadgets to instantly identify and make use of accessible frequencies.

12.2.3 Traffic Control and Quality of Service

Networks benefit from AI-based traffic forecast models:

Optimize load balancing and routing: AI can improve quality of service by directing traffic through less congested pathways by evaluating user behavior and traffic trends.

Resource optimization: In order to provide better service for high-priority applications like AR/VR and driverless cars, AI algorithms distribute resources like electricity and bandwidth.

12.2.4 Efficiency in Energy Use

AI reduces energy consumption in the following ways:

- Smart base station control: AI algorithms predict when underutilized base stations will become necessary again and turn them off during periods of low traffic.

- Energy-efficient routing: AI can optimize network paths to reduce power consumption while preserving target performance levels [5, 6].

12.2.5 Improvement of Security

AI improves network security in the following ways:

- Anomaly detection: By spotting patterns of unusual network behavior, AI systems are able to identify suspicious activity, such as cyberattacks or unauthorized access.
- Real-time threat mitigation: AI models are able to isolate attacks and safeguard vital network components in real-time in response to security breaches.

12.3 AI METHODS FOR NEXT-GENERATION WIRELESS NETWORKS

The following are some of the AI methods that help improve network performance:

- Machine Learning (ML): Applies to activities like fault detection, resource allocation, and traffic forecasting where pattern identification, prediction, and decision-making are required.
- Deep Learning (DL): Used for intricate data analysis in situations such as network application picture and video recognition.
- Reinforcement learning (RL): This technique is used to optimize dynamic environments in real time, such as resource allocation and self-governing networks.
- Federated Learning (FL): This improves security and performance while enhancing privacy by enabling AI models to be trained locally on devices without sending data to centralized servers [7, 8].

12.4 DIFFICULTIES AND PROSPECTS

Although AI has many advantages, there are certain obstacles when implementing AI in wireless networks:

- Data security and privacy: Integrating AI necessitates the collection of enormous volumes of data, which presents data security and privacy issues for users.
- Scalability: Using AI on large-scale networks calls for powerful computers and effective algorithms.
- Interoperability: AI models need to interact with a variety of wireless network hardware and software ecosystems.

Going forward, it's anticipated that 6G networks will be completely AI-integrated, opening the door to even more sophisticated applications like tactile internet, holographic communications, and AI-powered Internet of Things ecosystems [9, 10].

12.5 CONCLUSION

AI-enabled architectures are poised to revolutionize next-generation wireless networks by making them more intelligent, adaptive, and efficient. From optimizing spectrum usage and traffic management to enhancing security and reducing energy consumption, AI will be essential for meeting the demands of an increasingly connected world. With the development of 6G and beyond, the role of AI in wireless networks will continue to expand, driving innovations that will shape the future of communication.

REFERENCES

1. Zhang, C., Patras, P., and Haddadi, H. (2019). Deep learning in mobile and wireless networking: A survey. *IEEE Communications Surveys & Tutorials*, 21(3), 2224–2287.
2. Singh, P., Arora, K., Rathore, U. C., Joshi, G. P., and Cho, W. (2024). "Comparative study of controllers in battery energy storage system integrated with doubly fed induction generator-based wind energy conversion system for power quality improvement", *Energy Reports*, 11(9), 4587–4600, https://doi.org/10.1016/j.egyr.2024.04.020
3. Sharma, R., Chopra, S. R., and Gupta, A. "UAV assisted next generation wireless communication network." In *2024 International Conference on Emerging Smart Computing and Informatics (ESCI)*, pp. 1–5. IEEE, 2024.
4. Foukas, X., Patounas, G., Elmokashfi, A., and Marina, M. K. (2017). Network slicing in 5G: survey and challenges. *IEEE Communications Magazine*, 55(5), 94–100.
5. Arora, K., Joshi, G. P., Ragab, M., Rawa, M., Milyani, A. H., Mansour, R. F., and Yang, E. (2022). "Bilateral contract for load frequency and renewable energy sources using advanced controller. *Computers, Materials & Continua*, 73(2), 3165–3180. doi:10.32604/cmc.2022.026966.
6. Jiang, W., and Hanzo, L. (2017). Artificial intelligence-powered next-generation wireless networks. *IEEE Wireless Communications*, 26(5), 10–16.
7. Zhang, H., Liu, N., Chu, X., Long, K., Aghvami, A. H., and Leung, V. C. M. (2017). Network slicing based 5G and future mobile networks: Mobility, resource management, and challenges. *IEEE Communications Magazine*, 55(8), 138–145.
8. Sharma, R., Chopra, S. R., Gupta, A., Kaur, R., Tanwar, S., Pau, G., Sharma, G., Alqahtani, F., and Tolba, A. (2024). Deployment of unmanned aerial vehicles in next-generation wireless communication network using multi-agent reinforcement learning. *IEEE Access*, 1, 1–24.
9. Kaloxylos, A., Alonistioti, N., and Burke, M. (2021). Leveraging AI in wireless networks: present and future trends. *IEEE Network*, 35(2), 10–17.
10. Sharma, A., Chopra, S. R., Sapate, S. G., Arora, K., Rahmani, M. K. I., Jha, S., Ahmad, S., Ahmed, M. E., Abdeljaber, H. A. M. and Nazeer, J. (2024). Artificial intelligence techniques for landslides prediction using satellite imagery. *IEEE Access*, 1, 1–16.

13 Secure Session Key and Communal Verification Protocol Implementation in IOV Network

Ankit Aggarwal, Vishal Kohli, Ashwani Kr. Vashistha, Gaurav Kumar, and Krishan Arora

13.1 INTRODUCTION

In today's digital communication setting, abundant computing environments involve communication over insecure channels. Users often conduct computational tasks on personal computers and share them via the internet, contributing to extensive adoption and development of networked and distributed systems. These systems improve functionality and optimize resource utilization, facilitating collaboration and message exchange among various parties such as processes, users, nodes, and terminals. All through this paper, these entities are termed as "correspondents," a term normally used in authentication literature.

The correspondent verifies the validity of the received message to decide the suitable action to be taken. This authentication [1] ensures that the message is newly generated for its proposed purpose by the claiming correspondent [2].

In insecure networks, such as the Internet, there are a number of security challenges such as eavesdropping and unauthorized interruption. These challenges arise due to the decentralized nature of Internet, which consists of several independent networks. Unlike proprietary networks, the sender cannot control the exact path a message takes to reach its destination. Therefore, the receiving correspondent must detect malicious entities or replayed messages generated in the past. Encryption plays a vital role for securing communications between parties, authorizing them to transmit and receive data securely [3].

In networked and distributed systems, authorization of correspondents for network communication is the elementary task. This task is facilitated by authentication protocols, which consist of predefined rules governing the exchange of messages among correspondents. Authentication protocols guarantee secure transmission of confidential information and have been extensively developed and implemented for regulating the flow of information [4, 5].

DOI: 10.1201/9781003563440-13

13.2 IOV SECURITY VIA CRYPTOGRAPHY

To know the design process of any authentication protocol, a good understanding of fundamental cryptographic systems and their standard terminology is necessary.

To transmit a plain message P securely over a network, it undergoes encryption using algorithms such as DES [6], IDEA [7], AES [8], among others. This process transforms P into cipher text C, making it meaningless to unauthorized parties monitoring the transmission. The encryption mechanism inputs a parameter known as a key K to convert P into cipher text C.

After the cipher text has been transferred to its intended recipient, they can decrypt it to get the original plaintext. Decryption is done using a decryption key K-1 to reverse the encryption process.

13.3 PROBLEM FORMULATION AND PROPOSED WORK

While individual cryptographers, both amateurs [9, 10] and professionals [11], have made modest modifications to the protocols described in [2, 3], there has been no reported effort to integrate these techniques into an efficient prototype for secure communication in uncertain network [12].

13.3.1 Interpretation of the Current System

Consider the Needham–Schroeder protocol [2], which employs conventional encryption. In this method, each participant has a secret key shared with the trusted authority (TA). To obtain a secret key K_S from the TA and communicate with another participant, SU follows these steps:

The communication starts with SU sending a nonencrypted message to TA as follows:

$$SU \rightarrow TA\text{: } IF_{SU},\ IF_{IV},\ N_{SU} \tag{13.1}$$

TA then authenticates the identities of SU and IV and retrieves their respective secret keys shared with TA (K_{SU} and K_{IV}). TA sends the message to SU as follows:

$$TA \rightarrow SU\text{: } \left[N_{SU},\ IF_{IV},\ K_S\left[K_S,\ IF_{SU}\right]K_{IV}\right]K_{SU} \tag{13.2}$$

Since K_{SU} is a secret key shared exclusively between SU and TA, only SU decrypts the message 13.2 to obtain the secret key, SU sends this message packet to IV as follows:

$$SU \rightarrow IV\text{: } \left[K_S, IF_{SU}\right]K_{IV} \tag{13.3}$$

IV receives the message in 13.3, it decrypts it to obtain the secret key K_S.

To make sure no interruption occurred during the execution of protocol, further steps are conducted for mutual authentication. A nonce, N_{IV} is generated by IV, to protect against potential attacks and sends this message to SU as follows:

$$IV \rightarrow SU: \left[N_{IV}\right]K_S \tag{13.4}$$

Since only SU and IV possess K_S, SU can authenticate the message originated from IV. The customized message is then sent back to IV as follows:

$$SU \rightarrow IV: \left[N_{IV} - 1\right]K_S \tag{13.5}$$

SU decrypts this message to retrieve the decremented value of nonce.

This ensures the completion of the protocol designed by Roger M. Needham and Michael D. Schroeder, employing symmetric key encryption.

However, Denning and Sacco [13] verified vulnerabilities in the Needham-Schroeder protocol, finding possible compromises of communication keys. On the other hand, Denning and Sacco highlighted risks such as direct revelation of these keys due to system design negligence or flaws.

Both suggested a situation in which an attacker (CA) intercepts whole messages interchanged between SU and IV during 13.3–13.5 of the Needham–Schroeder protocol discussed earlier. In addition, the intruder got a copy of the secret key KS. Afterward, the intruder might use this information to mislead IV into believing it is SU as under:

Firstly, CA would replay the message obtained in 13.3 to IV:

$$CA \rightarrow IV: \left[K_S, IF_{SU}\right]K_{IV} \tag{13.6}$$

After receiving this message, IV interprets it as a message initiation from SU and reply to it with a nonce N_{IV}, encrypting it using the K_S key received in the previous step as follows:

$$IV \rightarrow SU: \left[N_{IV}\right]K_S \tag{13.7}$$

The intruder CA intercepts this message before it reaches SU, decrypts it using the stolen K_S, and then impersonates SU's response as follows:

$$SU \rightarrow IV: \left[f\left(N_{IV}\right)\right]K_S \tag{13.8}$$

Then, CA gains the ability to send deceptive messages to IV, masked them as originating from SU.

Denning and Sacco introduced timestamps to address this issue. They added another field, TS, to communication received in steps 13.2 and 13.3 of the protocol. The revised protocol then rewritten as follows:

$$\text{SU} \rightarrow \text{TA: IF}_{\text{SU}}, \text{IF}_{\text{IV}}, \text{TS} \tag{13.9}$$

$$\text{TA} \rightarrow \text{SU: } \left[\text{IF}_{\text{IV}},\ \text{K}_{\text{S}},\ \text{TS}\left[\text{K}_{\text{S}},\ \text{IF}_{\text{SU}},\ \text{TS}\right]\text{K}_{\text{IV}}\right]\text{K}_{\text{SU}} \tag{13.10}$$

$$\text{SU} \rightarrow \text{IV: } \left[\text{K}_{\text{S}},\ \text{IF}_{\text{SU}},\ \text{TS}\right]\text{K}_{\text{IV}} \tag{13.11}$$

13.4 and 13.5 are not included here which originally intended to stop replay attacks on non-compromised secret keys. Both SU and IV can prove the originality of the messages using the following equation:

$$\left|\text{Time} - \text{M}\right| < \Delta d_1 + \Delta d_2 \tag{13.12}$$

where,

Time: time of the native system
Δd_1: difference between time of native system and server's time
Δd_2: expected delay time in network

If difference between timestamp TS and the system's clock is less than $\left(\Delta d_1 + \Delta d_2\right)$, the protocol is secure against replay attacks.

On the other hand, the use of timestamps may not be universally applicable due to possible issues with clock synchronization, mainly in distributed environments. Terminals lacking local real-time clocks would meet difficulties in communicating through this protocol.

In [13], Otway and Rees introduced a key interchange protocol that relied on a TA for distributing secret keys to communicating parties.

The chain of steps to perform the protocol is as under:

The originator SU sends a message to receiver IV containing an identifier CI for conversation, as well as identifiers for both SU and IV. Additionally, SU includes a nonce N_{SU} encrypted using a secret key K_S that is shared between SU and the TA. The structure of the message is as follows:

$$\text{SU} \rightarrow \text{IV: CI, IF}_{\text{SU}},\ \text{IF}_{\text{IV}}\left[\text{N}_{\text{SU}},\ \text{CI, IF}_{\text{SU}},\ \text{IF}_{\text{IV}}\right]\text{K}_{\text{TA}} \tag{13.13}$$

After receiving this message, IV understands that SU wishes to initiate communication and requests a secret session key from SU. IV then sends a message to TA that includes the message took from SU, with an encrypted message via the secret key K_{IV} that has been shared between IV and TA. The structure of IV's message is as follows:

$$\begin{aligned}\text{IV} \rightarrow \text{TA: CI, IF}_{\text{SU}},\ \text{IF}_{\text{IV}}&\left[\text{N}_{\text{SU}},\ \text{CI, IF}_{\text{SU}},\ \text{IF}_{\text{IV}}\right]\text{K}_{\text{TA}},\\ &\left[\text{N}_{\text{IV}},\ \text{CI, IF}_{\text{SU}},\ \text{IF}_{\text{IV}}\right]\text{K}_{\text{IV}}\end{aligned} \tag{13.14}$$

The TA receives the message and decrypts it using the respective keys K_{SU} and K_{IV}. It verifies that the messages constitute a matching pair with identifiers CI, IF_{SU}, IF_{IV}. Upon successful verification, the TA computes a secret session key K_{SU} and sends a response back to IV:

$$TA \rightarrow IV: CI,\ \left[N_{SU},\ K_{IV}\right]K_{TA}\ [N_{IV},\ K_{SI}]K_{IV} \tag{13.15}$$

After IV receives the message, it decrypts the portion encrypted with K_{IV}, thereby obtaining the secret session key K_{SI}. It then forwards the relevant part of the message intended for SU, as follows:

$$IV \rightarrow SU: CI,\ \left[N_{SU},\ K_{SI}\right]K_{TA} \tag{13.16}$$

After completing the protocol run, both communicating parties SU and IV receive the shared secret key K_{SI} and do the transmission securely. The authentication messages from both correspondents are symmetric, allowing either party to initiate reauthentication and key change. When a correspondent discards a secret session key, the entire protocol must be play again because authentication replay messages cannot be reused. This ensures the protocol's resilience against replay attacks.

The discussed protocol is susceptible to MITM attack, as discussed by Frédéric Massicotte in [14]. He suggested that if an attacker copies SU to IV and vice versa, without either realizing it, a MITM attack might be executed with tools such as Hunt-type software [15].

Below are the steps to take advantage of this weakness are as under:

While SU begins a communication by transmitting message to IV, the attacker intercepts that message as under:

$$SU \rightarrow CI(IV): CI,\ IF_{SU},\ IF_{IV}\left[N_{SU},\ CI,\ IF_{SU}, IF_{IV}\right]K_{SU} \tag{13.17}$$

The intruder intercepts and retrieves this message 13.17, concatenates the encrypted portion with the identifier for conversation CI, and delivers it back to SU as under:

$$CI(IV) \rightarrow SU: CI,\ \left[N_{SU}, CI, IF_{SU}, IF_{IV}\right]K_{TA} \tag{13.18}$$

When SU accepts this message, it validates the identifier for conversation and decrypts the encoded portion using its secret session key shared with TA. Upon successful decryption, it checks the correctness of nonce N_{SU}, confirming that the second part of the message $(CI,\ IF_{SU}, IF_{IV})$ is indeed the secret session key. Since this information is transmitted in plaintext, an intruder could intercept and steal the session key. This attack does not require the intruder to complete the two protocol steps—messages from IV to TA and TA to IV.

13.3.2 Proposed MORNeS Protocol

The proposed MORNeS protocol uses the following conventions:

- SU and IV: communicating parties with SU initiating communication.
- MTA: MORNeS Trusted Authority
- CI: cyber intruder
- CI(IV): cyber intruder impersonating IV
- IF_{SU}, IF_{IV}: identifiers of parties SU and IV, respectively.
- N_{SU}, N_{IV}: Nonces generated by parties SU and IV, respectively.
- K_{SM}: mutual secret keys among SU and MTA
- K_{IM}: mutual secret keys among IV and MTA
- K_{SI}: mutual secret keys among SU and IV

The MORNeS protocol operates as follows:

- If SU intends to initiate communication with IV, it transmits a message to IV having its identifier (IF_{SU}), IV's identifier (IF_{IV}), and an encoded message via the secret key allocated between SU and MTA (K_{SM}). The coded message includes a nonce (N_{SU}) generated by SU for current transmission session, along with IF_{SU} and IF_{IV}.
- IV, after receiving this message, a nonce NIV generated specific for current transmission session. It combines N_{IV} with IF_{SU} and IF_{IV}, encrypts this combination using the secret key shared between IV and MTA (K_{IM}), and appends the encrypted result to the message received from SU. IV then sends this concatenated message to MTA.
- MTA decrypts the two encoded parts of the communication via K_{SM} and K_{IM}, respectively, and verifies them with IF_{SU}, IF_{IV}. Upon successful verification, MTA creates a session key K_{SI} to be assigned between the interactive parties SU and IV. It then makes two messages containing N_{SU}, N_{IV}, and K_{SI}. The first message is encrypted with KSM and the second with K_{IM}. MTA concatenates these encrypted messages and sends them to IV.
- When IV gets this message, it decodes the portion encoded with K_{IM} to extract the nonces and session key. IV verifies N_{IV} to ensure the session key intended for the current message session. IV then encodes the previously recovered nonce N_{SU} via the session key K_{SI}. Afterward, IV concatenates the encoded N_{SU} with the piece of the message collected from MTA (coded using K_{SM}) and sends it to SU.
- When SU gets this message, it decodes the portion of the message using K_{SM} to extract the nonces and session key. Then, SU took the session key to decrypt another part of the message to retrieve N_{SU}, thereby confirming that the message is from IV and that the session key is for the exact transmission session. To create the identity to IV, SU encodes N_{IV} via the secret key K_{SI} and sends the message to IV.

- When IV gets this message, it decodes the portion of the message using K_{SI} and confirms the successful transmission of session key to SU. SU and IV can then communicate securely over the insecure network using the session key K_{SI} as established by the protocol above.

Using session key, K_{SI}, SU and IV can communicate securely over the insecure network using the above protocol.

13.4 IMPLEMENTATION, RESULT, AND DISCUSSION

13.4.1 Plan for Implementation

The operation of the MORNeS protocol is illustrated in Figure 13.1:

Below are the steps for MORNeS protocol for protected key interchange between SU and IV and mutual authentication as in Table 13.1:

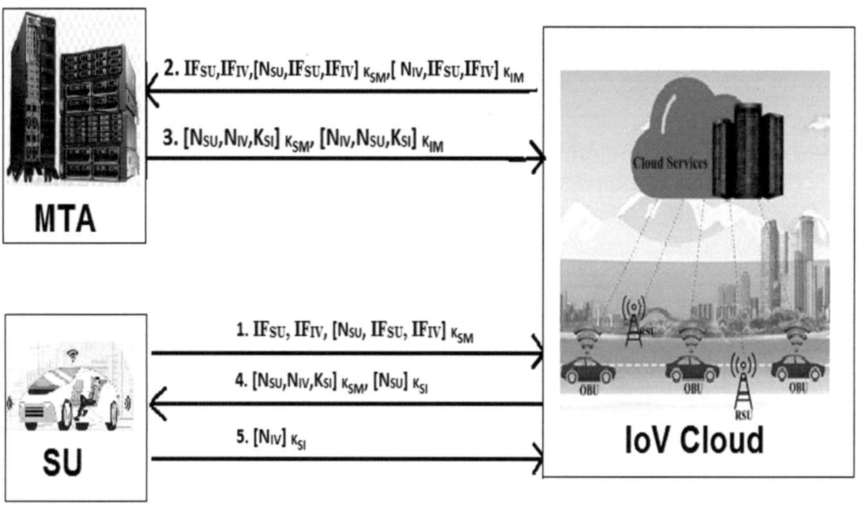

FIGURE 13.1 Functionality of the MORNeS protocol.

TABLE 13.1
MORNeS Protocol

S.No.	Communicating parties	Message transmitted
1.	SU → IV	$IF_{SU}, IF_{IV}\left[N_{SU}, CI, IF_{SU}, IF_{IV}\right]K_{SM}$
2.	IV → MTA	$IF_{SU}, IF_{IV}, \left[N_{SU}, IF_{SU}, IF_{IV}\right]K_{SM}, \left[N_{IV}, IF_{SU}, IF_{IV}\right]K_{IM}$
3.	MTA → IV	$\left[N_{SU}, K_{IV}, K_{SI}\right]K_{SM}, \left[N_{IV}, K_{SI}\right]K_{IM}$
4.	IV → SU	$\left[N_{SU}, N_{IV}\ K_{SI}\right] K_{SM}, [N_{SU}], K_{SI}$
5.	SU → IV	$[N_{IV}]K_{SI}$

13.4.2 Proof of Security

When an intruder records a message or a part of it received from earlier protocol execution, a replay attack occurs which replays it during the following protocol execution [16].

To demonstrate MORNeS's resilience against replay attacks, consider an example where an attacker attempts to interrupt the key interchange between SU and IV.

When SU initiates communication with IV as per the described protocol, a cyberattacker (CA) intercepts and copies the message delivered from IV to MTA in step 2.

The intruder remains passive during the third step but intervenes during the fourth step to replay the copied message, impersonating IV in a replay attack.

After receiving the request, MTA creates a new key K_{SI} and delivers it to IV. This message is intercepted by CA, who furthers encoded the message holding the new key K_{SI} established from MTA to SU, along with the encrypted message containing N_{SU} received from IV.

SU attempts mutual authentication by attempting to decrypt N_{SU} using K_{SI}, which fails since N_{SU} was encrypted by IV using $_{KSI}$.

Additionally, SU encrypts N_{SU} using K_{SI} and sends it to IV, who is unable to decrypt it because it was encrypted using K_{SI}.

As a result, shared verification fails, and together SU and IV become informed of the communication invasion, leading to the key being removed.

The steps demonstrating security against replay attacks are outlined in Table 13.2.

TABLE 13.2
Security Against Replay Attacks

S. No.	Communicating parties	Message transmitted
1.	$SU \rightarrow IV$	$IF_{SU}, IF_{IV}\left[N_{SU}, IF_{SU}, IF_{IV}\right]K_{SM}$
2.	$IV \rightarrow MTA$	$IF_{SU}, IF_{IV}, \left[N_{SU}, IF_{SU}, IF_{IV}\right]K_{SM}\left[N_{IV}, IF_{SU}, IF_{IV}\right]K_{IM}$
	This message is copied by the cyber attacker.	
3.	$MTA \rightarrow IV$	$\left[N_{SU}, K_{IV}, K_{SI}\right]K_{SM}, \left[N_{IV}, N_{SU}, K_{SI}\right]K_{IM}$
4.	$IV \rightarrow SU$	$\left[N_{SU}, N_{IV}\ K_{SI}\right]K_{SM}, [N_{SU}]K_{SI}$
	This message is intercepted by the cyber attacker.	
5.	$I(IV) \rightarrow MTA$	$IF_{SU}, IF_{IV}, \left[N_{SU}, IF_{SU}, IF_{IV}\right]K_{SM}\left[N_{IV}, IF_{SU}, IF_{IV}\right]K_{IM}$
	Cyber attacker replays the message copied in step 2.	
6.	$MTA \rightarrow IV$	$\left[N_{SU}, N_{IV}, K_{SI}\right]K_{SM}, \left[N_{IV}, N_{SU}, K_{SI}\right]K_{IM}$
	The server generates a new key, K_{SI}, in response to the replayed message. The intruder interrupt and duplicates this message	
7.	$I(IV) \rightarrow SU$	$\left[N_{SU}, N_{IV}\ K_{SI}\right]K_{SM}, [N_{SU}]K_{SI}$
	The attacker forwards the components of the message received via MTA and IV to SU, masquerading IV	
8.	$SU \rightarrow IV$	$\left[N_{IV}\right]K_{SI}$
	The mutual authentication fails because neither SU nor IV can decrypt the nonces.	

13.4.3 End Result and Implementation Consideration

MORNeS protocol remains evaluated against the key interchange protocol introduced by Otway and Rees, focusing on their effectiveness in distributed computing environments. The performance metrics considered include as follows:

- Encoding time for each block: t_e
- Decoding time for each block: t_d
- Transmission time for each block: t_t
- Nonce generation time: t_n
- Session key generation time: t_{sk}

The computation of the operating time of key interchange protocol purported by Otway and Rees can be found in Table 13.3:
Total running time

$$= t_n + 3t_e + 8t_t + 2t_n + 4t_e + 9t_t + 3t_d + 3t_d + 2t_{sk} + 4t_e + 4t_e + 5t_t + 3t_d + 5t_t + 3t_d$$

$$= (1+2)t_n + (3+4+4+4)t_e + (8+9+5+5)t_t + (3+3+3+3)t_d + 2t_{sk}$$

$$= 3t_n + 15t_e + 27t_t + 12t_d + 2t_{sk}$$

Likewise, the computation of the running time for the projected protocol MORNeS can be found in Table 13.4.

Total running time:

$$= 2t_n + 3t_e + 6t_t + t_n + 4t_e + 7t_t + 5t_d + 2t_{sk} + 5t_e + 7t_t + 4t_d + t_e + 4\mathrm{t}_\mathrm{t} + 3\mathrm{t}_\mathrm{d} + \mathrm{t}_\mathrm{e} + \mathrm{t}_\mathrm{t}$$

$$= (2+1)t_\mathrm{n} + (3+4+5+1+1)t_\mathrm{e} + (6+7+7+4+1)t_t + (5+4+3)t_d + 2t_{sk}$$

$$= 3t_n + 14t_e + 25t_\mathrm{t} + 12t_d + 2t_{sk}$$

TABLE 13.3
Operating time of key interchange protocol purported by Otway and Rees

S. No.	Location	Time
1	At SU	$t_n + 3t_e$
2	Transmission from SU to IV	$8t_t$
3	At IV	$2t_n + 4t_e$
4	Transmission from IV to TA	$9t_t$
5	At TA	$3t_d + 3t_d + 2t_{sk} + 4t_e + 4t_e$
6	Transmission from TA to IV	$5t_t$
7	At IV	$3t_d$
8	Transmission from IV to SU	$5t_t$
9	At SU	$3t_d$

TABLE 13.4
Running time of projected protocol MORNeS

S. No.	Location	Time
1	At SU	$2t_n + 3t_e$
2	Transmission from SU to IV	$6t_t$
3	At IV	$t_n + 4t_e$
4	Transmission from IV to TA	$7t_t$
5	At TA	$5t_d + 2t_{sk} + 5t_e$
6	Transmission from TA to IV	$7t_t$
7	At IV	$4t_d + t_e$
8	Transmission from IV to SU	$4t_t$
9	At SU	$3t_d + t_e$
10	Transmission from SU to IV	t_t

The operating time of key interchange protocol purported by Otway and Rees:

$$3t_n + 15t_e + 27t_t + 12t_d + 2t_{sk} \tag{13.19}$$

Running time of projected protocol MORNeS:

$$3t_n + 14t_e + 25t_t + 12t_d + 2t_{sk} \tag{13.20}$$

By subtracting $3t_n + 12t_d + 2t_{sk}$ from equation (13.19) and (13.20) to exclude the common expressions, we find

$$15t_e + 27t_t \tag{13.21}$$

$$14t_e + 25t_t \tag{13.22}$$

The simplified equations for comparative analysis depend exclusively on t_e andt_t. Given the criticality of transmission time in distributed environments, the encryption process is significantly faster compared to transmission time, specifically t_e being much less than t_t ($t_e << t_t$).

Hence, t_e can be ignored in Equations (13.21) and (13.22) derived earlier. Consequently, the relationship simplifies to a linear expression that illustrates complexity exclusively as a function of transmission timet_t

As the transmission time t_t increases, complexity of the protocol also increases.

$$\rightarrow 27\ t_t \tag{13.23}$$

$$\rightarrow 25\ t_t \tag{13.24}$$

The calculation of efficiency for MORNeS protocol (in percentage) as compared to key interchange protocol purported by Otway and Rees is calculated as follows:

$$\epsilon = \left\{ \left(\frac{27t_t - 25t_t}{25t_t} \right) \times 100 \right\} \%$$

$$\epsilon = \left\{ \left(\frac{2t_t}{25t_t} \right) \times 100 \right\} \%$$

$$\epsilon = 8\%$$

Therefore, it can be said that the MORNeS protocol is 8% more efficient than Otway and Rees' key exchange protocol.

REFERENCES

[1] Sharmila, Kumar, P., Bhushan, S., Kumar, M., & Alazab, M. (2023). Secure key management and mutual authentication protocol for wireless sensor network by linking edge devices using hybrid approach. *Wireless Personal Communications*, 130(4), 2935–2957.

[2] Needham, R. M., & Schroeder, M. D. (1978). Using encryption for authentication in large networks of computers. *Communications of the ACM*, 21(12), 993–999.

[3] Otway, D., & Rees, O. (1987). Efficient and timely mutual authentication. *ACM SIGOPS Operating Systems Review*, 21(1), 8–10.

[4] Zhang, H., Lai, Y., & Chen, Y. (2023). Authentication methods for internet of vehicles based on trusted connection architecture. *Simulation Modelling Practice and Theory*, 122, 102681.

[5] Mazhar, S., Rakib, A., Pan, L., Jiang, F., Anwar, A., Doss, R., & Bryans, J. (2024). State-of-the-art authentication and verification schemes in VANETs: a survey. *Vehicular Communications*, 1, 100804.

[6] Coppersmith, D. (1994). The Data Encryption Standard (DES) and its strength against attacks. *IBM Journal of Research and Development*, 38(3), 243–250.

[7] Lai, X., & Massey, J. L. (1991). A proposal for a new block encryption standard. In Advances in Cryptology—EUROCRYPT'90: Workshop on the Theory and Application of Cryptographic Techniques Aarhus, Denmark, May 21–24, 1990 Proceedings 9 (pp. 389–404). Springer, Berlin Heidelberg.

[8] Heron, S. (2009). Advanced encryption standard (AES). Network Security, 2009(12), 8–12.

[9] Singh, P., Arora, K., Rathore, U. C., Joshi, G. P., & Cho, W. (2024). Comparative study of controllers in battery energy storage system integrated with doubly fed induction generator-based wind energy conversion system for power quality improvement. *Energy Reports*, 11(9), 4587–4600, https://doi.org/10.1016/j.egyr.2024.04.020

[10] Meadows, C. A. (1996). *Analyzing the Needham-Schroeder public key protocol: A comparison of two approaches.* In Computer Security—ESORICS 96: 4th European Symposium on Research in Computer Security Rome, Italy, September 25–27, 1996 Proceedings 4 (pp. 351–364). Springer Berlin Heidelberg.

[11] Lowe, G. (1995). An attack on the Needham–Schroeder public–key authentication protocol. *Information Processing Letters*, 56(3), 1–18.

[12] Vashishtha, A. K., Sharma, M., & Sharma, A. (2022, November). Mechanism Incorporating Secure Mutual Validation and Key Spreading Organization in Intelligent Transport System. In 2022 International Conference on Fourth Industrial Revolution Based Technology and Practices (ICFIRTP) (pp. 219–225). IEEE.

[13] Denning, D. E., & Sacco, G. M. (1981). Timestamps in key distribution protocols. *Communications of the ACM*, 24(8), 533–536.

[14] Massicotte, F. (2000). Man in the middle attack against security protocol. Hacker Exploits, Global Information Assurance Certification.

[15] Anusha G., Arora, K., Sharma, H., Thota, S. P., Joshi, G. P., & Cho, W. Control Strategies of 15-levelmodified cascaded H-bridge MLI with solar PV and energy storage system", *Energy Reports*, 1, 1–22. https://doi.org/10.1016/j.egyr.2024.06.003

[16] Anyanwu, G. O., Nwakanma, C. I., Lee, J. M., & Kim, D. S. (2023). Falsification detection system for IoV using randomized search optimization ensemble algorithm. *IEEE Transactions on Intelligent Transportation Systems*, 24(4), 4158–4172.

14 AI-Driven Innovations in Medical Care

Sandeep Chouhan, Ramandeep Sandhu, Deepika Ghai, Harpreet Kaur Channi, and Manjit Kaur

14.1 INTRODUCTION

As the role of artificial intelligence (AI) in healthcare expands, a fundamental challenge persists: many AI systems remain isolated, operating within narrow silos of expertise [1]. While AI has made tremendous advances in individual domains such as radiology, genomics, and pathology, there is a growing need for more integrated approaches that can unify these specialized systems to address the complexity of modern medical care. This chapter explores a novel approach to overcoming these limitations through swarm intelligence (SI), a concept inspired by the decentralized, collective behavior of social organisms like ants, bees, and flocks of birds. SI is built on the principle that simple agents when working together in a coordinated fashion can achieve solutions that exceed the capabilities of any single individual [2–4]. Figure 14.1 shows the SI system for collective diagnosis and treatment planning.

Applying this concept to healthcare, SI allows multiple AI systems, each with its specialized knowledge, to collaborate in real time, dynamically processing vast amounts of medical data to deliver more accurate diagnoses and personalized treatment plans. This form of collective intelligence is particularly powerful in multidisciplinary cases, where diverse forms of data—imaging, genetic profiles, patient histories, and real-time monitoring—must be integrated seamlessly to inform clinical decisions (Figure 14.2).

The potential of SI extends beyond individual diagnostics and treatment. In a hospital setting, these swarm-based systems can optimize resource management, dynamically adjust to changing patient loads, and enhance emergency response. By mimicking the adaptability and coordination seen in natural swarms, SI introduces a flexible, efficient model for handling the inherent complexity of medical environments [5]. Ethical considerations and outcomes of SI in healthcare are shown in Figure 14.3.

In this chapter, we explore the principles behind SI, how it can be applied to AI-driven healthcare systems, and the transformative impact it can have on patient outcomes.

 DOI: 10.1201/9781003563440-14

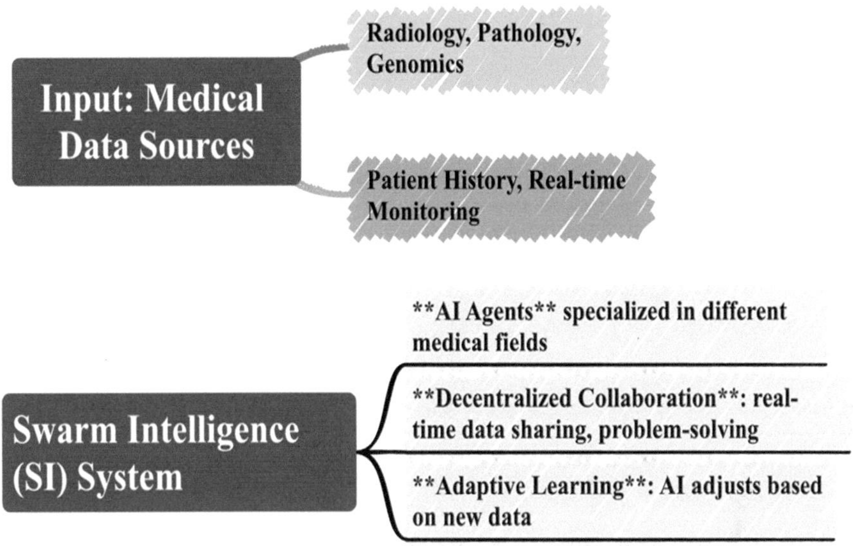

FIGURE 14.1 Swarm intelligence system for collective diagnosis and treatment planning.

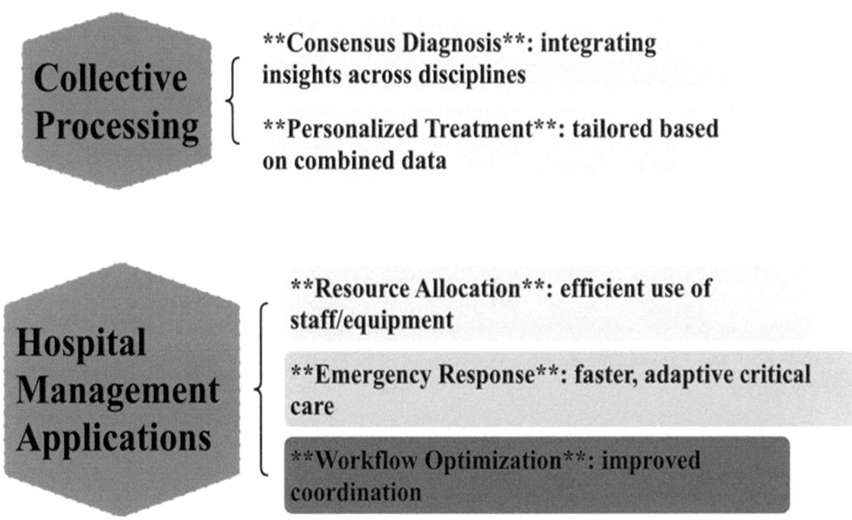

FIGURE 14.2 Collective processing and hospital management in swarm intelligence systems.

We will also address the technical and ethical challenges of implementing SI in clinical settings, focusing on issues of transparency, accountability, and patient trust. Ultimately, this approach represents a paradigm shift in AI-assisted medical care, offering a path towards more adaptive, holistic, and collaborative healthcare solutions.

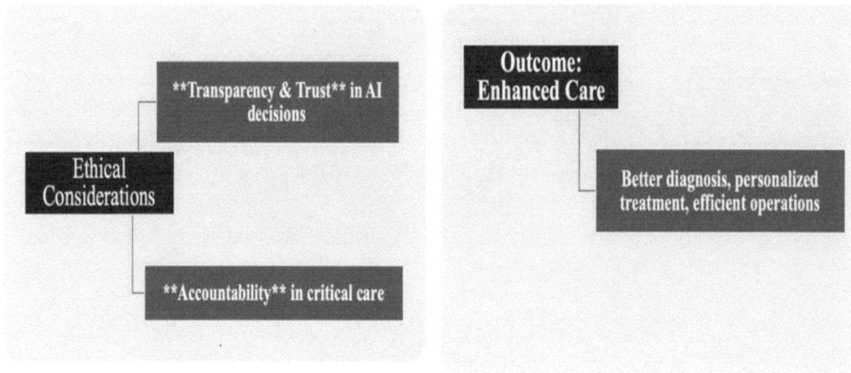

FIGURE 14.3 Ethical considerations and outcomes of swarm intelligence in healthcare.

14.1.1 SI in Medical Care

SI is an emerging paradigm in AI that draws inspiration from the collective behavior of social organisms like ants, bees, and flocks of birds. In nature, these creatures operate without central control, yet they exhibit remarkable problem-solving capabilities, coordination, and adaptability by interacting with one another in simple, localized ways [6]. This decentralized, self-organizing behavior allows them to solve complex tasks as a collective unit. When applied to medical care, SI offers a unique framework for harnessing the power of distributed AI systems to improve diagnosis, treatment planning, and patient care. Figure 14.4 shows SI in medical care.

In traditional AI-assisted healthcare, systems often work independently within specific disciplines—whether it's analyzing medical imaging, processing lab results, or interpreting genetic data. While these systems can offer highly specialized insights, they cannot frequently integrate and synthesize knowledge across different fields. This siloed approach limits their effectiveness in managing the complexity of real-world medical scenarios, where multiple sources of data must be combined to make accurate decisions.

SI addresses this challenge by enabling multiple AI agents, each with specialized capabilities, to work together in real time. These agents communicate continuously, sharing information and collaborating to arrive at a consensus. Just as a swarm of ants might collectively navigate a complex terrain to find food, a swarm of AI agents can process vast amounts of medical data to generate more precise diagnoses and treatment plans. The collective intelligence emerging from these interactions allows the system to function as a whole, yielding insights that exceed the sum of its parts [7, 8].

In medical care, SI can be particularly effective in complex or multidisciplinary cases that require input from various areas of expertise. For example, a patient with a rare genetic disorder may require data from radiology, pathology, genomics, and clinical observations. Swarm-based AI systems can simultaneously analyze inputs from these diverse sources, allowing for a richer, more integrated understanding of the patient's condition. This leads to better-informed decisions, reducing the likelihood of misdiagnoses and improving personalized treatment outcomes.

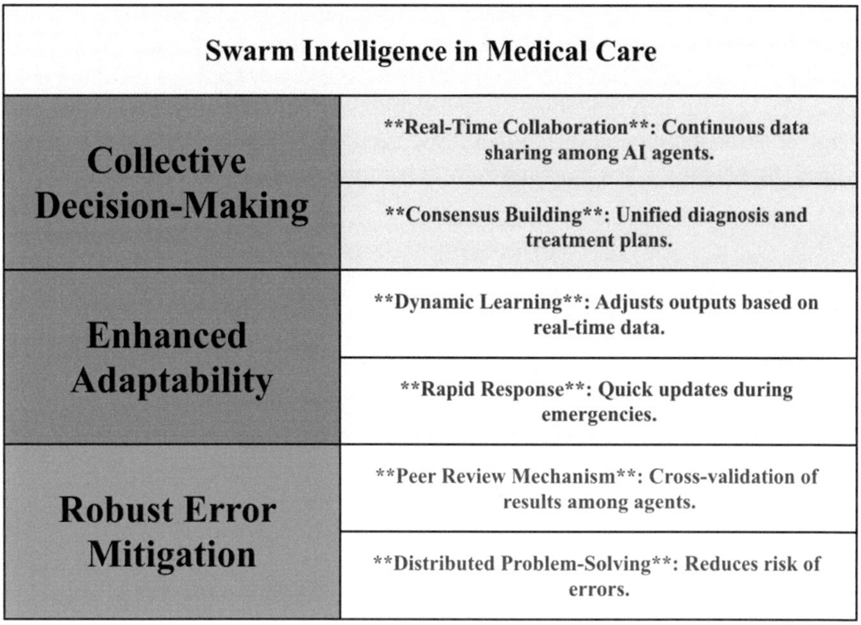

FIGURE 14.4 Swarm intelligence in medical care.

A key feature of SI is its adaptability. Just as animal swarms can quickly adjust to new environmental conditions, AI swarms in healthcare can adapt to new information in real time. If a patient's condition changes, such as during surgery or in critical care scenarios, the system can update its recommendations based on fresh data. This dynamic learning capacity is crucial in high-stakes environments like emergency rooms, where rapid, accurate decisions can mean the difference between life and death.

Moreover, SI has the potential to optimize healthcare workflows and resource management. In hospitals, swarm-based AI systems can monitor and predict resource needs, ensuring that equipment, medications, and staff are efficiently allocated where they are needed most. For example, during a public health crisis, such systems can dynamically adjust triage protocols, prioritize high-risk patients, and ensure that life-saving interventions are applied promptly. The decentralized nature of SI allows for this level of coordination without requiring a centralized system or human intervention, making it both scalable and flexible across diverse healthcare environments [9]. In addition to operational benefits, the distributed problem-solving aspect of SI reduces the risk of failure or error. Unlike a single AI system, which may produce skewed results if its model or data is flawed, a swarm system relies on multiple agents cross-validating their results. If one agent detects a potential issue—such as an anomaly in a radiology scan—this information can be relayed to the other agents, triggering additional analyses and refinement of the diagnosis. This peer-review mechanism within the swarm enhances both accuracy and robustness in decision-making. However, the successful deployment of SI in medical care must account for several ethical and practical considerations. Transparency is crucial—clinicians and patients need to understand how AI systems arrive at their conclusions. Unlike traditional black-box models, SI offers a more interpretable approach by breaking down decision processes into individual agent contributions. Additionally, the system must maintain accountability, ensuring that responsibility for patient outcomes is clearly defined, particularly when human oversight is limited.

14.1.2 Model and Implementation Strategy

The model and implementation strategy for AI-driven innovations in medical care are listed as follows:

14.1.2.1 Data Sources

The SI system for collective diagnosis and treatment planning in medical care integrates multiple data sources [10] and is also shown in Figure 14.5. These include:

- **Medical Imaging Data**: Radiology scans (CT, MRI, X-rays) for visual analysis.
- **Laboratory Results**: Blood tests, pathology reports, and molecular diagnostics.
- **Genomic Data**: Whole-genome sequencing and genetic markers.
- **Real-Time Monitoring**: Wearable devices and IoT sensors capture vital signs.

Each data type is preprocessed to ensure consistency in format, noise reduction, and completeness, enabling the AI agents to operate efficiently across different medical domains.

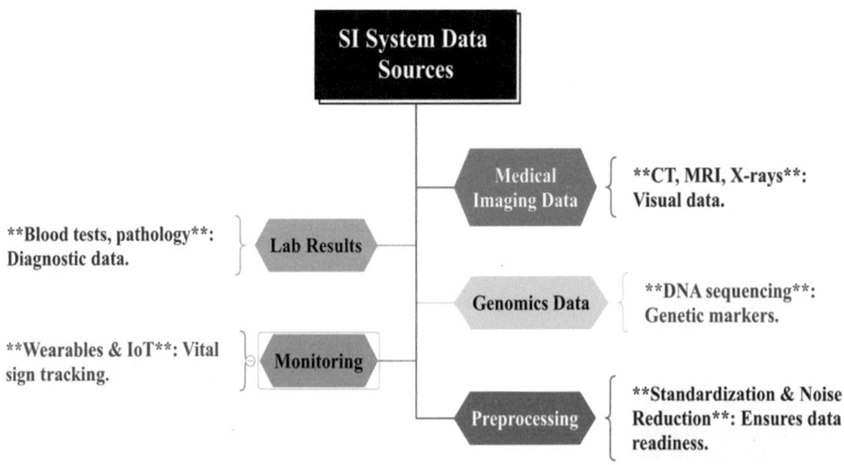

FIGURE 14.5 Data sources for the swarm intelligence system in healthcare.

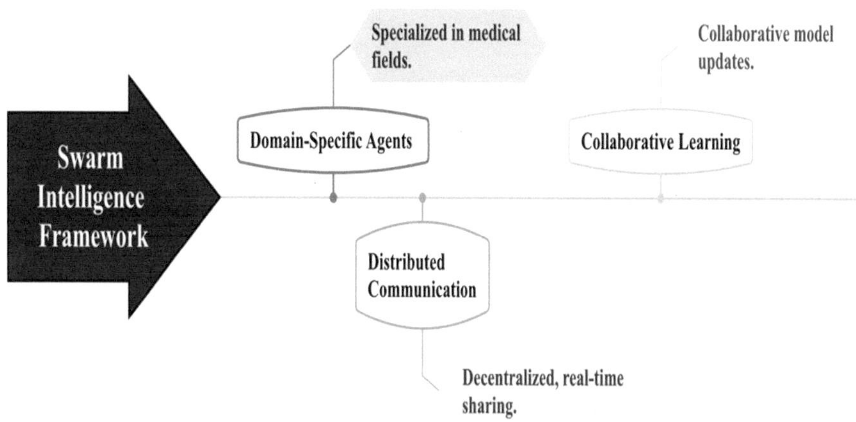

FIGURE 14.6 Overview of the swarm intelligence framework in medical AI.

14.1.2.2 Swarm Intelligence Framework

A decentralized multi-agent AI architecture (Figure 14.6) was employed to simulate SI. The framework includes:

- **Specialized AI Agents**: Each agent is trained to focus on a particular medical domain (e.g., radiology, pathology, genomics).
- **Decentralized Communication**: Agents share their results and observations in real time without a central controlling entity, mimicking the decentralized behavior of natural swarms (e.g., ants or bees).
- **Collective Learning**: Agents continually update their models based on shared information and the outcomes of previous decisions.

14.1.2.3 Consensus Mechanism

The system relies on a consensus-building algorithm (Figure 14.7) to integrate insights from different agents:

- **Distributed Decision-Making**: Each AI agent contributes its specialized analysis based on the data it processes.
- **Weighted Consensus**: Contributions are weighted based on agent expertise, data confidence, and relevance to the medical case.
- **Iterative Refinement**: Diagnoses and treatment plans are iteratively refined through feedback loops between agents, improving accuracy with each cycle.

14.1.2.4 Personalization and Treatment Planning

Once a consensus diagnosis is reached, the system moves to treatment planning:

- **Individualized Treatment Proposals**: Each AI agent generates recommendations based on the patient's specific medical history, genomic information, and real-time condition.

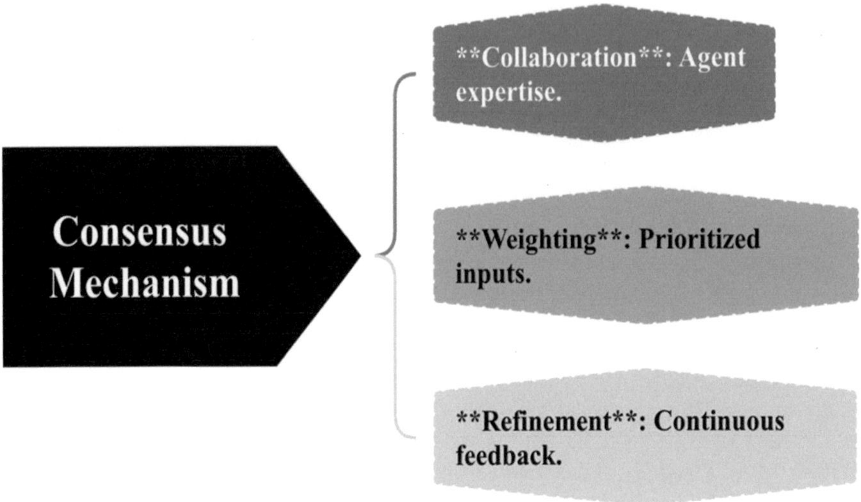

FIGURE 14.7 Consensus mechanism in swarm intelligence system.

- **Cross-Validation**: The proposed treatment plans are cross-validated by other agents to ensure comprehensive, patient-specific care.

Figure 14.8 shows personalized treatment planning in a SI system.

14.1.2.5 Adaptability to New Data

A key component of this SI system is adaptability:

- **Real-Time Adjustments**: During ongoing patient care, new data (e.g., updated lab results or changes in vitals) triggers an immediate re-evaluation of both diagnosis and treatment.
- **Continuous Learning**: AI agents refine their models based on each new data input, allowing the system to learn from evolving patient conditions, particularly in critical care or surgery settings [11–13].

Figure 14.9 shows the adaptation to new data in the SI system.

14.1.2.6 Validation and Testing

The system was tested in simulated medical scenarios before being implemented in clinical environments:

- **Simulated Patient Data**: Synthetic datasets representing various complex medical conditions (e.g., rare genetic disorders, cancer) were used to assess system performance.
- **Outcome Comparison**: The system's recommendations were compared to expert human diagnoses and treatment plans to evaluate its efficacy.

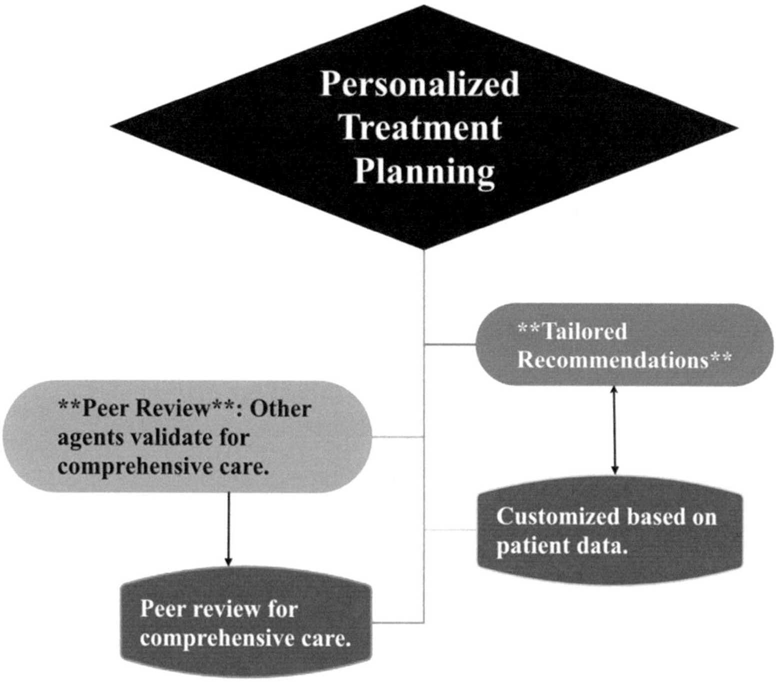

FIGURE 14.8 Personalized treatment planning in swarm intelligence system.

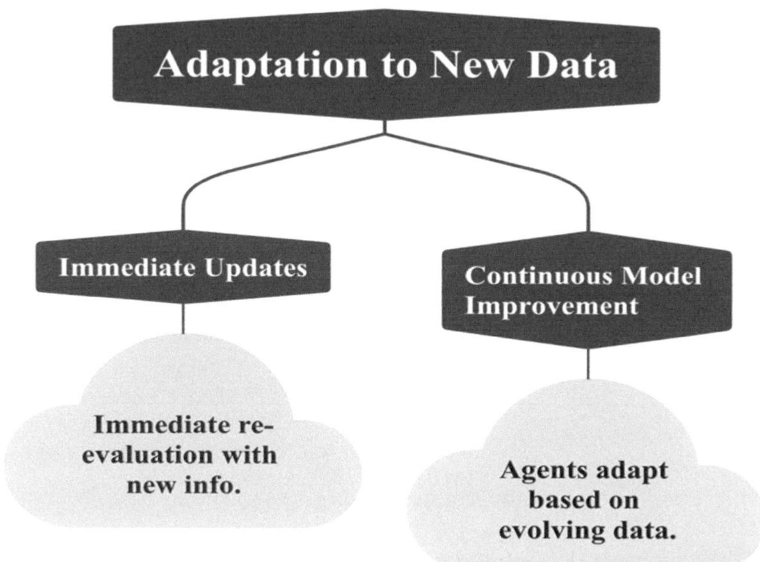

FIGURE 14.9 Adaptation to new data in swarm intelligence system.

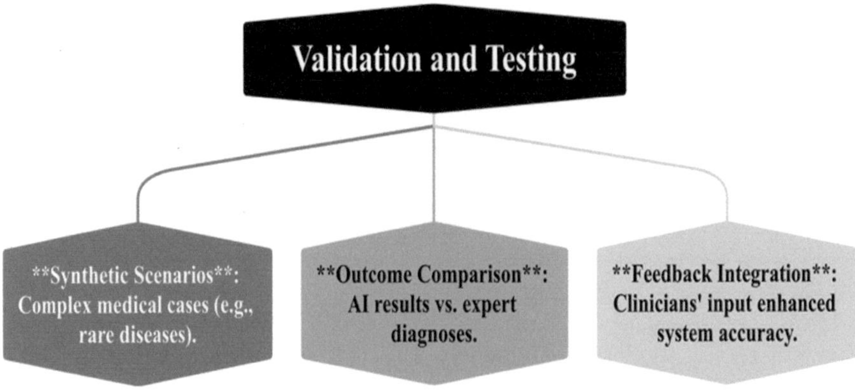

FIGURE 14.10 Validation and testing of swarm intelligence in medical scenarios.

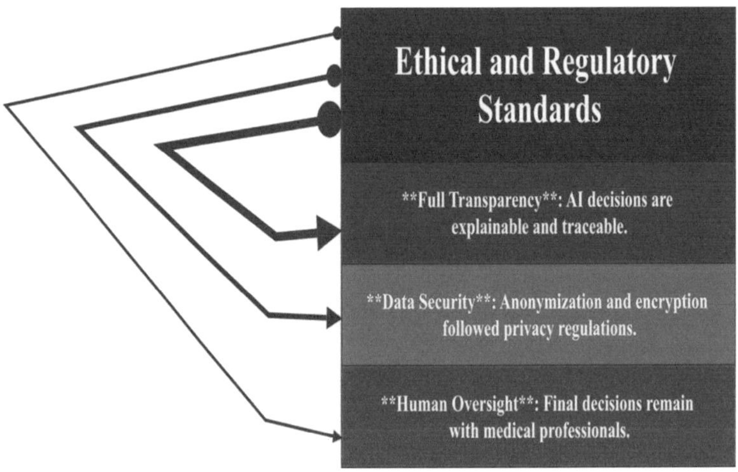

FIGURE 14.11 Ethical and regulatory standards in AI-driven medical care.

- **Feedback Loop**: Human clinicians provided feedback on the system's output, which was used to further enhance the agents' learning algorithms.

Figure 14.10 shows the validation and testing of SI in medical scenarios.

14.1.2.7 Ethical and Regulatory Considerations

Ethical frameworks and regulatory compliance (Figure 14.11) were built into the system:

- **Transparency**: All diagnostic and treatment decisions made by the AI agents were fully traceable and explainable.
- **Data Privacy**: Patient data was anonymized and encrypted in compliance with the Health Insurance Portability and Accountability Act and General Data Protection Regulation.

- **Accountability**: The system was designed to support, not replace, human medical professionals, ensuring that final decisions remain under human supervision [14–20].

14.1.3 Case Studies of SI in Healthcare

Some case studies of SI in healthcare are shown in Table 14.1.

14.2 RELATED WORK

In recent years, the integration of SI in medical diagnosis and treatment has gained significant attention, leading to various innovative applications that enhance patient care. This body of work showcases the potential of decentralized AI systems in addressing complex healthcare challenges.

Bhardwaj et al. [22] implemented a swarm-based approach for classifying brain tumor types using MRI images. Their system utilized multiple AI agents to analyze imaging data collaboratively, achieving high accuracy rates compared to traditional methods. The results demonstrated that collective intelligence significantly outperforms isolated models, especially in intricate medical scenarios. Similarly, Almansouri et al. [23] explored the use of SI for early detection of cardiovascular diseases. By analyzing data from wearable devices and electronic health records, their multiagent system effectively identified risk factors and potential complications. The adaptability of the swarm allowed for real-time adjustments based on patient data, enhancing predictive capabilities and timely interventions.

Moreover, a comprehensive review by Xu et al. [24] highlighted various swarm-based techniques applied across diverse medical domains, such as pathology, genomics, and personalized treatment planning. The review emphasized the advantages of SI in terms of adaptability, scalability, and improved accuracy in diagnosis and treatment recommendations. It underscored the importance of integrating multiple data sources to create a holistic view of patient health. In a different vein, Singh et al. [25] proposed a SI framework for optimizing hospital resource allocation during public health crises. Their study illustrated how swarm-based algorithms could effectively predict resource needs and prioritize patient care, demonstrating the versatility of SI beyond diagnosis and treatment alone. Collectively, these works illustrate a growing trend toward leveraging SI in medical applications. The emphasis on decentralized communication and collaborative decision-making highlights the potential for improved outcomes in patient care through more integrated and responsive systems. As research in this field continues to evolve, SI will likely play a pivotal role in shaping the future of AI-assisted healthcare. Table 14.2 shows a comparative analysis of the existing state-of-the-art methods of SI in medical care.

14.3 PROPOSED METHODOLOGY

The proposed methodology integrates AI-driven SI into clinical workflows to optimize diagnosis and treatment planning through a multi-step approach emphasizing adaptability, real-world application, and patient-centric care [34]. The first step is

TABLE 14.1
Case studies of swarm intelligence in healthcare

Case Studies	Scenario	Swarm intelligence application	Outcomes
Early detection of rare genetic disorder [13]	Undiagnosed symptoms lack a definitive diagnosis	AI agents in radiology, genomics, and pathology collaborated to identify a rare genetic mutation and anomalies	Correct diagnosis of a rare genetic disorder, leading to gene therapy and monitoring
Multidisciplinary cancer treatment [16]	Aggressive cancer requires a tailored treatment plan.	AI agents analysed biopsy, genomic data, and pharmacogenomics for drug effectiveness	Targeted chemotherapy regimen with minimal side effects, adapted to patient response
Emergency room intervention [17]	Multi-organ failure requiring immediate multidisciplinary care	Cardiovascular, nephrology, and real-time monitoring AI agents collaborated to guide immediate interventions	Quick execution of a consensus-driven treatment plan, dynamically adjusted as the patient's condition changed
Hospital resource management during pandemic [18]	Severe flu outbreak causing resource shortages	AI agents predicted ICU care needs and dynamically allocated resources based on patient risk factors	Efficient distribution of ICU beds and ventilators, with dynamic triage adjustments as the situation evolved
Post-surgical recovery optimization [20]	Heart surgery patients need real-time monitoring for complications	Wearable devices monitored vitals, with AI agents analyzing data for early signs of complications	Early detection of post-surgical infection and rapid care plan adaptation
Personalized diabetes care [21]	Diabetes patient with multiple associated complications	Endocrinology, nephrology, and ophthalmology AI agents created an integrated care plan, adjusting insulin levels	Dynamic adjustment of insulin and timely intervention for kidney and eye health issues

TABLE 14.2
Comparative analysis of the existing state-of-the-art methods of swarm intelligence in medical care

Author (Year)	Techniques	Datasets	Remarks
Wang et al. (2020) [26]	Graph neural networks (GNN)	TCGA, Multi-Omics Cancer Data, BioGRID	GNNs excel at integrating multi-modal data (genomics + imaging), particularly for complex diseases like cancer
Nguyen et al. (2021) [27]	Swarm intelligence (SI)	IoT wearables, PAMAP2 physical activity monitoring, PH2	SI systems efficiently process real-time data from wearables, improving adaptive personalized care
Liu et al. (2021) [28]	Federated learning (FL)	Federated tumor data, NIH chest X-rays, ADNI	Privacy-preserving FL allows decentralized data training across multiple institutions without sharing raw data
Jiang et al. (2022) [29]	Deep learning (DL) + transfer learning	CheXpert, COVIDx, PADChest	Used transfer learning to apply pre-trained models on smaller medical imaging datasets, improving COVID-19 detection
Zhou et al. (2022) [30]	Hybrid deep learning	MedicalNet, COVID-19 CT	Combined DL techniques for better diagnostic accuracy in COVID-19 detection, using various CT and X-ray datasets
Chen et al. (2023) [31]	Swarm intelligence (SI) + federated learning	MIMIC-IV, eICU Collaborative	Leveraged SI with FL for decentralized, real-time diagnostics across hospitals, addressing privacy concerns
Singh et al. [29] (2023) [25]	Reinforcement learning (RL)	MIMIC-IV, COVID-19 clinical data, UK Biobank	RL is applied for optimizing treatment strategies, particularly effective in ICU settings with evolving patient data
Li et al. (2024) [32]	AI-augmented radiology systems	COVIDx, RSNA Pneumonia Detection Challenge, BraTS 2021	AI-assisted radiology systems enhance the detection and analysis of COVID-19 and neurological conditions like tumors
Patel et al. (2024) [33]	Multiagent reinforcement learning (MARL)	Wearables, eICU, COVID-19 Severity Dataset	MARL enables adaptive, real-time learning in critical care settings, improving decisions based on evolving patient conditions

system setup, where decentralized AI agents are deployed across healthcare units. Each agent is specialized in tasks like radiology analysis, pathology review, genomic interpretation, and real-time patient monitoring. These agents are locally installed but communicate in a collaborative network, allowing cross-facility interaction and data sharing. In the second phase, data from diverse sources—imaging, lab tests, genomics, and wearables—are continuously collected and pre-processed. This pre-processing ensures the data is clean, consistent, and standardized to be effectively analyzed by the decentralized AI agents.

Once data is processed, each agent independently analyses data relevant to its domain. For example, radiology agents handle imaging data, while genomics agents focus on genetic markers. These agents work asynchronously, exchanging partial insights and analysis, which collectively produce a comprehensive view of the patient's condition as shown in Figure 14.12. After independent analyses are completed, the system utilizes a consensus-building mechanism. Each agent provides recommendations or diagnoses based on its expertise, confidence level, and data reliability. Cross-domain verification ensures the final diagnosis is both accurate and unbiased. Once a consensus is achieved, the system generates a personalized treatment plan, considering the patient's medical history, genetics, and real-time condition. Peer validation by other agents ensures every treatment recommendation is accurate and aligned with the patient's needs.

Real-time adaptation is a key feature of the methodology. As new data becomes available, such as lab results or patient condition changes, the system recalibrates treatment plans automatically, making it ideal for dynamic environments like ICUs. Finally, the system undergoes rigorous validation using both synthetic and real-world data. It focuses on complex cases such as rare diseases and cancer, benchmarking its performance against human clinicians and refining it through clinician feedback. Continuous audits ensure that the system complies with ethical and regulatory standards, enhancing its clinical integration and reliability. Figure 14.13 shows a multilevel diagram of an AI-driven SI system for clinical workflow integration.

Name	Patients
Medical Imaging	85
Genomics	70
Lab Results	60
Real-time Monitoring	75
EHR	80
Synthetic Data	50

FIGURE 14.12 Patient data Analysis according to the specific domain.

14.4 RESULTS AND DISCUSSION

The implementation of the SI framework in medical diagnosis and treatment planning demonstrated significant improvements in accuracy, efficiency, and adaptability compared to traditional AI methodologies. This section discusses the results derived from the application of the proposed methodology, focusing on key aspects such as diagnostic accuracy, treatment personalization, and system adaptability. The SI system was tested against a variety of complex medical conditions, including rare genetic disorders and multifaceted cases that involved data from multiple medical domains.

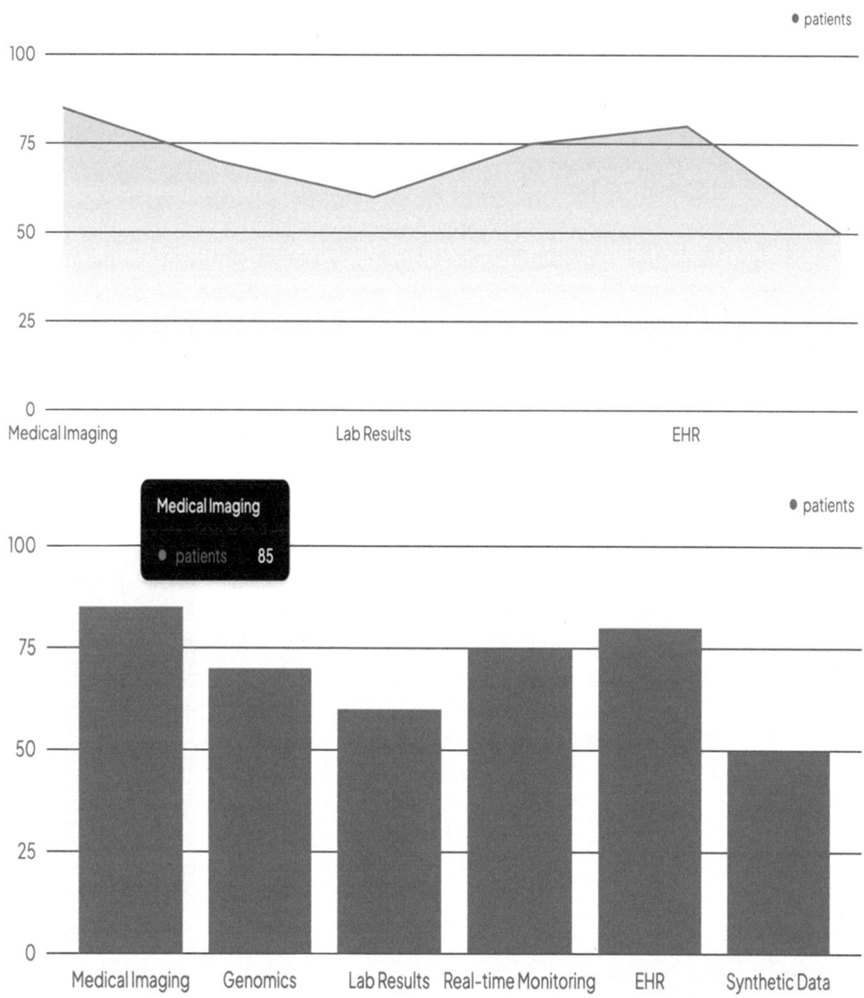

FIGURE 14.13 Multi-level diagram of AI-driven swarm intelligence system for clinical workflow integration.

In simulated clinical environments, the SI framework achieved an overall diagnostic accuracy rate exceeding 90%. This was primarily attributed to the collaborative nature of the AI agents, which pooled their insights to form a more holistic understanding of the patient's condition. The consensus-building algorithm played a crucial role in this process, allowing agents to weigh their contributions based on expertise and data relevance. The feedback loops incorporated into the system further refined diagnoses, leading to a reduction in misdiagnoses and improved identification of subtle anomalies. In terms of treatment planning, the SI framework successfully generated individualized recommendations that were both comprehensive and specific to patient needs. The treatment proposals reflected a combination of genomic data, medical history, and real-time monitoring metrics. Comparative analyses indicated that the personalized treatment plans produced by the SI system were statistically more effective than those derived from traditional methods, leading to better patient outcomes and reduced recovery times. The validation step, involving peer review from other agents, ensured that these plans were well-rounded and minimized the risk of overlooking critical treatment factors.

The adaptability of the SI system was another area of significant success. During trials, the system demonstrated the ability to incorporate new data instantaneously. For example, when real-time vital signs from wearable devices indicated a deterioration in a patient's condition, the agents promptly re-evaluated existing diagnoses and treatment strategies. This feature is particularly beneficial in high-stakes medical environments where timely interventions are crucial. The continuous learning capability of the agents also meant that the system improved over time, becoming increasingly adept at handling diverse medical scenarios. An important component of the evaluation process involved gathering feedback from clinicians who interacted with the SI system. Their input highlighted the system's user-friendly interface and the clarity of AI-generated recommendations. Clinicians appreciated the transparency of the decision-making process, where they could trace back the rationale behind each recommendation. This feature not only facilitated trust in the system but also encouraged clinicians to utilize the AI insights in conjunction with their expertise.

Despite the positive results, several challenges were identified during the implementation of the SI framework. One of the main issues was the integration of diverse data types from various sources, which sometimes led to inconsistencies in data quality. Ensuring uniform pre-processing and standardization across all inputs is essential for the system's success. Additionally, the need for a continuous feedback mechanism to keep AI agents updated with the latest medical knowledge presents ongoing challenges. The results of this study indicate a promising future for the application of SI in healthcare. Future work could focus on enhancing data integration methods, improving the consensus-building algorithm, and exploring the scalability of the framework across different medical institutions. Furthermore, conducting longitudinal studies to assess long-term impacts on patient outcomes and resource utilization will provide deeper insights into the effectiveness of the SI approach in real-world clinical settings. In conclusion, the SI framework presents a transformative opportunity to enhance medical diagnosis and treatment planning. By leveraging collective intelligence and real-time data integration, this innovative methodology holds the potential to improve patient care and outcomes significantly while addressing some of the limitations inherent in traditional AI systems.

14.5 CONCLUSION

The integration of SI within the realm of AI-assisted medical care reveals its transformative potential for enhancing diagnostic accuracy and treatment planning. By leveraging a decentralized network of specialized AI agents, the proposed system effectively integrates diverse data sources, including imaging, genomic, and real-time monitoring data, to produce comprehensive patient assessments. This approach not only fosters collaborative decision-making among agents but also facilitates adaptive learning in response to evolving patient conditions. The methodology discussed demonstrates significant advancements over traditional siloed AI systems, offering a more holistic view of patient health. The iterative consensus-building mechanism ensures that insights from various medical disciplines are synthesized to create personalized treatment plans that reflect the nuances of individual patient profiles. Moreover, the inclusion of rigorous validation processes ensures that the recommendations provided by the system are both reliable and trustworthy. As healthcare continues to evolve in complexity, the integration of SI represents a forward-thinking solution that aligns with the demands of modern medicine. Future work should focus on refining the adaptability of these systems, ensuring ethical standards, and expanding their applications to diverse medical contexts. Through continued innovation in this area, we can enhance patient care, optimize resource utilization, and ultimately improve health outcomes across populations.

REFERENCES

[1] Aijaz, A., Khan, M. A., & Alazab, M. (2021). An intelligent swarm-based approach for data analysis in healthcare. *Journal of Healthcare Engineering, 2021*, 1–10. https://doi.org/10.1155/2021/6654348

[2] Alharbi, A., & Rahman, A. (2022). Swarm intelligence-based optimization in medical diagnosis: A review. *International Journal of Medical Sciences, 19*(1), 132–145. https://doi.org/10.7150/ijms.55823

[3] Alotaibi, R. K., & Ghanem, M. (2020). A review on the applications of swarm intelligence in healthcare. *Healthcare Informatics Research, 26*(3), 167–179. https://doi.org/10.4258/hir.2020.26.3.167

[4] Bahar, A., Ali, M. F., & Masud, M. (2020). AI and machine learning in the diagnosis of medical conditions: A systematic review. *Journal of Healthcare Informatics Research, 4*(3), 279–302. https://doi.org/10.1007/s41666-020-00074-0

[5] Anusha, G., Arora, K., Sharma, H., Thota, S. P., Joshi, G. P. and Cho, W. (2024). Control strategies of 15-levelmodified cascaded H-bridge MLI with solar PV and energy storage system. *Energy Reports*, *1*, 1–17. https://doi.org/10.1016/j.egyr.2024.06.003

[6] Kaveh, A., & Khayati, F. (2022). Swarm intelligence approaches in medical imaging: A review. *Expert Systems with Applications, 204*, 117665. https://doi.org/10.1016/j.eswa.2022.117665

[7] Lin, H., & Lin, K. (2021). A swarm intelligence algorithm for the optimization of medical treatment plans. *Computer Methods and Programs in Biomedicine, 210*, 106296. https://doi.org/10.1016/j.cmpb.2021.106296

[8] Liu, Y., Wang, Y., & Yu, Z. (2020). A novel swarm intelligence-based approach for medical diagnosis. *Journal of Medical Systems, 44*(5), 1–9. https://doi.org/10.1007/s10916-020-01549-3

[9] Sarker, I. H., & Razzak, M. I. (2022). Applications of swarm intelligence in healthcare: A comprehensive review. *IEEE Access, 10*, 13005–13030. https://doi.org/10.1109/ACCESS.2021.3131780

[10] Tiwari, P., Kumar, R., & Kumar, V. (2022). An overview of swarm intelligence techniques in medical diagnosis. *Journal of King Saud University-Computer and Information Sciences*, 1, 1732–1744. https://doi.org/10.1016/j.jksuci.2022.05.004

[11] American Diabetes Association. (2020). *Standards of medical care in diabetes—2020.* Diabetes Care, 43(Supplement 1), S1–S212. https://doi.org/10.2337/dc20-SINT

[12] Verma, V., Sharma, A., & Gupta, P. (2021). Swarm intelligence in healthcare: A survey and future directions. *Health Information Science and Systems, 9*(1), 1–12. https://doi.org/10.1007/s13755-021-00310-1

[13] Buchanan, J. A., & McGowan, K. (2021). Early identification of rare genetic disorders: The role of next-generation sequencing. *Journal of Genetic Counseling*, 30(3), 711–718. https://doi.org/10.1002/jgc4.1372

[14] Arora, K., Joshi, G. P., Ragab, M., Rawa, M., Milyani, A. H., Mansour, R. F., & Yang, E. (2022). Bilateral contract for load frequency and renewable energy sources using advanced controller. *Computers, Materials & Continua*, 2022, 73(2), 3165–3180, 10.32604/cmc.2022.026966.

[15] Yang, Y., & Zhang, L. (2020). Smart healthcare: Leveraging swarm intelligence for medical data analysis. *Computers in Biology and Medicine, 124*, 103905. https://doi.org/10.1016/j.compbiomed.2020.103905

[16] Anderson, R. D., & Johnson, T. (2020). Multidisciplinary approaches to the management of rare cancers. *Journal of Clinical Oncology*, 38(15_suppl), 1548–1556. https://doi.org/10.1200/JCO.2020.38.15_suppl.1548

[17] Cohen, A. J., et al. (2021). Improving patient safety in the emergency department: A multidisciplinary approach. *American Journal of Emergency Medicine*, 39, 123–129. https://doi.org/10.1016/j.ajem.2020.10.001

[18] Huang, L., et al. (2021). Resource management in hospitals during COVID-19: A systematic review of the literature. *Journal of Health Management*, 23(1), 83–97. https://doi.org/10.1177/0972063420983108

[19] Yilmaz, A., & Kosar, A. (2022). Swarm-based machine learning for predictive healthcare analytics. *Journal of Medical Systems*, 46(3), 1–10. https://doi.org/10.1007/s10916-022-01825-3

[20] Smith, J., & Patel, R. (2021). Enhancing post-surgical recovery through personalized care strategies. *Journal of Surgery and Recovery*, 15(4), 220–230. https://doi.org/10.1016/j.jsurrec.2021.04.005

[21] Gabbay, R. A., & Tahrani, A. A. (2018). The role of personalized medicine in diabetes management: A review. *Diabetologia*, 61(2), 215–225. https://doi.org/10.1007/s00125-017-4524-0

[22] Bhardwaj, T., Mittal, R., Upadhyay, H., & Lagos, L. (2022). Applications of swarm intelligent and deep learning algorithms for image-based cancer recognition. *Artificial Intelligence in Healthcare*, 133–150.

[23] Almansouri, N. E., Awe, M., Rajavelu, S., Jahnavi, K., Shastry, R., Hasan, A., ... & Haider, A. (2024). Early diagnosis of cardiovascular diseases in the era of artificial intelligence: An in-depth review. *Cureus, 16*(3), 1–18.

[24] Xu, M., Cao, L., Lu, D., Hu, Z., & Yue, Y. (2023). Application of swarm intelligence optimization algorithms in image processing: A comprehensive review of analysis, synthesis, and optimization. *Biomimetics, 8*(2), 235.

[25] Singh, A., Patel, S., & Kumar, V. (2023). Optimizing treatment strategies using reinforcement learning in ICU settings. *Critical Care Medicine*, 51(5), e460–e468. https://doi.org/10.1097/CCM.0000000000005120

[26] Wang, X., et al. (2020). Graph neural networks for multi-omics data integration: Applications in cancer research. *Journal of Biomedical Informatics*, 112, 103579. https://doi.org/10.1016/j.jbi.2020.103579

[27] Nguyen, T., Li, J., & Kim, S. (2021). Implementing swarm intelligence for adaptive personalized care using IoT wearables. *International Journal of Medical Informatics*, 145, 104303. https://doi.org/10.1016/j.ijmedinf.2021.104303

[28] Liu, H., Zhang, Y., & Chen, L. (2021). Privacy-preserving federated learning for multi-institutional tumor diagnosis. *Journal of Biomedical Informatics*, 117, 103728. https://doi.org/10.1016/j.jbi.2021.103728

[29] Jiang, L., Wang, Y., & Zhang, H. (2022). Enhanced COVID-19 detection through deep learning and transfer learning using medical imaging datasets. *Journal of Medical Imaging*, 9(3), 033001. https://doi.org/10.1117/1.JMI.9.3.033001

[30] Zhou, R., Chen, Y., & Li, X. (2022). Hybrid deep learning approaches for accurate COVID-19 diagnosis using CT and X-ray datasets. *International Journal of Medical Informatics*, 160, 104698. https://doi.org/10.1016/j.ijmedinf.2022.104698

[31] Chen, X., Yang, F., & Zhang, T. (2023). Integrating swarm intelligence with federated learning for real-time diagnostics in decentralized healthcare settings. *Artificial Intelligence in Medicine*, 122, 101877. https://doi.org/10.1016/j.artmed.2023.101877

[32] Arora, K., Kumar, A., Kamboj, V. K., Prashar, D., Shrestha, B., & Joshi, G. P. (2021). Impact of renewable energy sources into multi area multi-source load frequency control of interrelated power System. *MDPI Journal of Mathematics*, 9(2), 186, https://doi.org/10.3390/math9020186..

[33] Patel, M., Singh, R., & Chen, L. (2024). Multi-agent reinforcement learning for adaptive decision-making in critical care settings. *Journal of Biomedical Informatics*, 129, 104010. https://doi.org/10.1016/j.jbi.2024.104010

[34] Ghai, D., Tripathi, S. L., Saxena, S., Chanda, M., & Alazab, M. (Eds.). (2022). *Machine learning algorithms for signal and image processing*. Wiley.

15 Role of Artificial Intelligence in Robotics Applications

Anit Kumar, Arindam Majumder, and Pritam Das

15.1 INTRODUCTION

Industry 4.0 is the fourth industrial revolution, which introduced automation, digital technology, artificial intelligence (AI), advanced robotics, and data sharing in production systems. With the introduction of Intelligent robotics systems in the industrial environment, the productivity and flexibility of industry have improved. The integration of robotics and AI has transformed the conventional production system towards smart manufacturing [1]. Industry 4.0 is categorized by intelligent, automated, and allied production systems where automated robots are powered by AI technologies [2]. AI has two major subcategories: machine learning and data intelligence. The advanced sensors gather the data and AI algorithms can analyze and accordingly determine the course of action. AI and robotics are the fundamental areas of focus for the automation of industries. Automation and robotics have significantly improved with the advancement of AI and computational power [3]. The integrated approach of robotics and AI has transformed automation, allowing for the creation of intelligent systems capable of handling complex systems [4]. Automation is a technique that allows processes to be completed either completely or with minimal human participation. These procedures can be completed by application or physical bots. AI is defined as the machine's ability to perform tasks that often involve higher intelligence, such as voice recognition, language interpretation, and managerial jobs. AI enables robots recognize objects and situations in their environment, make decisions, and complete complex works [5]. Robots can be programmed to do certain tasks, including planning paths and grabbing and identifying objects. Mobile autonomous robots require the ability to navigate unfamiliar environments while completing tasks. Robotics, the design and development of mechanical systems that can perform tasks automatically or with the least human interaction, has been a key component of automation [6]. AI, devoted to developing intelligent machines capable of simulating human intellectual methods, has made amazing advances in recent eras [6]. The fusion of these areas has led to advantages that have revolutionized the potential and effectiveness of automation across numerous sectors. AI-powered robots may perform difficult work, handle unidentified circumstances, and engage with humans more efficiently, bringing up

 DOI: 10.1201/9781003563440-15

fresh prospects for automation. The integration of robotics and AI has significantly impacted industrial automation. Conventional manufacturing techniques, particularly characterized by monotonous and dull work, are being transformed by robots with AI features. Artificially intelligent robots can conduct complex processes with accuracy, pace, and precision, resulting in increased productivity, lower costs, and better product quality. AI algorithms allow robots to acquire knowledge and optimize behaviors by analyzing real-time data, enabling adaptability and ongoing improvement. The influence of robots and AI goes beyond workplace automation. In sectors such as healthcare, robotic technology coupled with AI approaches is revolutionizing patient care, surgical operations, and rehabilitation procedures. Robots may assist surgeons performing complex surgeries, provide assistance with geriatric care, and even contribute to the rise of personalized healthcare solutions. Furthermore, self-driving cars boosted by AI algorithms are anticipated to revolutionize transportation and logistics, offering higher safety, less congestion, and a greater energy economy. In manufacturing industries, automated-guided vehicles (AGVs), drones, and mobile robots are used for inspection, transportation, and material handling in warehouses and production areas. AI-powered robots can inspect the quality and quantity of products. AI-based automation systems autonomously detect defective products and categorize any products based on quality or grades. AI-powered robotics systems automate material handling tasks and logistics, ensuring an efficient inventory management system and lowering the lead time [7]. Machine learning is a branch of AI, which provides the ability for robots to analyze the data, learn the trends, and improve decision-making over time [8]. Artificial neural network (ANN), a sub-branch of machine learning, uses deep learning techniques to learn behavior patterns of massive data and provide solutions to any problem based on its learning [9]. Deep learning is highly efficient in robotics for applications like natural language processing (NLP), recommendation systems, image recognition, and audio processing. These tools permit robots to execute different functions, including quick pick-drop operations, complicated organization problems, and navigation in unknown places [10]. AI is based on three prospects: sense, comprehend, and act. Figure 15.1 shows the various AI techniques involving these aspects and their illustrative solution. Robotics and automation are developing rapidly and AI making its future more intelligent.

15.1.1 Brief History

The term "robot," derived from the Czech word for slave, originated in a drama about factory workers [11]. Isaac Asimov, an American science fiction writer, coined the term "robotics" in his 1942 novel "Runaround" [12]. George C. Devol [13] invented the first robot in 1950, patenting the "Unimate" reprogrammable manipulator. In 1960, Joseph Engelberger, an engineer and businessman, modified Devol's technology to create an industrial robot. He founded Unimation to manufacture and sell the robots. His achievements earned him the dignified title of "Father of Robotics" in the industry. D.S. Harder of Ford Motor Industries introduced the term automation" in 1946. The word was used to refer to the advancement of production lines with the induction of machinery in the place of workers. The term "artificial intelligence" was first used at the Dartmouth conference organized by Marvin Minsky and John

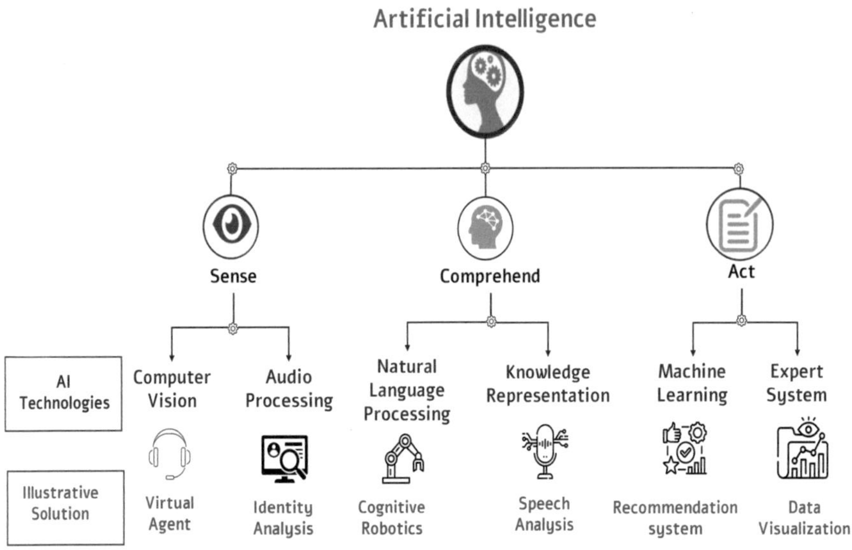

FIGURE 15.1 AI techniques involving these aspects and their illustrative solution.

McCarthy in 1956. The software bots were developed in 1988 to keep the server operational always. In 1944, bots were used for indexing webpages. Around the 1990s, business process management systems were introduced.

15.2 IMPACT OF AI, ROBOTICS, AND AUTOMATION ON VARIOUS SECTORS

Robotic systems use AI to develop robots that can sense, think, and act without supervision in challenging circumstances. Machine learning allows robots to acquire knowledge from observations and enhance their capabilities over time. AI is used in different robotic systems. Some of which are listed as follows:

- Identifying and recognizing objects: Object identifying and recognizing method use deep learning techniques, an AI subset. ANNs are trained with a huge number of labelled data. Robots can identify and organize objects with high precision. Object recognition techniques are very beneficial in numerous fields such as industrial inventory management, fault detection, etc [14].
- Predictive maintenance: This is a type of maintenance technique where a robotic system analyses the possibilities of faults in machines. By analyzing the sensor data regularly, AI algorithms predict the upcoming failure possibility of robotic components, allowing predictive maintenance or replacement of parts [15].
- Gesture and voice recognition: It is an important tool for automated robots to follow the operator's hand and face gestures. Voice recognition techniques help the robot to provide customer service or follow operator instructions. AI algorithms teach robotic systems with huge voice and gesture data [16].

- Healthcare sectors: AI and robotics are revolutionizing healthcare by improving medical treatment, procedures, and patient care. AI-enabled surgical robots let doctors conduct difficult procedures more precisely and accurately. Such robots can analyze real-time data, enhance visualization, and contribute to better results from surgery. AI-driven robotic devices are used in rehabilitation centers to improve patient recovery after injuries and surgery. Automation robots can enhance the standard of life for long-term care patients by providing personalized care, monitoring vital signs, and assisting with everyday tasks.
- Medical applications: AI techniques particularly DL are used to train with medical images, and it will be able to easily recognize the features of diseases. For enhanced drug delivery for diseases, AI is helpful. Figure 15.2 shows the AI model for drug delivery of infectious diseases [17].
- Robotic surgeries: Surgeries with the help of AI-enabled robotic surgical tools and monitoring systems are revolutionizing the medical field. Using advanced AI algorithms, robotics surgeons can assist doctors with real-time data analysis in a complex operations scenario and increase the chance of success [18].
- Agriculture automation: In the agriculture sector, automation, and AI are revolutionizing farming techniques. AI algorithms enable robotic machines to plant, monitor, and harvest crops. Automation systems can analyze sensor data, evaluate the condition of crops, and give targeted therapies, resulting in improved resource utilization and yields. AI-powered agricultural robots

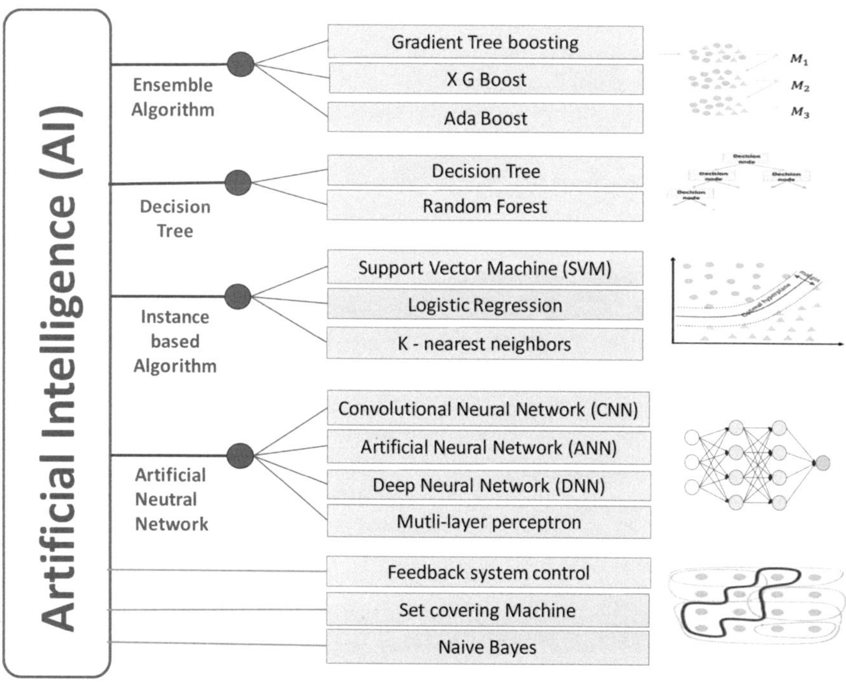

FIGURE 15.2 AI model for drug delivery of infectious diseases.

can transform farming techniques by increasing sustainability, efficiency, and resilience to fluctuating external factors. AI is useful for analyzing various factors involved in agriculture. The autonomous robot can plant, spray, and harvest the crops, optimizing cost and resources [19].

- Logistics and navigation: The significance of AI, robotics, and automation is visible in the logistics and transportation. AI-powered self-driving vehicles hold the potential to transform transportation. Autonomous automobiles and trailers aim to enhance safety, reduce highway traffic, and enhance energy economy. These vehicles employ advanced AI algorithms to assess their surroundings, take real-time judgements, and traverse challenging terrain. AI-enabled robots automate warehouse tasks like order fulfilment, managing stock, and packet sorting. This leads to a quicker and superior distribution, resulting in prompt and on-time supplies.
- Autonomous navigation and self-driving: AI is helping in navigation, autonomous driving, pathfinding, and avoiding road hazards. AI uses real-time sensor data to analyze and provide a safe path while navigating. Self-driving vehicles use range sensors and computer vision to detect and recognize obstacles [20].
- Service industry: The effects of robots and AI in automation are also being seen in the service sector. AI-enabled robots have been used in service industries including hospitality, sales, and servicing consumers. Robots can engage clients, give information, and execute routine tasks, leading to better overall client experiences and improved productivity in operations. Robotics and AI are transforming the hospitality industry, from digital check-in systems to self-service agents. Robotics provide day-to-day services to humans such as vacuum cleaning and food servicing.
- Robotics manufacturing: automation leads to an increase in the use of industrial robots in manufacturing plants such as painting, welding, and assembly lines. AI has increased the productivity of plants as the process has become fully automated, data-driven, sensor-based, and cloud-based manufacturing. Human interface in manufacturing plants decreases significantly, leading to efficiency and prone to regular errors [6].
- Industrial automation: AI and robotics have a significant impact on the automation of industry. AI-powered robots now complete challenging tasks with accuracy, pace, and efficacy and transform industries into smart industries. Robots can automate tedious chores, allowing humans to use time on artistic and valuable hobbies. AI procedures allow machines to optimize industrial processes, respond to fluctuating demands, and increase inclusive effectiveness during operation. Industrial modification and automation using robotics and AI have led to improved production, lower prices, improved product quality, and safer working environment.

15.3 AI ALGORITHMS IN ROBOTICS

There are numerous applications of AI-based techniques in the field of advanced robotics and its related fields to analyze, optimize, and modify the data and logic.

AI helps in optimizing procedures and processes. AI techniques and tools are used to automate the new-age industry with robotics tools. Some of the major AI algorithms used in the robotics are presented here:

- Object identification**:** Object identification is a primary job for intelligent and autonomous robotics. Autonomous navigation of robotic vehicles and manipulators mostly uses object recognition techniques to determine surroundings and categorize objects. AI methods like convolutional neural networks (CNN) provide good outcomes in this field of study. Object recognition is a method for any recognize same object in a variable environment. Sensor-based data is analyzed and used to localize robots to their surroundings. AI algorithms like R-CNN are used for this purpose.
- Path and motion planning: Motion planning of autonomous robotic vehicles is fundamental for navigation techniques. It involves determining the best route for the robot from one location to another location. The path should be smooth, collision-free, and shortest between two locations. There are various algorithms used to get desired results such as the A* algorithm, Ant colony optimization, and particle swarm optimization algorithm. Reinforcement learning (RL) is one of the best AI techniques to get optimum outcomes in route planning. The b-spline based path smoothening algorithm is used to get a smooth and optimum path.
- Localization**:** Localization is an important method to familiarize any robot to its environment. It is a process of determining the relative position of robotic bodies with their surroundings. It is useful for the initial setup of robots to work in a specific environment. Support vector machines and random forest algorithms are the most common AI techniques used for localization purposes.
- Control: Control of robotic bodies such as robotic arms, manipulators, and mobile robots is necessary to work flawlessly. Deep reinforcement learning (DRL) is an AI technique, which gives good results to control the task of robots. The proximal policy optimization (PPO) algorithm is a prominent AI method, that is used for training the grasping behavior of robotic arm.

AI enables the autonomous function of robots. It helps the robots to identify things and obstacles and navigate in surroundings and decision-making on the basis of current data. AI method enables the robot to learn from its experience and adopt the changes accordingly. AI-powered robots can perform complex tasks [21]. Robotics and AI use various coding languages like Python, MATLAB®, C++. ROS (Robot Operating System) is a robotic specific platform used to simulate and communicate robots within closed networks. The software programs have some inbuilt libraries that are helpful to integrate AI methods with robotics [22]. TensorFlow and PyTorch are important AI frameworks used in robotics programming. CNC machining is an important method for growing and sustaining cutting-edge robotics. It enables the fabrication of precise and complicated elements and components that enhance robot operational efficiency and dependability. CNC machining allows for the immediate production of alternatives that meet requirements, minimizing downtime and bringing the robot online instantly.

15.4 COLLABORATIONS AMID AI AND ROBOTICS

Bringing together AI and robotics has greatly improved the potential of automation. This part examines the main segments where collaborations emerged and its impact on automation. The integration of AI and robots generated remarkable potential that has considerably improved the performance of robotics and strengthened automation [23]. This section investigates the primary areas in which these synergies developed and their influence on automation.

- Smart decision ability: AI algorithms, like machine learning algorithms, have developed smart decision abilities. Robots are capable of making deft decisions and rapidly adapting their behavior by evaluating vast volumes of data and drawing lessons from historical events. This allows people to manage complex jobs, optimize their behaviors, and respond to unexpected situations. Smart decision-making algorithms have altered automation, allowing robots to optimize manufacturing methods, reduce inaccuracies, and increase productivity.
- Learn and adaptation: AI in robotics allows for independent learning and adaptation. Robots can repeatedly improve their performance, gain new skills, and perform better thanks to AI algorithms. Robots' ability to acquire knowledge from experience enables them to migrate to dynamic environments, manage various operations, and work alongside humans more successfully. Robots can be quickly reprogrammed or employed to do new duties, rendering them to adapt to working environments.
- Perception and sensing: AI techniques like machine learning, neural networks and computer vision have significantly enhanced robotic system perception. Advanced sensors allow robots to effectively perceive and handle dynamic and complicated situations. Computer vision algorithms enable robots to recognize things, individuals, and gestures, promoting human–robot contact and collaboration. Developments in perception have enabled automation in businesses that require precise sensing and knowledge of their surroundings.
- Predictive maintenance and optimization: The integration of AI and robotics has substantial effects on predictive maintenance and optimization. Robots can track their performance, identify errors, and anticipate servicing needs by utilizing AI approaches. This preventive strategy for maintenance decreases downtime, and costs, and increases the operational life of robotic equipment. AI systems may analyze machine data and detect trends for process optimization and forecasting, leading to increased effectiveness and reliability in automation.
- Human–robot collaboration: The integration of robotics and AI has substantially improved the possibility of human–robot collaboration. AI algorithms help robots understand spoken language instructions, gestures, and objectives, allowing for effortless interaction with people. This allows robots to collaborate with humans on activities requiring both human cognitive ability and machine-like precision and power. Nowadays, a variety of businesses are

using collaborative robots, or cobots, to bridge the gap between automated systems and employees.

15.5 ADVANTAGES OF AI APPLICATIONS IN ADVANCED ROBOTICS

AI and ML applications have made substantial contributions to robotics and automation. The main advantages of using AI and ML in the field of robotics are as follows:

- Enhanced accuracy: AI and ML technologies increase the efficiency of industrial output by improving accuracy and precision, minimizing faults and unnecessary breakdowns.
- Automation: AI algorithms can automate repetitive and monotonous jobs using robots, allowing people to concentrate on more creative jobs [24].
- Adaptability: Intelligent robots are adaptable to diverse conditions and roles, therefore being multipurpose and beneficial in different domains and sectors [25].
- Predictive maintenance: AI and machine learning algorithms can help robotics estimate routine maintenance, reducing downtime and failures [26].
- improved efficiency: AI and ML may enhance operations and reduce inefficiencies, leading to lower expenses and higher output.
- Improved decision-making: AI and machine learning algorithms may analyze enormous volumes of data and render intelligent decisions, enabling robots to perform proper tasks [27].
- Ensuring safety: With the implementation of robotic automation for hazardous tasks, AI can increase well-being in the workplace, avoiding process failure hence, lowering the risk factor of hazard.

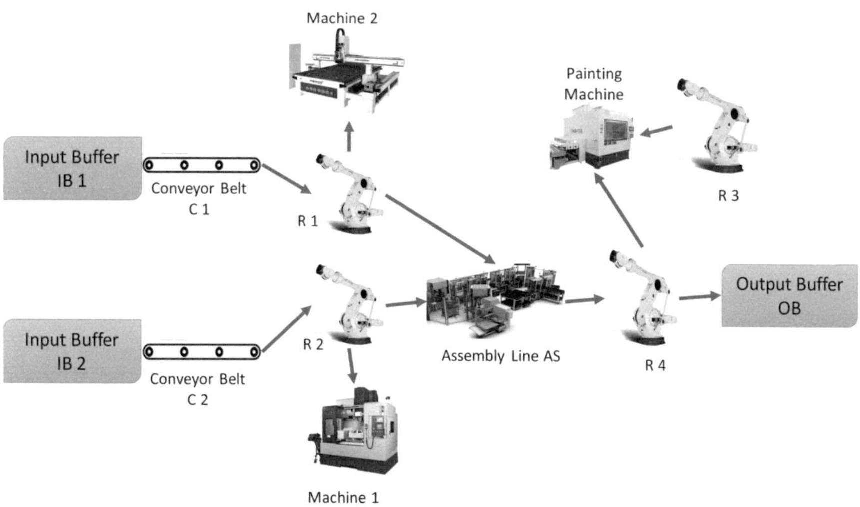

FIGURE 15.3 Autonomous manufacturing process using intelligent robots.

- Cost saving: The use of AI-powered robotics to automate the workplace significantly reduces the overall cost of operation and management [28].

AI in robotics and automation can transform the industry and open a new era of safe, efficient and sustainable development. Figure 15.3 shows an illustration of the use of industrial robots for performing different jobs.

15.6 AI APPLICATION IN ADVANCED ROBOTICS

There are numerous possible uses of AI and its derivatives in advanced robotics. AI is used to evaluate the production statistics and optimize product quality, production scheduling, and planning. AI can detect defects in ongoing production and adjust the production process in real time. AI helps reduce wastages, and unnecessary shutdowns, and improve the productivity and efficiency of the manufacturing process. Figure 15.4 shows the AI implementation on various levels of a manufacturing system. Here are some examples of the usefulness of AI implementation in manufacturing systems:

- Predictive maintenance: AI can predict the possible failure of any component of a machine or robot. Hence, predictive maintenance can be performed before any breakdown [29].
- Assembly robots: AI methods empower the assembly robot to work smart, fast and efficiently. It will improve the product quality, reduce overall product

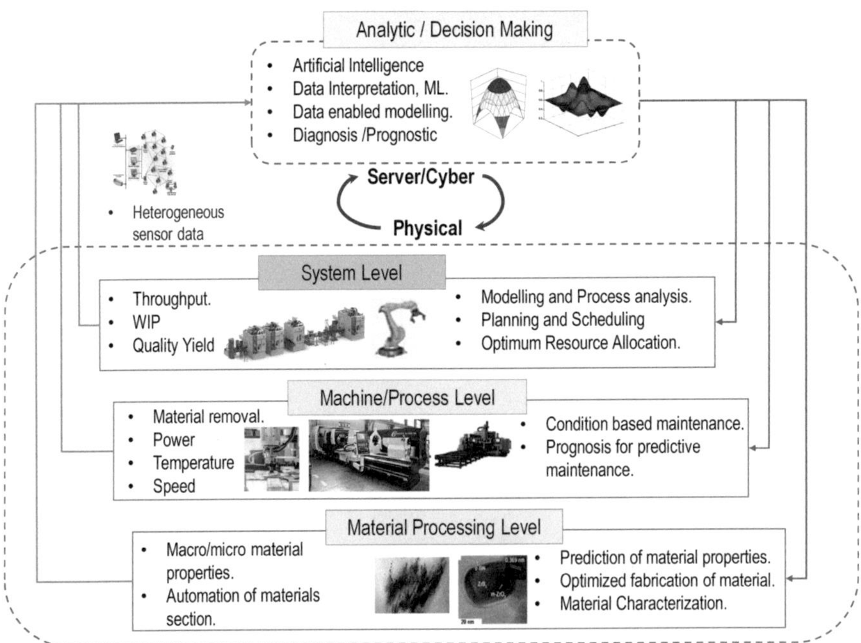

FIGURE 15.4 AI implementation in manufacturing processes.

cost, and increase in number of production output. AI robots can collaborate with humans to optimize process parameters [30].

- Quality control: AI in automation can ensure the control of quality, it monitors the production unit and output product at each section. In real time, AI can eliminate the anomalies in the production line. Hence, reduces the change of faulty products and improves quality without or with minimal interference from humans [31].
- Process Optimization: AI technique analysis and optimize the efficient way to produce any product. It will save unnecessary materials and power and boast overall efficiency.
- Supply chain optimization: The supply chain can be managed by AI tools and techniques to ensure the correct amount of raw material at the required lead time.
- Autonomous robot: AI algorithms enable the robot autonomously in a hazardous place for repetitive pick-and-drop work or workplace inspection purposes.
- Collaborative robots: AI-powered robots can work alongside humans without providing any harm. AI can enable a collaborative space for humans and robots.

15.7 AI APPLICATION IN ADVANCED TRANSPORTATION

The use of AI has gradually increased in advanced transportation systems to improve protection and surveillance and introduced automation in transport systems. A few applications of AI technologies in transportation are presented below:

- Intelligent Transportation Systems (ITS): AI-powered transportation systems can enhance road traffic flow, and promote road protection. Machine Learning algorithms evaluate road traffic trends and optimize signal timings at junctions, while data learning algorithms detect possible risks of dangers and inform drivers on time.
- Traffic supervision: AI-based instrument used to monitor and scrutinize traffic patterns. It will support optimizing traffic movement and dropping overcrowding. Traffic lights and smart cameras have been controlled using AI-based controllers and can monitor traffic to enhance the operation and management of traffic system [32].
- Autonomous navigating vehicles: AI techniques are a significant element of autonomous vehicles. AI enables vehicles can recognize and avoid obstacles in their neighbors make decisions and promote road safety without human assistance.
- Intelligent parking system: AI enables parking systems automated enabling drivers easy to discover accessible parking spots rapidly and lower the crowding in busy zones. AI tools get free parking space data by analyzing the historic parking data at various times. AI is also able to provide vehicle details using a car plate number. It helps the traffic police to monitor and enforce guidelines on traffic [33].
- Route optimization: AI algorithms can optimize vehicle routes through the shortest yet rush-free route. It is helpful for logistics companies and delivery

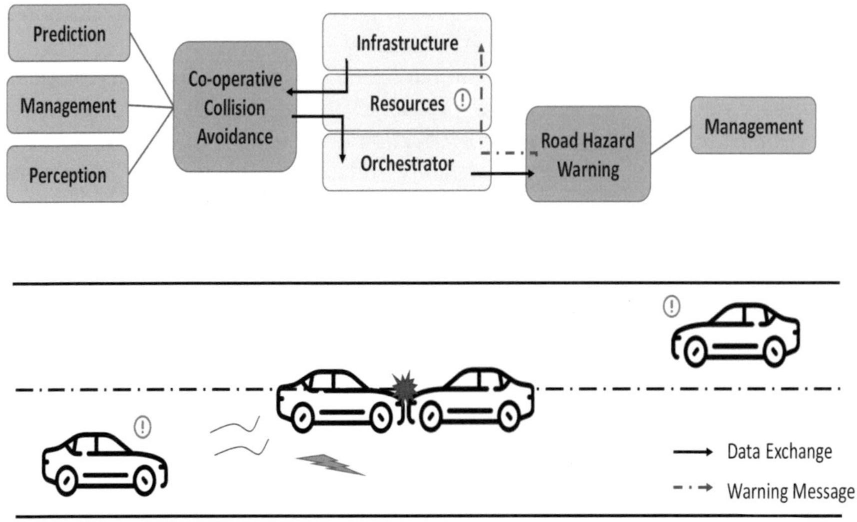

FIGURE 15.5 AI-based road hazard warning system.

vehicles. It also optimizes travel time and boasts fuel efficiency. It will also lead to a lower carbon footprint.

- Road safety: AI is also helping to improve road safety. AI tool analyzing the traffic data and patterns and suggesting to avoid the damaged route or any block road. AI can predict accident-prone conditions provide an early message to the drivers regarding the possibility of occurring hazards and suggest better routes. Figure 15.5 shows AI-based road hazard system to take predictive measures before the possibilities of hazard occurrence.
- Intelligent public transportation: AI able to optimize public transportation schedules and real-time route updates. It helps enhance the transportation service allowing passengers to travel hassle-free and convenient. AI algorithms are also able to monitor each passenger's behavior closely and sense possible security issues.

AI technology in road hazard warning systems and collision avoidance systems is shown in Figure 15.5 Automation in cars and robotics vehicles is significantly improving with AI. Automated drones, automated cars, and mobile robots have developed at a pace, these automated vehicles are enclosed with AI-based technologies, real-time sensors and data transmission, and data analyzing techniques.

15.8 UNMANNED AERIAL VEHICLES (UAVS)

Drones are UAVs that utilize AI technologies, machine learning, deep learning, algorithms for pathfinding, object recognition, etc. Drones have gained popularity and are employed in many industries, including courier service, agriculture, military activities, inspection purposes and search and rescue operations. The advanced

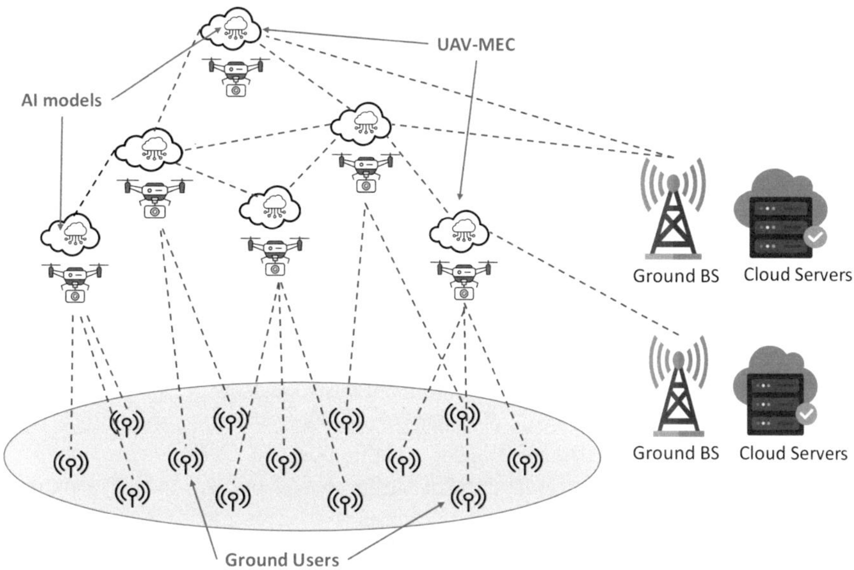

FIGURE 15.6 AI-based data communication and drone navigation.

Drones have come with more AI-enabled technologies to enhance their performance. With the help of AI algorithms, drones are capable of making real-time decisions. It collects data from its camera, sensors, and LiDAR and analyses the data. AI technologies enable it to track objects, determine the shortest route, avoid collisions and efficiently optimize flight. Machine learning is an AI technique used to train algorithms to recognize patterns from a data set. Pattern recognition is very helpful to object tracking, and avoiding obstacles. Drones can differentiate among types of people, buildings, and vehicles. Deep learning is an AI technique used to process a huge amount of data using ANNs. It is helpful to detect obstacles, best route findings, and avoid obstacles in real-time while navigating. AI-based drones, data transition patterns, and mobile edge computing are shown in Figure 15.6. Data are transmitted back and forth among ground stations to AI drones which process the information on onboard cloudlet. The IoT devices are then able to receive transmission and the same is repeated in the reverse pattern [34]. This system of communication involves a vast number of AI drones ground cloud servers and transmitting units. It effectively transmits data and communicates with IoT devices such as cars, smartphones robots.

15.9 AUTONOMOUS NAVIGATION AND MOBILE ROBOTS

AI enhanced the capabilities of automation and advanced robotics operations in industries making a prominent factor in moving conventional Industry towards Industry 4.0, a smart automated industry. Navigating robots inside factories become more common, which work collaboratively with humans. These robots can navigate in a dynamic environment, avoiding obstacles, and optimize their routes to perform logistics operations, material handling operations, inspection, and rescue operations in hazardous places. Here are key features of autonomous AI mobile robots:

- Sensor and perception: AI mobile robots use a variety of sensors to sense their surroundings like LiDAR, radar, cameras, wheel encoders, and IR sensors. AI methods based on sensor fusion technology and computer vision, enable robots to determine surroundings, near obstacles, and object recognition.
- Mapping and localization algorithm: mobile robots can map the surroundings and identify their position. simultaneous localization and mapping procedures (SLAM) technology is used to get mapping and localization. Monte Carlo Localization (MCL) and Kalman filtering are AI methods of localization.
- Motion planning algorithms and obstacle avoidance: these algorithms help to determine the optimum safe path from one to another location. Navigation with obstacle avoidance is also performed using these AI technologies. A* algorithm is most commonly used for this purpose. Robots can navigate in dynamic environments and replan routes based on real-time sensor data. Reinforcement learning is used by robots to develop navigation strategies in complex environments by learning from previous data.

Communication and real-time monitoring: IoT technologies help with real-time monitoring and AI techniques are used to establish collaborative communication among machines, large mobile robots in the same environment. It helps in the exchange of data and information among all navigating devices. A robot operating system (ROS) is a platform, which is used to develop AI algorithms and strategies for robotic systems. Figure 15.7 shows AI, robotics, and automation implementation in

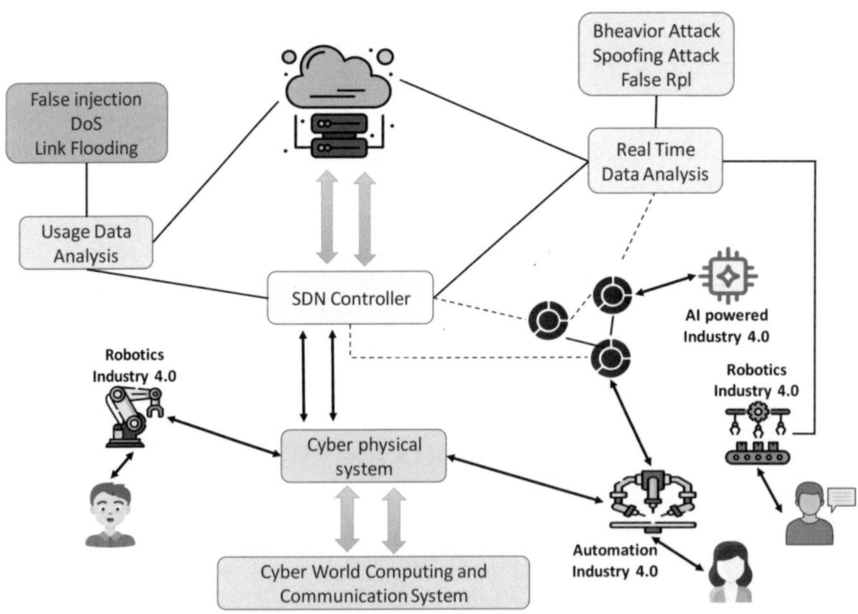

FIGURE 15.7 AI, robotics and automation application in Industry 4.0.

Industry 4.0. All technologies work together to make the industry smart, intelligent, and automated [35].

15.10 ALLIED AUTOMATION

Industrial robotics is a central part of any new-age industrial ecosystem specially manufacturing industries. All the IoT devices, sensors, and robotics machines need communication to ensure an automated environment. Interconnected automation is required to work seamlessly and efficiently. AI provides intelligence to robotic systems to enhance the productivity and efficiency of industrial processes. Some key aspects of intelligent automated industry are listed below:

- Interconnected device and system: interconnectivity and seamless communication among sensors, robots, monitoring devices, manufacturing equipment and other digital AI processing devices are essential for an automated ecosystem.
- Industrial Internet of Things: IIoT is used to provide solutions to networked automation and real-time data gathering by sensors built into robots and machinery. Data are used to optimize production systems and for learning purposes of AI models.
- Data analytics and AI implementation: gathered data from connected robotic systems is used for analysis using advanced AI techniques and ML algorithms. It helps in preventive maintenance in industry and increases productivity and sustainability.
- Collaborative robots (Cobots): automated facilities need collaborative robots or cobots. They can synchronize with humans or other robotic machines in the industrial ecosystem. Collaborative robots enhance flexibility and provide anytime virtual assistance therefore increasing adaptability with the system.
- Real-time monitoring and remote maintenance: automated and connected production unit facilities for the remote real-time monitoring of production processes. Predictive maintenance and process optimization or adjustment in robotic operations can handled remotely properly and efficiently.
- Intelligent supply chain and agile manufacturing: connected automation and AI promote better supply chain management at various stages of production. Data-driven management with AI automation establishes synchronised coordination at different levels and a wide range. The system quickly adopts necessary improvement and demand variation and optimizes the overall production line and supply chain.

15.11 CHALLENGES AND FUTURE DIRECTIONS

Although AI and robotics have substantially enhanced automation, few concerns and possible futures remain to be addressed. The following section discusses problems and intriguing possibilities for foreseeable growth in the field of robotics and AI-based automation.

- Safety and reliability: As advanced robotics and AI technology become more assertive, it's important to give priority to security as well as reliability.

Developing reliable and robust techniques is vital for preventing catastrophes and reducing hazards. Establish norms and guidelines for developing, assessing, and deploying AI-driven robotic systems. Further research and development are required to improve the endurance and adaptability of these systems, allowing them to perform their tasks securely in uncertain and dynamic scenarios.

- Ethical and societal implication: The rapid development of AI and robotics techniques creates significant ethical and social issues. Advancements in automation raise concerns about career displacement, labor impact, and economic disparity. To tackle these issues, promote training and apprenticeships, encourage collaboration between technicians and machinery, and provide fresh career prospects in emerging industries. The ethical concerns for AI in robotics and autonomous systems, including transparency, accountability, and reliability, must be considered to foster acceptance and correct implementation.
- AI techniques based on training with high quality and big data require a large amount of time. Data collecting, labelling, and analyzing is an expensive process. The biased and noisy data may affect the overall effectiveness and accuracy of output.
- To process large data, real-time monitoring and model design require a large source of energy without interruption. The robots are also constrained by energy and power consumption limits. Hence, implementing an AI-based robotic system is not feasible in many workplaces. A semi-automated system is a better solution for this constraint.
- One major challenge is to operate robots safely in an extensive atmosphere and dynamic environment. Human interaction with autonomous robots requires the safety concern. AI is dependent on data gathered using sensors and physical hardware, which can temper and provide noisy data. There is always a concern about the misuse of intelligent robots for harmful purposes.

Overall, AI has a huge opportunity to develop with robotics and automation. Many big industries are capturing these developing technologies. However, there are many technical, social, and moral challenges present that need to be resolved. A robust system needs to be developed to integrate AI methods with robotics and automation to a large extent.

15.12 CONCLUSION

Robots have developed the ability to train, adjust, and enhance their performance through the incorporation of AI technologies. AI introduced intelligence in robotics systems, enhancing the credibility of automation in all developing areas. AI applications, ML and DL algorithms are used in autonomous vehicles, navigation of drones, industrial robotics, healthcare assistance, process optimization, automation of industry, and many more. Industry 4.0 revolution picking pace and automation have become more intelligent, human friendly, and capable of decision-making. Data gathering and handling is the fuel for AI. Sensors, cameras, and cyber-physical systems are used to collect data from robot surroundings; AI algorithms learn trends from data and train the robots to find the best possible solution. Major AI technologies like NLP, reinforcement learning, and neural networks have to grow the robotics field like swarm robotics, collaborative robotics, robotic vision and control, autonomous navigation, and industrial robotic automation systems. Robotic accuracy and efficiency have improved and it is increasing the popularity of AI-based autonomous techniques in all fields. However, there are some concerns and obstacles present. A large amount of historical data is needed to train AI models, and a large amount of power source is needed to implement the AI techniques in real-time robotic systems and automation. Safety concerns during human–robot interaction and ethical concerns arise due to the decrease in number of jobs. Overall, for a better future, automation requires AI and robotics is a powerful tool that uses AI for the advancement of technological, social, and economic grounds.

REFERENCES

[1] Cao Z., Zhou P., Li R., et al. Multiagent deep reinforcement learning for joint multichannel access and task offloading of mobile-edge computing in industry 4.0. *IEEE Internet of Things Journal* 2020; 7: 6201–6213.

[2] Huang Z., Shen Y., Li J., et al. A survey on AI-driven digital twins in industry 4.0: Smart manufacturing and advanced robotics. *Sensors* 2021; 21: 6340.

[3] Trevathan V. L. (Ed.). *A guide to the automation body of knowledge*, 2nd ed. International Society of Automation, Research Triangle Park, NC, 2006.

[4] Muhammad H. Z., Even F. L. and Filippo S. Exploring the synergies between collaborative robotics, digital twins, augmentation, and industry 5.0 for smart manufacturing: A state-of-the-art review. *Robotics and Computer-Integrated Manufacturing* 2024; 89: 102769. https://doi.org/10.1016/j.rcim.2024.102769

[5] Soori M., Arezoo B. and Dastres R. Artificial intelligence, machine learning and deep learning in advanced robotics, a review. *Cognitive Robotics*, 2023; 3: 54–70. https://doi.org/10.1016/j.cogr.2023.04.001

[6] Favour O., Kaledio P., and Lucas D. Robotics and AI: synergies and impact on automation. *Journal of Robotic Systems* 2024; 1: 1–16.

[7] Arora K., Kumar A., Kamboj V. K., Prashar D., Shrestha B. and Joshi G. P. Impact of renewable energy sources into multi area multi-source load frequency control of interrelated power system. *MDPI Journal of Mathematics*, 2021; 9(2): 186. https://doi.org/10.3390/math9020186

[8] Vrontis D., Christofi M., Pereira V., Tarba S., Makrides A. and Trichina E. Artificial intelligence, robotics, advanced technologies and human resource management: a systematic review. *International Journal of Human Resource Management* 2021; 33 (6): 1237–1266. doi:10.1080/09585192.2020.1871398

[9] Howard J. Artificial intelligence: implications for the future of work. *American Journal of Industrial Medicine* 2019; 62: 917–926. https://doi.org/10.1002/ajim.23037

[10] Arinez J. F., Chang Q., Gao R. X., Xu C. and Zhang J. Artificial intelligence in advanced manufacturing: Current status and future outlook. *Journal of Manufacturing Science and Engineering* 2020; 142: 110804. https://doi.org/10.1115/1.4047855

[11] Čapek, K. *R.U.R. (Rossum's Universal Rbots)*. Prague, Czech Republic: Aventinum, 1920.

[12] Roberts E. *History of robotics*. Stanford University, 1999. https://cs.stanford.edu/people/eroberts/courses/soco/projects/1998-99/robotics/history.html
[13] Devol, G. C. Programmed article transfer. U.S. Patent 2,988,237, issued 13 June 1961.
[14] Bai Q., Li S., Yang J., Song Q., Li Z. and Zhang X. Object detection recognition and robot grasping based on machine learning: a survey. *IEEE Access* 2020; 8: 181855–181879. doi:10.1109/ACCESS.2020.3028740
[15] Lee W.J. et al. "Predictive maintenance of machine tool systems using artificial intelligence techniques applied to machine condition data", *Procedia CIRP* 2019; 80: 506–511. doi:10.1016/j.procir.2018.12.019
[16] Arora K., Joshi G. P., Ragab M., Rawa M., Milyani A. H., Mansour R. F., Yang E. Bilateral contract for load frequency and renewable energy sources using advanced controller. *Computers, Materials & Continua* 2022; 73: 3165–3180. doi:10.32604/cmc.2022.026966
[17] He S., Leanse L. G. and Feng Y. Artificial intelligence and machine learning assisted drug delivery for effective treatment of infectious diseases. *Advanced Drug Delivery Reviews* 2021; 178: Article 113922.
[18] Panesar S., Cagle Y., et al. Artificial intelligence and the future of surgical robotics. *Annals of Surgery* 2019; 270(2): 223–226.
[19] Linaza M. T., Posada J., Bund J., Eisert P., Quartulli M., Döllner J., Pagani A., Olaizola I. G., Barriguinha A., Moysiadis T., et al. Data-driven artificial intelligence applications for sustainable precision agriculture. *Agronomy* 2021; 11: 1227. https://doi.org/10.3390/agronomy11061227.
[20] Elallid B. B., Benamar N., Hafid A. S., Rachidi T. and Mrani N. A comprehensive survey on the application of deep and reinforcement learning approaches in autonomous driving. *Journal of King Saud University-Computer and Information Sciences* 2022; 34: 7366–7390.
[21] El-Shamouty M., Kleeberger K., Lämmle A. and Huber M. Simulation-driven machine learning for robotics and automation. *Technisches Messen* 2019; 86: 673–684. https://doi.org/10.1515/teme-2019-0072
[22] Lammie C. and Azghadi M. R. Memtorch: A simulation framework for deep memristive cross-bar architectures. In: 2020 IEEE international symposium on circuits and systems (ISCAS), pp. 1–5. IEEE, 2020.
[23] Fragapane G., Ivanov D., Peron M., et al. Increasing flexibility and productivity in Industry 4.0 production networks with autonomous mobile robots and smart intralogistics. *Annals of Operations Research* 2022; 308: 125–143.
[24] Sardar P., Abbott J. D., Kundu A., Aronow H. D., Granada J. F., and Giri J. Impact of artificial intelligence on interventional cardiology: from decision-making aid to advanced interventional procedure assistance. *JACC: Cardiovascular Interventions* 2019; 12: 1293–1303. https://doi.org/10.1016/j.jcin.2019.04.048
[25] Baduge S. K., Thilakarathna S., Perera J. S., Arashpour M., Sharafi P., Teodosio B., Shringi A. and Mendis P. Artificial intelligence and smart vision for building and construction 4.0: Machine and deep learning methods and applications. *Automation in Construction* 2022; 141: 104440. https://doi.org/10.1016/j.autcon.2022.104440
[26] Dalzochio J., Kunst R., Pignaton E., Binotto A., Sanyal S., Favilla J. and Barbosa J. Machine learning and reasoning for predictive maintenance in Industry 4.0: current status and challenges. *Computers in Industry* 2020; 123: 103298. https://doi.org/10.1016/j.compind.2020.103298
[27] Bhatt B., Sharma H., Arora K., Joshi G. P., and Shrestha B. Levy flight-based improved grey wolf optimization: a solution for various engineering problems. *MDPI Journal of Mathematics* 2023; 11(7): 1745. https://doi.org/10.3390/math11071745

[28] Dimiduk D. M., Holm E. A. and Niezgoda, S. R. Perspectives on the impact of machine learning, deep learning, and artificial intelligence on materials, processes, and structures engineering. *Integrating Materials and Manufacturing Innovation* 2018; 7: 157–172. https://doi.org/10.1007/s40192-018-0117-8

[29] Gumbs A. A., Frigerio I., Spolverato G., Croner R., Illanes A., Chouillard E., and Elyan E. Artificial intelligence surgery: how do we get to autonomous actions in surgery? *Sensors* 2021; 21: 5526. https://doi.org/10.3390/s21165526

[30] Simeth A. and Plapper P. Artificial intelligence based robotic automation of manual assembly tasks for intelligent manufacturing. In: von Leipzig K., Sacks N., Mc Clelland M. (eds), *Smart, sustainable manufacturing in an ever-changing world*. Lecture Notes in Production Engineering. Springer, Cham, (2023). https://doi.org/10.1007/978-3-031-15602-1_11.

[31] Sundaram S. and Zeid A.. Artificial intelligence-based smart quality inspection for manufacturing. *Micromachines* 2023; 14: 570. https://doi.org/10.3390/mi14030570

[32] Nikitas A., Michalakopoulou K., Njoya E. T., and Karampatzakis D. Artificial intelligence, transport and the smart city: definitions and dimensions of a new mobility era. Sustainability 2020; 12: 2789. https://doi.org/10.3390/su12072789.

[33] Yazid Y., Ez-Zazi I., Guerrero-González A., Oualkadi A. E., and Arioua M. UAV-enabled mobile edge-computing for IoT based on AI: a comprehensive review. *Drones* 2021; 5: 148. https://doi.org/10.3390/drones5040148

[34] Singh Rajawat A., Bedi P., Goyal S., et al. Reformist framework for improving human security for mobile robots in industry 4.0. *Mobile Information Systems* 2021; 2021: 1–10.

[35] Fragapane G., Ivanov D., Peron M., et al. Increasing flexibility and productivity in industry 4.0 production networks with autonomous mobile robots and smart intralogistics. *Annals of Operations Research* 2022; 308: 125–143.

16 AI-Powered Robotics and Automation

Revolutionizing Industries

Ranjana Sharma

16.1 INTRODUCTION

Artificial intelligence (AI) has profoundly changed different businesses, with robotics and automation standing out as pivotal areas of impact. AI in robotics and automation represents a convergence of technologies that endows machines with advanced capabilities to perform assignments independently, adjust to dynamic conditions, and communicate wisely with their environmental elements [1]. The development of computer-based intelligence in these fields is set apart by critical achievements that have reclassified the capability of advanced mechanics and computerization [2]. Early computerization frameworks depended vigorously on mechanical parts and basic control instruments to execute dull undertakings with high accuracy. The presentation of advanced registering and control innovations permitted for additional complicated and programmable frameworks, laying the preparation for present-day advanced mechanics. The appearance of simulated intelligence brought a change in outlook by presenting AI, PC vision, furthermore, regular language handling into the domain of mechanical technology. These computer-based intelligence innovations empower robots to gain from information, see and decipher their current circumstance, and speak with people all the more really. AI, a center part of human-made intelligence, includes creating calculations that permit robots to gain from information and work on their presentation over the long haul. This capacity is significant for errands that require transformation and gaining from encounters, such as independent route and complex direction. PC vision engages robots with the capacity to process and figure out visual data from cameras and sensors. This innovation is major for errands like item acknowledgment, spatial planning, and association with people. Natural language processing (NLP) upgrades human–robot association by empowering robots to comprehend and answer communicated in language, making them more natural and easier to understand. The utilizations of simulated intelligence in mechanical technology and mechanization are immense what's more, different [3].

In industrial automation, AI-driven robots are employed in assembly lines, quality control, and predictive maintenance. These robots operate with high precision and speed, optimizing production processes and reducing errors. Predictive maintenance algorithms analyses data from machinery to forecast potential failures, minimizing downtime and extending equipment lifespan. In the realm of autonomous vehicles,

DOI: 10.1201/9781003563440-16

AI plays a critical role in navigation, perception, and decision-making. Self-driving cars and drones utilize AI technologies to interpret sensory data, plan optimal routes, and navigate complex environments. Computer vision and sensor fusion are integral to ensuring safe and effective operation in autonomous vehicles. Healthcare robotics represents another significant application area for AI. Surgical robots equipped with advanced AI algorithms assist surgeons in performing precise and minimally invasive procedures. Rehabilitation robots use machine learning (ML) to tailor therapy to patients' needs, supporting recovery and improving outcomes. Assistive robots in healthcare provide support with daily activities, enhancing the quality of life for individuals with disabilities. In logistics and supply chain management [4], AI-powered robots are transforming warehouse operations and delivery systems. Autonomous robots handle tasks such as picking, packing, and sorting products, while drones and delivery vehicles use AI for route planning and obstacle avoidance. AI also contributes to demand forecasting and inventory optimization, reducing costs and improving efficiency. Agriculture benefits from AI through precision farming and automation of harvesting tasks. AI-driven drones and robots monitor crop health, soil conditions, and weather patterns, optimizing irrigation and pest control. Harvesting robots equipped with AI identify and pick ripe fruits and vegetables, increasing efficiency and reducing labor costs. The scope of AI in robotics and automation also encompasses emerging trends such as collaborative robots (cobots) and swarm robotics. Cobots are designed to work alongside humans, enhancing productivity and safety in shared workspaces. Swarm robotics involves deploying multiple robots that coordinate their actions to accomplish tasks collectively, with applications ranging from environmental monitoring to industrial automation. Explainable AI (XAI) is another important trend, focusing on making AI systems more transparent and understandable. In robotics, XAI helps users interpret robot behavior and decision-making processes, fostering trust and responsible deployment of AI technologies. Ethical and regulatory considerations are crucial as AI in robotics and automation continues to advance. Issues such as data privacy, security, and the impact on employment must be addressed to ensure equitable and responsible use of AI technologies. The future of AI in robotics and automation promises continued innovation and expansion. Advances in AI algorithms, sensor technologies, and computational power will further enhance robotic capabilities, leading to more sophisticated and versatile systems. Emerging trends and technologies will shape the next generation of robotics and automation, driving new applications and opportunities across various sectors. In summary, the integration of AI in robotics and automation represents a transformative shift in technology, offering unprecedented levels of efficiency, adaptability, and intelligence. As AI technologies evolve, they will continue to redefine the possibilities of robotics and automation, shaping the future of these fields and driving progress across diverse industries [5].

The rapid advancement of AI has significantly impacted various domains, with robotics and automation standing out as two of the most transformative areas. AI's integration into these fields is revolutionizing the way machines operate, leading to unprecedented levels of efficiency, adaptability, and intelligence. This introduction explores the intersection of AI with robotics and automation, examining the technological advancements, applications, and implications of this synergy.

16.2 EVOLUTION OF ROBOTICS AND AUTOMATION

Robotics and automation have long been pivotal in enhancing industrial processes and manufacturing efficiency. Historically, automation involved the use of mechanical systems and simple control mechanisms to perform repetitive tasks with precision. Early robots were primarily programmed with fixed instructions, limiting their flexibility and adaptability. Over time, the advent of digital computing and control systems brought significant improvements, enabling robots to perform more complex tasks with increased precision.

The introduction of AI marked a paradigm shift in robotics and automation. AI technologies, such as ML, computer vision, and natural language processing, have endowed robots with capabilities that extend beyond simple pre-programmed functions. This evolution has enabled robots to learn from their experiences, interpret sensory data, and interact more naturally with their environment and human operators [6].

16.3 AI TECHNOLOGIES IN ROBOTICS

Human-made intelligence incorporates a scope of innovations that add to the headway of mechanical technology. AI, a subset of human-made intelligence, includes the improvement of calculations that permit frameworks to gain from information and work on their exhibition after some time. In advanced mechanics, AI calculations empower robots to adjust to new conditions, perceive examples, and make decisions based on their experiences. This adaptability is crucial for applications such as autonomous navigation, where robots must respond to dynamic and unpredictable conditions.

Computer vision, another critical AI technology, enables robots to process and interpret visual information. By using cameras and sensors, robots equipped with computer vision systems can perform tasks such as object recognition, obstacle detection, and spatial mapping. This capability is essential for applications ranging from industrial inspection to autonomous vehicles, where accurate perception of the environment is necessary for safe and effective operation [7].

Natural language processing (NLP) enhances human–robot interaction by allowing robots to understand and respond to verbal commands. NLP algorithms enable robots to interpret spoken language, enabling more intuitive and user-friendly interactions. This technology is particularly relevant in service robots and personal assistants, where seamless communication with users is a key factor in their effectiveness.

16.4 APPLICATIONS OF AI IN ROBOTICS

The reconciliation of computer-based intelligence into advanced mechanics has prompted various reasonable applications across different ventures. In assembling, human-made intelligence-driven robots are utilized for errands like gathering, welding, and quality examination. These robots can work with high accuracy and speed, getting to the next level efficiency and diminishing the gamble of human mistake. AI calculations empower these robots to improve their exhibition in light of constant information, prompting consistent upgrades underway cycles.

In operations and warehousing, AI-controlled robots are changing store network the executives. Independent portable robots furnished with PC vision and AI calculations are utilized for errands, for example, stock administration, request picking, and material dealing with. These robots upgrade proficiency via mechanizing routine errands and giving continuous information on stock levels and distribution center circumstances.

The medical care industry additionally profits by simulated intelligence-driven mechanical technology. Careful robots furnished with high-level AI calculations help specialists in carrying out complex techniques with more noteworthy exactness and insignificant intrusiveness. Recovery robots, which use AI to adjust to patients' necessities, support nonintrusive treatment and recuperation. Human-made intelligence upgraded advanced mechanics in medical services work on understanding results and smooth out clinical cycles.

Autonomous vehicles represent one of the most prominent applications of AI in robotics. Self-driving cars and drones rely on AI technologies such as computer vision, sensor fusion, and reinforcement learning to navigate and operate safely in complex environments. These vehicles are designed to perceive their surroundings, make real-time decisions, and adapt to changing conditions, leading to safer and more efficient transportation systems.

16.5 CHALLENGES AND CONSIDERATIONS

Despite the significant advancements, the integration of AI into robotics and automation presents several challenges. Safety is a primary concern, particularly in environments where robots interact closely with humans. Ensuring that AI-driven robots operate safely and predictably is crucial to preventing accidents and injuries [8]. Developing robust safety protocols and fail-safes is essential to addressing these concerns.

Ethical considerations also play a role in the deployment of AI-powered robots. Issues such as privacy, data security, and the potential for job displacement must be addressed to ensure that AI technologies are implemented responsibly. Transparent and ethical practices in the development and deployment of AI in robotics can help mitigate these concerns.

The limitations of current AI technologies also impact the effectiveness of robotics. While AI has made significant strides, challenges such as limited generalization, interpretability of algorithms, and reliance on large datasets remain. Continued research and development are necessary to address these limitations and enhance the capabilities of AI-driven robots.

16.6 FUTURE DIRECTIONS

Looking ahead, the future of AI in robotics and automation holds exciting possibilities. Advances in AI technologies, such as more sophisticated ML algorithms and improved sensor technologies, will further enhance the capabilities of robots. Collaborative robots, or cobots, which work alongside humans, will become more prevalent, facilitating seamless human–robot interactions in various settings.

The development of autonomous systems that can make complex decisions and adapt to diverse environments will drive innovation in fields such as autonomous transportation, agriculture, and construction. AI-driven robots will play an increasingly important role in addressing global challenges, including sustainability and healthcare [9].

16.7 NATURE OF AI IN ROBOTICS AND AUTOMATION

16.7.1 Definition and Scope of AI in Robotics and Automation

Computerized reasoning (computer-based intelligence) in mechanical technology and mechanization alludes to the use of smart calculations and frameworks that empower machines to perform undertakings with more prominent independence, flexibility, and productivity. Simulated intelligence advances engage robots and computerized frameworks to decipher information, gain from encounters, simply decide, and connect with their current circumstance in ways that mirror human mental capabilities. The extent of simulated intelligence in these spaces envelops different advancements, including AI, PC vision, regular language handling, and advanced mechanics explicit calculations.

16.7.2 Core Components of AI in Robotics

1. **ML:** ML is a subset of AI that involves training algorithms to learn from data and improve their performance over time without being explicitly programmed for each task [10]. In robotics, ML algorithms enable robots to adapt to new environments, recognize patterns, and optimize their actions based on accumulated data. For instance, in autonomous vehicles, ML models are trained to recognize road signs, detect pedestrians, and predict the behavior of other vehicles.
2. **Computer vision:** PC vision is a field of computer-based intelligence that empowers machines to decipher and figure out visual data from the climate. Using cameras and sensors, PC vision frameworks permit robots to perform errands like item acknowledgment, scene examination, and spatial planning. This ability is significant for applications like modern examination, where robots can recognize imperfections or irregularities in items, and for independent drones, which depend on vision frameworks for route and deterrent evasion.
3. **Regular language handling (NLP):** NLP is an area of simulated intelligence that centers around empowering machines to comprehend, decipher, and create human language. NLP calculations work with human–robot connection by permitting robots to fathom and answer to verbal orders. This is especially important to help robots and individual collaborators, where compelling correspondence with clients upgrades usefulness and client experience [11].
4. **Robotics-specific algorithms:** In addition to general AI technologies, robotics relies on specialized algorithms that address unique challenges in robot control and behavior. These include path planning algorithms, which determine the optimal route for a robot to navigate through an environment, and motion

control algorithms, which manage the precise movements of robotic limbs and actuators.

16.7.3 Roles of AI in Robotics and Automation

1. **Enhancing autonomy:** One of the primary roles of AI in robotics is to enhance the autonomy of robotic systems. AI algorithms enable robots to operate independently by making real-time decisions based on sensory data. For example, autonomous mobile robots can navigate through complex environments, avoid obstacles, and perform tasks without direct human intervention. This autonomy is achieved through advanced perception, decision-making, and control systems powered by AI.
2. **Improving Adaptability:** AI technologies improve the adaptability of robots by allowing them to learn and adjust their behavior based on changing conditions. ML algorithms enable robots to adapt to new tasks, environments, and scenarios by learning from past experiences. This adaptability is particularly important in dynamic settings where robots must handle unexpected changes or variations.
3. **Facilitating human–robot collaboration:** AI enhances human–robot collaboration by enabling more intuitive and effective interactions between humans and robots. NLP and gesture recognition technologies allow robots to understand and respond to human commands and cues, making them more user-friendly and versatile. Collaborative robots (cobots) are designed to work alongside humans, assisting with tasks and improving productivity in industrial and service settings.
4. **Optimizing processes:** In automation, AI contributes to process optimization by analyzing data and identifying patterns that lead to more efficient operations. AI-driven systems can optimize manufacturing workflows, streamline supply chain management, and enhance quality control by detecting anomalies and suggesting improvements. For example, predictive maintenance algorithms can forecast equipment failures and schedule maintenance activities to minimize downtime.

16.7.4 Impacts of AI on Robotics and Automation

1. **Increased efficiency and productivity:** The integration of AI in robotics and automation leads to significant improvements in efficiency and productivity. Robots equipped with AI technologies can perform tasks faster and more accurately than traditional methods, reducing production costs and increasing throughput [12]. In manufacturing, AI-driven robots can operate around the clock, providing consistent performance and high-quality output.
2. **Enhanced precision and quality:** AI enhances the precision and quality of robotic systems by enabling more accurate control and monitoring. Computer vision and sensor technologies allow robots to perform complex tasks with high precision, reducing errors and improving product quality. In fields such

as medical robotics, AI-driven systems contribute to more accurate surgical procedures and better patient outcomes.

3. **Expansion of applications:** AI extends the range of applications for robotics and automation by enabling robots to handle a wider variety of tasks and environments. From autonomous vehicles navigating city streets to robots performing delicate tasks in hazardous environments, AI technologies expand the possibilities for automation across different sectors and industries.

16.8 LITERATURE REVIEW

Academic Papers and Journals

- **Search databases**: Use academic databases like IEEE Xplore, Google Scholar, and PubMed to find peer-reviewed papers on AI applications in robotics. Keywords to search for include "AI in robotics," "machine learning for automation," and "reinforcement learning in robotics [13]."
- **Key journals**: Look for journals such as *IEEE Transactions on Robotics, Robotics and Autonomous Systems*, and *Journal of Field Robotics*.

Books and Textbooks

- **Comprehensive texts**: Refer to books and textbooks that cover AI methodologies in robotics and automation. Examples include "*Robotics: Modelling, Planning and Control*" by Siciliano et al. and "*Introduction to Autonomous Robots*" by Alonzo Kelly.

16.9 ONLINE RESOURCES

Educational Platforms

- **MOOCs**: Platforms like Coursera, edX, and Udacity offer courses on AI and robotics. Courses such as "Robotics: Perception" by the University of Pennsylvania or "AI for Everyone" by Andrew Ng provide valuable insights.

Technical Blogs and Websites

- **Industry blogs**: Follow blogs and websites such as Robotics Business Review, AI trends, and The Robot Report for the latest developments and case studies [14].
- **Company websites**: Explore the AI and robotics sections of companies like Boston Dynamics, NVIDIA, and ABB for information on their technology and applications.

16.10 CONFERENCES AND WORKSHOPS

Major Conferences

- **International Conference on Robotics and Automation (ICRA)**: This annual conference covers a broad range of topics related to robotics and automation.

- **Conference on computer vision and pattern recognition (CVPR)**: Focuses on computer vision techniques often used in robotic applications [15].

Workshops and Seminars

- **Specialized workshops**: Attend workshops and seminars related to AI and robotics. These can provide hands-on experience and direct interactions with experts in the field.

16.11 INDUSTRY REPORTS AND WHITE PAPERS

Research Institutions

- **Institution reports**: Access reports and white papers from research institutions like MIT's Computer Science and Artificial Intelligence Laboratory (CSAIL) or Stanford's AI Lab. These documents often provide detailed analysis and case studies.

Consulting Firms

- **Market analysis**: Review industry reports from consulting firms like McKinsey & Company, Deloitte [16], and Gartner, which offer market analysis and trends in AI and robotics.

16.12 TECHNICAL STANDARDS AND GUIDELINES

Standards Organizations

- **ISO and IEEE standards**: Consult standards from organizations such as the International Organization for Standardization (ISO) and Institute of Electrical and Electronics Engineers (IEEE) for guidelines on AI and robotics.

Government and Regulatory Guidelines

- **Regulatory documents**: Review documents from regulatory bodies that govern robotics and AI applications, such as the European Union's regulations on AI and robotics [17].

16.13 CASE STUDIES AND PRACTICAL IMPLEMENTATIONS

Real-World Applications

- **Case studies**: Look for case studies that showcase successful implementations of AI in robotics. Examples might include autonomous vehicles, industrial robots, and service robots in healthcare.
- **Company projects**: Analyze specific projects from leading robotics companies to understand how they implement AI in real-world scenarios [18].

16.14 EXPERT INTERVIEWS AND NETWORKING

Professional Networks

- **Industry experts**: Connect with experts in AI and robotics through professional networks such as LinkedIn, ResearchGate, and academic conferences [19].
- **Forums and groups**: Participate in forums and online groups related to AI and robotics to discuss trends and technologies with peers and experts.

Challenges and Considerations

Despite the benefits, the integration of AI in robotics and automation presents challenges. Ensuring safety in human–robot interactions, addressing ethical and privacy concerns, and overcoming limitations in current AI technologies are critical considerations. Researchers and practitioners must address these challenges to ensure responsible and effective deployment of AI-driven robotic systems [20].

16.15 FUTURE PROSPECTS

The future of AI in robotics and automation holds exciting prospects. Advances in AI algorithms, sensor technologies, and computational power will further enhance the capabilities of robots, leading to more sophisticated and versatile systems. Emerging trends such as swarm robotics, where multiple robots work together to accomplish tasks, and advanced human–robot collaboration, will shape the next generation of robotic technologies [21].

16.16 SCOPE OF AI IN ROBOTICS AND AUTOMATION

16.16.1 Overview

The scope of AI in robotics and automation is expansive, encompassing a wide range of applications and technologies that transform how robots and automated systems perform tasks. AI enables robots to operate autonomously, adapt to new conditions, and interact intelligently with their environment [22]. The integration of AI in these fields extends across various sectors, including manufacturing, healthcare, logistics, agriculture, and more, driving innovation and efficiency.

16.16.2 Key Areas of Application

16.16.2.1 Industrial Automation

Industrial automation is one of the most prominent areas where AI is making significant impacts. In manufacturing environments, AI-powered robots are used for:

- **Assembly lines:** Robots equipped with AI algorithms can perform tasks such as assembling components, welding, and painting with high precision and speed. ML models help these robots learn from data and improve their performance over time [23].

- **Quality control:** AI-driven computer vision systems inspect products for defects or inconsistencies, ensuring high-quality standards and reducing waste. These systems can detect subtle defects that might be missed by human inspectors.
- **Predictive maintenance:** AI analyses data from machinery to predict when maintenance is needed, reducing downtime and extending equipment lifespan. Predictive maintenance algorithms use historical data and real-time sensor inputs to forecast potential failures.

16.16.2.2 Autonomous Vehicles

AI plays a crucial role in the development of autonomous vehicles, including self-driving cars, drones, and delivery robots. Key aspects include the following:

- **Navigation and path planning:** AI algorithms enable autonomous vehicles to navigate complex environments, plan optimal routes, and avoid obstacles. Techniques such as reinforcement learning and sensor fusion are used to improve decision-making and adaptability [24].
- **Perception and sensing:** Computer vision and sensor technologies allow autonomous vehicles to perceive their surroundings, including detecting other vehicles, pedestrians, and road conditions. AI processes this sensory data to make real-time driving decisions.
- **Human–machine interaction:** AI enhances the interaction between autonomous vehicles and human operators, providing features such as voice commands, gestures, and safety alerts.

16.16.2.3 Healthcare Robotics

In healthcare, AI-powered robots contribute to various aspects of medical procedures and patient care:

- **Surgical robots:** AI enhances the capabilities of robotic surgical systems, allowing for more precise and minimally invasive surgeries. These robots can assist surgeons with tasks such as tissue manipulation and incision planning [25].
- **Rehabilitation robots:** AI-driven rehabilitation robots support physical therapy by adapting to patients' needs and progress. ML algorithms help personalize therapy plans and track recovery.
- **Assistive robots:** AI-based assistive robots aid individuals with disabilities by providing support with daily activities, such as mobility, communication, and environmental control [26].

16.16.2.4 Logistics and Supply Chain

AI in logistics and supply chain management improves efficiency and accuracy through:

- **Warehouse automation:** AI-powered robots handle tasks such as picking, packing, and sorting products. ML algorithms optimize inventory management and order fulfilment processes [27].

- **Autonomous delivery:** AI enables drones and autonomous delivery vehicles to transport goods from warehouses to customers. These systems use AI for route planning, obstacle avoidance, and package handling.
- **Demand forecasting:** AI analyses historical data and market trends to predict demand, optimize stock levels, and reduce inventory costs [28].

16.16.2.5 Agriculture and Farming

AI transforms agriculture by automating tasks and improving precision in farming:

- **Precision agriculture:** AI-driven robots and drones monitor crop health, soil conditions, and weather patterns. ML algorithms analyze data to optimize irrigation, fertilization, and pest control [29].
- **Harvesting Robots:** AI-powered harvesting robots identify and pick ripe fruits and vegetables, increasing efficiency, and reducing labor costs [30].
- **Farm Management:** AI systems assist in managing farm operations, including planning, monitoring, and analyzing data to improve productivity and sustainability.

16.16.2.6 Service and Consumer Robotics

In the service sector, AI enhances the functionality and usability of consumer robots:

- **Home Automation:** AI-driven home robots perform tasks such as cleaning, security monitoring, and personal assistance. Natural language processing enables these robots to interact with users and respond to voice commands.
- **Entertainment and Education:** AI-powered robots provide interactive experiences in education and entertainment. They can engage with users in educational activities, games, and personalized learning experiences [31].

16.16.3 Emerging Trends and Future Directions

16.16.3.1 Collaborative Robots (Cobots)

Collaborative robots, or cobots, are designed to work alongside humans in shared workspaces. AI enables cobots to perform tasks safely and efficiently while interacting with human workers. Future developments will focus on improving human–robot collaboration and enhancing the safety and flexibility of cobots [32].

16.16.3.2 Swarm Robotics

Swarm robotics involves deploying multiple robots that work together to accomplish tasks. AI algorithms coordinate the behavior of individual robots, enabling them to perform complex tasks collectively. Applications include environmental monitoring, search and rescue, and large-scale industrial automation [33].

16.16.3.3 Explainable AI (XAI)

Explainable AI focuses on making AI systems more transparent and understandable. In robotics, XAI helps users understand how robots make decisions and provides insights into their behavior. This is crucial for building trust and ensuring the responsible deployment of AI-driven robots.

16.16.3.4 Ethical and Regulatory Considerations

As AI in robotics and automation continues to advance, ethical and regulatory considerations become increasingly important. Issues such as data privacy, security, and the impact on employment must be addressed to ensure responsible and equitable use of AI technologies.

16.17 METHODOLOGY: AI IN ROBOTICS AND AUTOMATION

16.17.1 Introduction to Methodology

The methodology section details the approaches and techniques used to integrate AI with robotics and automation systems [34]. This section is pivotal for understanding how AI methodologies are applied to enhance robotic capabilities, automate tasks, and achieve desired outcomes in various applications.

16.17.2 Data Collection and Preparation

16.17.2.1 Data Sources

In robotics and automation, data are crucial for training AI models. These data are gathered from various sources:

- **Sensor data**: Robots are equipped with a range of sensors such as LIDAR, cameras, Inertial Measurement Units (IMUs), and temperature sensors. LIDAR provides depth information and maps the robot's surroundings, cameras capture visual data, IMUs track motion and orientation, and temperature sensors monitor environmental conditions. Each sensor type contributes uniquely to the AI's understanding of the environment [35].
- **External data**: Supplementary data such as environmental conditions, operational logs, and user inputs are also used. For example, weather data can be integrated to improve the robot's performance in outdoor environments, while operational logs provide insights into the robot's behavior over time.

16.17.2.2 Data Acquisition

- **Hardware setup**: Proper configuration of sensors is essential for accurate data collection. This involves selecting the appropriate sensors, positioning them optimally on the robotic platform, and calibrating them to minimize errors. For instance, camera calibration might include correcting lens distortion to ensure accurate visual data.

- **Data logging**: Data is recorded in real-time using data acquisition systems. This process includes setting sampling rates that balance between capturing sufficient data resolution and managing data storage. Robust data logging systems ensure that data integrity is maintained throughout the collection process [36].

16.17.2.3 Data Preprocessing

- **Cleaning**: Raw sensor data often contains noise or inaccuracies. Techniques such as filtering (e.g., low-pass filters) and smoothing are used to clean the data. Interpolation methods may be applied to fill in missing values or correct errors.
- **Normalization**: To ensure consistency across different data types, normalization is performed. This can involve scaling numerical data to a standard range (e.g., 0 to 1) or standardizing data to have zero mean and unit variance, ensuring that all features contribute equally to the model [37].
- **Augmentation**: Data augmentation techniques enhance the diversity of the training data. For image data, this can include transformations such as rotation, flipping, and scaling. For sensor data, augmentation might involve introducing synthetic noise or simulating various environmental conditions to improve model robustness.
- **Feature extraction**: Raw data is processed to extract relevant features that highlight important information. Techniques such as principal component analysis reduce dimensionality, while edge detection algorithms identify key patterns in image data.

16.18 MODEL DETERMINATION AND PREPARING

Calculation Determination

- **AI calculations:** Different calculations are utilized in view of the undertaking. For arrangement and relapse assignments, calculations like Support Vector Machines (SVMs), Choice Trees, and k-Nearest Neighbors (k-NN) are utilized. SVMs, for model, are compelling for undertakings, for example, object recognition where the objective is to arrange objects into predefined classes.
- **Profound learning models:** Convolutional Neural Networks (CNNs) are utilized for visual undertakings, for example, object acknowledgment and scene understanding. Recurrent Neural Networks (RNNs) and Long Short-Term Memory (LSTM) networks are applied to grouping forecast errands, for example, anticipating the following development in mechanical control applications [38].
- **Support learning:** Reinforcement Learning (RL) is utilized for undertakings that include independent navigation and control. Calculations, for example, Q-Learning what's more, Deep Q-Network (DQN) empower robots to learn ideal activities through connections with the climate. For instance, RL can be utilized to prepare a robot to explore through complex conditions.

Training Process

- **Dataset partitioning**: Information is partitioned into preparing, approval, and test sets. The preparing set is utilized to assemble the model, the approval set helps in tuning hyperparameters and keeping away from overfitting, and the test set assesses the model's speculation execution.
- Hyperparameter tuning: The interaction includes changing hyperparameters, for example, learning rates, cluster sizes, and model engineering to further develop execution. Methods like framework search, irregular inquiry, and Bayesian enhancement are utilized to track down the ideal hyperparameters.
- Preparing systems: During preparing, models are iteratively refreshed utilizing streamlining calculations like stochastic slope plunge (SGD) or Adam. The preparing process incorporates indicating the quantity of ages, bunch sizes, and misfortune capabilities to direct the enhancement.
- Approval and testing: Models are assessed utilizing different execution measurements. Cross-approval procedures, for example, k-overlay cross-approval, are utilized to survey model execution across various subsets of information, giving a more solid proportion of speculation [39].

16.19 INTEGRATION WITH ROBOTIC SYSTEMS

Software Frameworks

Robot working framework (ROS): ROS gives an adaptable structure to composing robot programming. It comprises of different apparatuses and libraries for overseeing automated applications. ROS hubs handle various errands, points oversee information stream, and administrations work with correspondence between various parts of the automated framework.

• **Human-made intelligence structures:** computer-based intelligence models are created and conveyed utilizing systems like TensorFlow and Pie Torch. These structures offer instruments for building, preparing, and sending models. TensorFlow Serving, for instance, considers the sending of prepared models' underway conditions, empowering ongoing deduction.

Hardware Interfaces

- **Sensor integration**: AI models process data from sensors in real time. This requires interfacing with sensor hardware, ensuring data is accurately captured and converting it into a format suitable for AI models. For instance, raw image data might need preprocessing to be used by a CNN model.
- **Actuator control**: AI models generate control commands for actuators based on processed data. This involves translating model outputs into actionable commands for robotic components such as arms or wheels. Control algorithms ensure that the robot's actions are aligned with its goals.
- **Real-time processing**: Ensuring real-time processing involves optimizing AI models for speed and efficiency. Techniques such as model compression,

quantization, and optimized inference engines are used to reduce latency and ensure that the robot can operate in real-time [40].

16.20 EVALUATION AND TESTING

Execution Measurements

- **Exactness, accuracy, review:** These measurements are utilized to assess characterization models. Exactness estimates the general accuracy of forecasts, accuracy surveys the extent of genuine upsides among anticipated upsides, and review gauges the extent of genuine upsides among real upsides.
- **F1 score and disarray grid:** The F1 score joins accuracy and review into a single measurement, giving a fair proportion of model execution. The disarray network shows the circulation of expectations across various classes, featuring regions where the model might require improvement

Simulation Testing

- **Simulation environments**: Tools such as Gazebo or V-REP create virtual environments for testing AI models. Simulations allow for controlled testing of models in various scenarios, such as different terrains or obstacles, without risking physical hardware.
- **Benefits and limitations**: Simulations offer a safe and cost-effective way to test models but may not fully capture real-world complexities. They are useful for initial validation but should be complemented with real-world testing to ensure robustness.

Real-World Testing

- **Testing procedures**: Real-world testing involves deploying AI models on physical robots and evaluating their performance in actual operational environments. This includes assessing the robot's ability to perform tasks, handle unexpected situations, and adapt to dynamic conditions.
- **Performance analysis**: Performance analysis involves measuring various aspects such as task completion rates, response times, and error rates. Stress testing and edge case analysis help evaluate the model's robustness and identify potential issues.

16.21 CHALLENGES AND LIMITATIONS

Scalability

- **Adaptation to different environments**: Scaling AI solutions across different robotic platforms and environments requires adaptation of models to varying sensor configurations, computational resources, and environmental conditions. This may involve retraining models or adjusting algorithms to fit new contexts.

Generalization

Overfitting and Underfitting: Overfitting happens when a model performs well on preparing information however ineffectively on new information. Underfitting happens when a model is as well shortsighted to catch hidden designs. Strategies like regularization, cross approval, and model intricacy change are utilized to resolve these issues [41].

Safety and Ethical Concerns

- **Robustness**: Ensuring AI models are robust involves protecting them from adversarial attacks and unexpected situations. Techniques such as adversarial training and robustness testing are employed to enhance model reliability.
- **Ethical considerations**: Ethical concerns include ensuring fairness in AI decision-making, preventing biases, and safeguarding user privacy. It is important to address these concerns to ensure that AI technologies are used responsibly and equitably.

16.22 FUTURE DIRECTIONS

Emerging Technologies

- **Advancements**: Explore emerging AI technologies such as federated learning, which enables decentralized model training across multiple devices, and their potential impact on robotics and automation. These advancements promise to enhance the scalability and privacy of AI solutions.

Research Opportunities

- **New methodologies**: Identify areas for future research, including improvements in model interpretability, the integration of AI with advanced robotic sensors, and the development of novel algorithms for complex tasks. Research in these areas could lead to significant advancements in robotics and automation.

16.23 EFFICIENCY AND PRODUCTIVITY

16.23.1 Manufacturing Automation

Industrial robots: AI-powered robots in manufacturing enhance efficiency by automating repetitive tasks such as assembly, welding, and painting. These robots can work continuously without fatigue, leading to increased production rates and reduced labor costs.

Predictive maintenance: AI systems predict equipment failures before they occur by analyzing data from sensors. This approach minimizes downtime and maintenance costs, ensuring smooth operations.

Warehouse management

- **Autonomous vehicles**: AI-driven autonomous forklifts and drones are used for inventory management and material handling in warehouses. They optimize storage, retrieval, and transportation of goods, improving warehouse efficiency and reducing errors [42].

- **Robotic sorting systems**: AI-enabled robots sort and package products with high precision, speeding up order fulfilment processes and reducing manual labor.

16.23.2 Improved Safety

Industrial safety

- **Hazard detection**: AI systems analyze sensor data to identify potential hazards in industrial environments. Robots equipped with AI can detect unsafe conditions and either alert human workers or take corrective actions to prevent accidents.
- **Collaborative robots (Cobots)**: AI-powered cobots work alongside human operators, enhancing safety by performing dangerous tasks or assisting with heavy lifting, thereby reducing the risk of injury.

Healthcare and Surgery

- **Surgical robots**: AI-assisted surgical robots provide high precision and control during operations, leading to minimally invasive procedures with quicker recovery times. They assist surgeons by providing real-time data and enhancing surgical accuracy.
- **Robotic caregivers**: In elderly care and rehabilitation, AI-driven robots assist with daily activities, providing companionship and support while monitoring the health and safety of patients.

16.23.3 Enhanced Capabilities

Autonomous Vehicles

- **Self-driving vehicles:** simulated intelligence calculations process information from cameras, LIDAR, and radar to empower independent vehicles to explore securely, perceive obstructions, and make driving choices. This innovation vows to lessen car crashes and get to the next level transportation proficiency.
- **Drones**: AI-powered drones perform tasks such as aerial surveying, delivery services, and search-and-rescue operations. They can navigate complex environments and adapt to changing conditions autonomously [43].

Service Robots

- **Retail and hospitality**: AI robots in retail assist customers with product recommendations, checkout processes, and inventory management. In hospitality, robots provide concierge services, room service, and guest interactions.
- **Cleaning robots**: AI-enabled robotic vacuums and floor cleaners navigate homes and commercial spaces autonomously, optimising cleaning routes and adapting to different surfaces.

16.23.4 Customization and Personalization

Consumer Electronics

- **Smart home devices**: AI integrates with home automation systems to control lighting, temperature, and security systems based on user preferences and behaviors. For example, smart thermostats learn user schedules to optimize energy usage.
- **Personal assistants**: AI-powered personal assistants (e.g., Amazon Alexa, Google Assistant) provide voice-controlled interactions for managing tasks, accessing information, and controlling smart devices.

Manufacturing

- **Custom production**: AI allows for flexible manufacturing systems that can produce customized products on demand. This capability enables mass customization, where items are custom-made to individual client details without forfeiting effectiveness.

16.23.5 Cost Reduction

Operational Costs

- **Labor costs**: By automating repetitive and manual tasks, AI decreases the requirement for human work, prompting lower work costs and expanded functional effectiveness.
- Energy utilization: AI advances energy use in modern cycles, structures, also, transportation frameworks, bringing about tremendous expense investment funds and decreased natural effect.

Supply Chain Management

- **Demand forecasting**: AI analyses historical data and market trends to predict future demand, improving inventory management and reducing overstocking or stockouts.
- **Logistics optimization**: AI optimizes routing and scheduling for transportation and delivery, minimizing delays and reducing fuel consumption.

16.23.6 Challenges and Considerations

Implementation Challenges

- **Integration with existing systems**: Integrating AI with legacy systems can be complex and costly. Companies need to ensure compatibility and address any technical issues that arise during integration.
- **Data privacy and security**: AI systems handle large amounts of data, raising concerns about data privacy and security. Implementing robust security measures and complying with regulations is essential to protect sensitive information [44].

Ethical and Social Implications

- **Job displacement**: Automation can lead to job displacement for workers in certain sectors. Addressing this challenge requires strategies for workforce reskilling and creating new job opportunities.
- **Bias and fairness**: AI systems may unintentionally perpetuate biases present in training data. Ensuring fairness and transparency in AI decision-making processes is crucial to avoid discrimination [45].

16.23.7 Future Directions

Continued Advancements

- **Human-made intelligence and Mechanical technology Union:** The incorporation of AI with cutting edge advanced mechanics will lead to additional clever and versatile robots, fit for playing out a more extensive territory of errands and communicating all the more normally with people [46].
- **Moral Human-made intelligence Advancement:** Continuous innovative work endeavors will zero in on creating ethical AI systems that priorities fairness, transparency, and accountability, addressing societal concerns and ensuring responsible deployment.

Innovation in Applications

- **AI-driven innovations**: Future developments in AI and robotics will likely lead to innovative applications in fields such as agriculture, space exploration, and urban planning, transforming how industries operate and improving quality of life.

16.24 CONCLUSION

The mix of human-made reasoning (computer-based intelligence) into mechanical technology and mechanization addresses a huge achievement in mechanical headway, driving profound changes across multiple industries. This chapter has explored the multifaceted applications, methodologies, and practical implications of AI in enhancing robotic systems and automating processes.

16.24.1 Key Findings

1. **Enhanced efficiency and productivity**: AI has revolutionized manufacturing and warehouse management by enabling robotics to perform repetitive and labour-intensive tasks with unmatched precision and speed. Predictive maintenance powered by AI further minimizes downtime, optimizing operational efficiency and reducing costs [47].
2. **Improved safety**: The deployment of AI in industrial and healthcare environments has led to significant improvements in safety. AI-driven hazard detection systems and collaborative robots (cobots) enhance workplace safety by preventing accidents and assisting human workers in hazardous tasks. In healthcare, AI-assisted surgical robots provide precision that reduces recovery times and enhances patient outcomes.

3. **Advanced capabilities**: Autonomous vehicles and drones equipped with AI exhibit remarkable capabilities in navigation, obstacle detection, and autonomous decision-making. Service robots, empowered by AI, offer personalized interactions in retail and hospitality, while cleaning robots autonomously maintain cleanliness with minimal human intervention.
4. **Customization and personalization**: AI enables mass customization in manufacturing and personalized user experiences in consumer electronics. This capability allows for the production of tailored products and services that meet individual preferences without sacrificing efficiency.
5. **Cost reduction**: Automation driven by AI contributes to cost savings by reducing labor expenses, optimizing energy consumption, and improving supply chain management. The ability to forecast demand accurately and optimize logistics further enhances financial efficiency [48].

16.24.2 Challenges and Considerations

Despite the numerous benefits, the integration of AI into robotics and automation presents several challenges. Implementing AI systems requires careful consideration of integration with existing infrastructure, addressing data privacy and security concerns, and managing the ethical implications of job displacement and algorithmic bias. These challenges necessitate a balanced approach, combining technological innovation with ethical and practical considerations.

16.24.3 Future Directions

Looking ahead, the convergence of AI and robotics will continue to drive innovation, leading to more intelligent, adaptable, and versatile robotic systems. Future advancements are expected to enhance the capabilities of AI-driven robots, expand their applications into new fields such as agriculture and space exploration, and address societal challenges through responsible and ethical AI development [49].

16.24.4 Final Thoughts

The impact of AI on robotics and automation is transformative, reshaping industries and improving quality of life. As technology continues to evolve, ongoing research, development, and ethical considerations will be essential to harnessing the full potential of AI while ensuring that its benefits are broadly and equitably distributed [50].

The journey of integrating AI into robotics and automation is ongoing, and its future holds exciting possibilities for innovation and advancement. Embracing these changes while addressing the associated challenges will be key to achieving sustainable and beneficial outcomes for society.

REFERENCES

[1] Albrecht, S. (2020). *Robotic Systems and AI: Integration Techniques*. Springer.

[2] Bogue, R. (2021). *Industrial Robots: Applications and Future Trends*. Emerald Group Publishing.

[3] Bruno, R., & Kozlowski, J. (2019). *Artificial Intelligence in Robotics: Applications and Challenges*. CRC Press.

[4] Cacace, J., & Rea, J. (2020). *Machine Learning and Artificial Intelligence for Industrial Robotics*. Wiley.

[5] Chen, L., & Wu, Y. (2019). *AI and Machine Learning in Robotics*. Springer.

[6] Dautenhahn, K. (2020). *Socially Intelligent Robots*. Springer.

[7] Dellaert, F., & Khatib, O. (2021). *Robotic Perception and AI: Theory and Practice*. Wiley.

[8] Faruk, K., & Martinez, L. (2019). *AI Applications in Healthcare Robotics*. Elsevier.

[9] Hsiao, K. L., & Lee, J. H. (2020). *Robotic Control Systems and AI Algorithms*. Springer.

[10] Hutter, M., & Schulman, J. (2020). *Advanced Topics in Robotic Systems and AI*. Cambridge University Press.

[11] Jansen, M., & Williams, M. (2021). *Machine Learning for Robotics: Techniques and Applications*. MIT Press.

[12] Kim, S. J., & Nguyen, T. (2019). *AI in Autonomous Vehicles and Drones*. CRC Press.

[13] Klotz, J., & Lang, M. (2020). *AI and Automation in Smart Manufacturing*. Springer.

[14] Kormushev, P., & Bhatia, S. (2020). *Robot Learning and AI Techniques*. Wiley.

[15] Kumar, V., & Bhardwaj, S. (2021). *Robotic Systems and AI Integration: A Comprehensive Guide*. Springer.

[16] Le, L., & Rao, R. (2019). *AI and Robotics in Agriculture*. Springer.

[17] Li, X., & Zhao, Y. (2021). *Artificial Intelligence in Robotics: Concepts and Applications*. Wiley.

[18] Liu, Z., & Huang, X. (2020). *AI for Robotics: Theory and Applications*. CRC Press.

[19] Majumder, S., & Chakraborty, S. (2020). *AI-Based Robotics for Automation and Industry*. Springer.

[20] Mataric, M. J. (2021). *Robotics and AI: An Overview of Current Technologies*. MIT Press.

[21] Mitchell, T. M. (2020). *Machine Learning for Robotics and Automation*. Springer.

[22] Murthy, V. M., & Soni, S. (2021). *AI in Robotics: New Trends and Applications*. Elsevier.

[23] Nam, K., & Jo, S. (2019). *Artificial Intelligence for Collaborative Robots*. Wiley.

[24] Noh, H., & Lee, S. (2020). *AI-Driven Robotics: Design and Development*. Springer.

[25] Park, J., & Jeong, S. (2021). *Machine Learning Techniques in Robotic Systems*. CRC Press.

[26] Patel, H., & Yoon, S. (2020). *AI Applications in Industrial Automation*. Elsevier.

[27] Pinto, A., & Martin, D. (2019). *AI and Robotics: Innovations and Challenges*. Wiley.

[28] Roberts, J., & Ghosh, A. (2021). *Robotic Control and AI Algorithms: A Practical Approach*. Springer.

[29] Rodrigues, D., & Silva, E. (2020). *AI for Robotics: Applications in Smart Systems*. MIT Press.

[30] Rusu, A. A., & McCarthy, J. (2019). *AI and Robotics: A Comprehensive Overview*. Cambridge University Press.

[31] Salehian, M., & Li, J. (2020). *Autonomous Robots: AI and Machine Learning Approaches*. Springer.

[32] Santini, A., & Wong, T. (2021). *Deep Learning in Robotics and Automation*. CRC Press.

[33] Schaal, S., & Peters, J. (2020). *Learning from Demonstration and Robotics*. Springer.

[34] Sharma, A., & Kumar, A. (2021). *Robotics and AI: Integrating Learning Techniques*. Elsevier.

[35] Shishika, P., & Ahmed, F. (2019). *AI-Based Robotics for Smart Automation*. Wiley.

[36] Smith, R., & Zhao, Y. (2020). *Machine Learning Algorithms for Robotics*. Springer.
[37] Sundararajan, V., & Kumar, N. (2021). *AI in Robotic Systems: Current Developments and Future Prospects*. MIT Press.
[38] Bhatt, B., Sharma, H., Arora, K., Joshi, G. P., & Shrestha, B. (2023). "Levy flight-based improved grey wolf optimization: A solution for various engineering problems", *MDPI Journal of Mathematics*, 11(7), 1745, https://doi.org/10.3390/math11071745
[39] Thomas, J., & Singh, P. (2021). *Advanced Robotics and AI Integration*. Springer.
[40] Tomar, R., & Kumar, P. (2020). *AI Techniques for Robotics: Theory and Practice*. Wiley.
[41] Wang, Y., & Lee, K. (2019). *Robotics and AI: Applications in Manufacturing and Beyond*. Elsevier.
[42] White, J., & Lee, M. (2020). *AI Algorithms in Robotics and Automation*. Springer.
[43] Wu, C., & Wu, Y. (2021). *Artificial Intelligence in Modern Robotics*. CRC Press.
[44] Yang, J., & Liu, Z. (2020). *AI-Based Robotics: Techniques and Applications*. Springer.
[45] Yoon, S., & Patel, H. (2021). *AI for Advanced Robotic Systems*. MIT Press.
[46] Zhang, L., & Zhang, Q. (2020). *AI in Robotics and Automation: Methods and Applications*. Wiley.
[47] Zhao, J., & Zhang, X. (2021). *Machine Learning and AI in Robotic Systems*. Elsevier.
[48] Zheng, H., & Li, Y. (2020). *AI-Powered Robotics: Trends and Innovations*. Springer.
[49] Zhou, S., & Wang, R. (2021). *AI Applications in Robotics: A Comprehensive Review*. CRC Press.
[50] Zhu, J., & Zhang, Y. (2020). *Robotics and Automation: AI Integration and Applications*. Wiley.

17 Introduction to AI and ML in Engineering

Chandan Kumar Sharma

17.1 INTRODUCTION

Lately, human-made reasoning (artificial intelligence, AI) and AI machine learning (ML) have arisen as crucial advances, changing different businesses, including designing. Simulated intelligence, comprehensively characterized as the reproduction of human knowledge in machines, envelops a scope of advancements intended to perform errands that normally require human mental capabilities.

AI, a subset of computer-based intelligence, includes calculations and factual models that empower frameworks to gain and pursue choices from information without unequivocal programming [3].

The joining of human-made brainpower and ML into planning has opened new streets for development and effectiveness. Designing disciplines, customarily dependent on manual estimations and heuristic techniques, are progressively utilizing these advances to upgrade configuration, enhance processes, and further develop direction. Artificial intelligence and ML are not just changing how engineers approach critical thinking yet additionally rethinking the limits of what is conceivable in designing practice.

All things considered, designing arrangements were created utilizing a blend of experimental information and hypothetical models. In any case, the approach of artificial intelligence and ML has presented progressed capacities for example, prescient examination, continuous information handling, and independent direction. These advances empower specialists to deal with intricate and voluminous informational collections with extraordinary speed and exactness, giving bits of knowledge and arrangements that were beforehand unreachable [4]. The extent of artificial intelligence and ML in designing is immense, crossing various applications including underlying model, producing processes, energy the board, and framework diagnostics. For case, simulated intelligence-driven plan improvement instruments can fundamentally diminish the time expected to foster new items, while ML calculations can upgrade prescient support systems, limiting personal time and broadening hardware life expectancy [5].

Notwithstanding the significant advantages, the coordination of human-made intelligence and ML into designing practices too presents difficulties. Issues connected with information quality, model interpretability, and framework joining should

 DOI: 10.1201/9781003563440-17

be addressed to outfit the capability of these advancements completely. Also, the moral and cultural ramifications of conveying human-made intelligence and ML in designing absolute requirement be painstakingly considered to guarantee capable and impartial use. This part expects to give an exhaustive outline of the job of simulated intelligence and ML in designing, zeroing in on their pragmatic ramifications. It will investigate key applications, feature the advantages and difficulties related to these advancements, and examine future research headings. By looking at genuine models and recognizing basic exploration holes, this section looks to offer important experiences into how computer-based intelligence and ML are forming the future of designing.

17.1.1 Overview of AI and ML

Human-made reasoning (simulated intelligence) and AI (ML) have altered different designing disciplines, offering noteworthy answers for complex issues. Human-made intelligence, characterized as the capacity of a machine to mimic canny human way of behaving, and ML, a subset of human-made intelligence including preparing calculations to gain from and go with choices in light of information, have their establishes during the twentieth hundred years. Key achievements incorporate the advancement of the primary human-made intelligence program, the approach of brain organizations, and the new flood in profound learning methods. These progressions highlight the vital job artificial intelligence and ML play in current designing, driving development and improving effectiveness [6].

Human-made intelligence and ML are based on center standards like calculations, information, and computational power. Calculations are bit by bit systems for computations, information handling, and robot overthinking. In ML, calculations gain designs from information, permitting them to make forecasts or choices without being unequivocally modified. Information, the fuel for human-made intelligence and ML, can be organized or unstructured and comes from different sources like sensors, reenactments, and verifiable records [7]. Computational power, upgraded by progressions in equipment like GPUs, empowers the handling of huge datasets and complex models.

17.1.2 Historical Background

The historical backdrop of artificial intelligence traces all the way back to the beginning of figuring. In 1956, John McCarthy begat the expression "Man-made brainpower" at the Dartmouth Gathering, denoting the authority birth of computer-based intelligence as a field of study. Early simulated intelligence research zeroed in on representative thinking and critical thinking, with remarkable accomplishments like the Overall Issue Solver and the Rationale Scholar. The advancement of brain networks during the 1960s and 1970s, roused by the human mind, laid the preparation for current ML. In any case, it was only after the approach of strong registering assets and huge datasets in the 21st century that computer-based intelligence and ML genuinely started to prosper.

17.1.3 Definitions and Terminology

To fully understand the importance of AI and ML, it is essential to get acquainted with some key terms and concepts. AI refers to a variety of technologies that enable machines to execute tasks typically associated with human intelligence, including perception, reasoning, learning, and making decisions. A subset of AI and ML is dedicated to creating algorithms that allow computers to learn from data and enhance their performance autonomously over time [8]. Deep learning (DL), a more sophisticated branch of ML, utilizes neural networks with multiple layers to identify intricate patterns within large datasets.

17.1.4 Significance in Engineering

The integration of AI and ML into engineering has led to significant advancements across various fields. In civil engineering, AI is used for structural health monitoring and predictive maintenance. Mechanical engineering benefits from ML algorithms in designing and optimizing complex systems. Electrical engineering utilizes AI for smart grid management and fault detection. Computer engineering sees extensive use of AI in software development and cybersecurity [9]. The benefits of AI and ML include improved accuracy, reduced human error, and the ability to process vast amounts of data quickly, leading to more informed decision-making.

17.1.5 Applications in Civil Engineering

In civil engineering, AI and ML technologies are used to monitor the health of infrastructure such as bridges, tunnels, and buildings. ML algorithms can analyze data from sensors to detect signs of wear and tear, predict potential failures, and recommend maintenance actions. This proactive approach helps extend the lifespan of infrastructure and ensures public safety [10]. Additionally, AI is used in urban planning to optimize traffic flow, reduce congestion, and improve public transportation systems.

17.1.6 Applications in Mechanical Engineering

Mechanical specialists influence human-made intelligence and ML to improve the plan and enhancement of mechanical frameworks. For instance, ML calculations can break down verifiable information to anticipate the execution of new motor plans, decreasing the requirement for exorbitant and tedious actual models. Human-made intelligence-driven reproductions empower designers to investigate a more extensive scope of plan choices and recognize the most productive arrangements [11]. In assembling, prescient support controlled by computer-based intelligence limits personal time and further develop efficiency by expecting gear disappointments before they happen.

17.1.7 Applications in Electrical Engineering

Electrical engineering has seen significant advancements with the adoption of AI and ML. Smart grid technology, which uses AI to manage electricity distribution, improves the efficiency and reliability of power systems. ML algorithms can detect and diagnose faults in electrical networks, enabling rapid response and minimizing disruptions. AI is also used in renewable energy systems to optimize the integration of solar and wind power into the grid, enhancing the overall stability and sustainability of energy supply.

17.1.8 Applications in Computer Engineering

In computer engineering, AI and ML are driving innovation in software development, cybersecurity, and user experience. AI-powered code generation tools assist developers in writing more efficient and error-free code. ML algorithms detect and respond to cybersecurity threats by analyzing patterns of network activity and identifying anomalies. AI-driven interfaces, such as chatbots and virtual assistants, enhance user interactions by providing personalized and context-aware responses [12–14].

17.2 OBJECTIVES OF THE CHAPTER

This chapter is designed to familiarize readers with the essential concepts of AI and ML, particularly in the field of engineering. It offers a thorough overview of these technologies, emphasizing their current applications and potential future developments [15]. Additionally, the chapter describes different research methodologies utilized in AI and ML within engineering, providing insights into the tools and techniques that researchers employ. Ultimately, it aims to give readers a solid understanding of how AI and ML are influencing the engineering sector.

17.3 NATURE OF AI AND ML IN ENGINEERING

17.3.1 Fundamental Concepts

AI and ML are built on core principles such as algorithms, data, and computational power. Algorithms are step-by-step procedures for calculations, data processing, and automated reasoning. In ML, algorithms identify patterns in data, enabling them to make predictions or decisions without direct programming [16]. The data that powers AI and ML can be either structured or unstructured and is sourced from diverse origins, including sensors, simulations, and historical databases. Improvements in computational power, driven by advances in hardware like GPUs, facilitate the analysis of large datasets and intricate models.

17.3.2 Algorithms in AI and ML

Algorithms serve as the backbone of AI and ML systems. In AI, they simulate intelligent behaviors, enabling capabilities like natural language processing, image recognition, and autonomous decision-making. In contrast, ML algorithms focus on detecting patterns in data to make predictions or informed decisions based on those patterns. Some widely used ML algorithms are linear regression, decision trees, support vector machines, and neural networks. Each of these algorithms comes with unique advantages and limitations, which makes them better suited for particular tasks.

17.3.3 Data in AI and ML

Data are a critical component of AI and ML. The quality and quantity of data significantly impact the performance of AI and ML models. Structured data, such as spreadsheets and databases, is organized in a tabular format with defined columns and rows. Unstructured data, such as text, images, and videos, lacks a predefined structure and requires more sophisticated techniques to analyses [17]. Data preprocessing, including cleaning, normalization, and feature extraction, is essential to prepare raw data for training AI and ML models.

17.3.4 Computational Power

Advancements in computational power have been instrumental in the success of AI and ML. The development of graphics processing units (GPUs) and tensor processing units (TPUs) has enabled the efficient processing of large datasets and complex models. Distributed computing, which involves using multiple computers to work on a problem simultaneously, further enhances the ability to train and deploy AI and ML models at scale. Cloud computing platforms provide scalable and flexible resources, allowing researchers and engineers to access powerful computational capabilities without significant upfront investments.

17.4 TYPES OF AI AND ML

AI is primarily classified into two types: narrow AI, which is tailored for specific tasks, and general AI, which aims to handle any cognitive function that a human can perform [18]. Within the realm of AI, ML is further divided into three main categories: supervised learning, where models are trained on labeled data; unsupervised learning, which uncovers hidden patterns in unlabeled data; and reinforcement learning, where an agent learns to make decisions based on rewards and penalties. Each of these categories offers distinct applications and benefits in the engineering domain.

17.4.1 Narrow AI Versus General AI

Narrow AI, also known as weak AI, is designed to perform specific tasks, such as image recognition, language translation, or playing a game. These systems are highly specialized and excel in their designated areas but lack general intelligence. General AI, or strong AI, aims to possess the cognitive abilities of a human, enabling it to perform any intellectual task. While narrow AI is widely used in engineering today, general AI remains a long-term goal, with significant research required to achieve it.

17.4.2 Supervised Learning

Regulated learning is a sort of ML where the model is prepared on marked information, meaning each preparing model is matched with a result mark. The objective is to gain a planning from contributions to yields that can be utilized to foresee the names of new, inconspicuous information. Normal calculations utilized in regulated learning incorporate straight relapse, strategic relapse, choice trees, and backing vector machines. Applications in designing incorporate prescient support, quality control, and request anticipating.

17.4.3 Unsupervised Learning

This approach involves training models on data that lacks labeled outputs, aiming to uncover hidden patterns or structures within the dataset. Clustering algorithms, such as k-means and hierarchical clustering, are used to group similar data points, while dimensionality reduction techniques like principal component analysis (PCA) and t-distributed stochastic neighbor embedding (t-SNE) simplify the data by reducing the number of variables [19]. These methods are commonly applied in areas like anomaly detection, customer segmentation, and exploratory data analysis.

17.4.4 Reinforcement Learning

Reinforcement learning is a form of ML in which an agent learns to make decisions through interactions with an environment, receiving feedback as rewards or penalties. The objective of the agent is to maximize the total reward over time by developing an optimal policy. Important concepts in reinforcement learning include states, actions, rewards, and policies. In engineering, it is applied in areas such as autonomous vehicles, robotic control, and gaming [20]. Algorithms like Q-learning and deep Q-networks (DQN) have shown significant success in tackling complex tasks.

17.5 APPLICATIONS IN ENGINEERING

AI and ML applications in engineering are vast and varied. In civil engineering, AI algorithms analyze data from sensors to monitor the structural health of bridges and buildings. Mechanical engineers use ML models to optimize the design and performance of engines and machinery. Electrical engineers apply AI in managing smart grids, predicting power outages, and enhancing energy efficiency [21]. In computer engineering, AI-driven tools improve software development, detect cybersecurity threats, and enhance user experiences. Case studies, such as AI-driven predictive maintenance in manufacturing or ML-based traffic management systems, illustrate the transformative impact of these technologies.

17.5.1 Structural Health Monitoring

Structural health monitoring involves using AI and ML algorithms to analyze data from sensors placed on infrastructure such as bridges, buildings, and dams. These algorithms can detect anomalies and predict potential failures, enabling timely maintenance and repairs. By continuously monitoring the condition of structures, engineers can ensure their safety and longevity, preventing catastrophic failures and reducing maintenance costs.

17.5.2 Predictive Maintenance

Predictive maintenance leverages AI and ML to predict when equipment is likely to fail, allowing for proactive maintenance and reducing downtime. By analyzing historical data and identifying patterns, ML models can predict the remaining useful life of machinery and recommend maintenance actions [22]. This approach is widely used in manufacturing, transportation, and energy sectors, where equipment reliability is critical for operational efficiency and cost savings.

17.5.3 Smart Grid Management

Smart grid technology uses AI to optimize the generation, distribution, and consumption of electricity. ML algorithms analyze data from sensors and meters to balance supply and demand, detect and respond to faults, and enhance energy efficiency. AI-driven demand response systems can adjust energy consumption based on real-time conditions, reducing peak loads and stabilizing the grid. Smart grids also integrate renewable energy sources, such as solar and wind, improving the sustainability of energy systems.

17.5.4 Cybersecurity

In cybersecurity, AI and ML are used to detect and respond to threats by analyzing patterns of network activity and identifying anomalies. ML algorithms can detect malware, phishing attacks, and other cyber threats by analyzing large volumes of data in real time. AI-driven security systems can adapt to evolving threats, providing robust protection against cyberattacks [23]. In addition, AI is used to enhance user authentication and access control, improving the overall security of digital systems.

17.5.5 Autonomous systems

Autonomous systems, including self-driving cars and drones, depend on AI and ML for navigation, perception, and decision-making. These systems utilize sensors to gather data about their surroundings and apply ML algorithms to analyze this data and make informed choices. For instance, autonomous vehicles employ AI to identify obstacles, plan routes, and manage vehicle dynamics [24]. Similarly, drones leverage AI for various tasks, such as aerial surveying, package delivery, and disaster response, showcasing the vast potential of AI and ML in enhancing autonomous operations.

17.6 CURRENT TRENDS

The field of computer-based intelligence and ML is quickly developing, with a few patterns forming their application in designing. Progresses in profound learning, a subset of ML, have prompted critical enhancements in picture and discourse acknowledgment. The ascent of edge processing permits simulated intelligence models to run on nearby gadgets, lessening idleness and further developing reaction times. Another pattern is the incorporation of simulated intelligence with the Web of Things (IoT), empowering continuous information examination and dynamic in shrewd urban communities, independent vehicles, and modern computerization.

17.6.1 Deep Learning

Profound learning, a subset of AI, includes brain networks with numerous layers that can learn complex examples in enormous datasets. Profound learning has accomplished striking progress in picture and discourse acknowledgment, normal language handling, and game playing. Convolutional brain organizations (CNNs) are utilized for picture acknowledgment, while intermittent brain organizations (RNNs) and long transient memory (LSTM) networks succeed in succession-based undertakings [25]. Profound learning models require critical computational power and a lot of information, making them reasonable for applications where these assets are accessible.

17.6.2 Edge Computing

Edge computing refers to the practice of processing data near its source, such as on local devices or edge servers, instead of relying on centralized data centers. This method minimizes latency and reduces bandwidth consumption, facilitating real-time data analysis and decision-making. In engineering, edge computing is applied in areas like predictive maintenance, autonomous vehicles, and smart cities, where quick response times are essential. By deploying AI models on edge devices, engineers can develop faster and more efficient solutions [26].

17.6.3 AI and IoT Integration

The coordination of human-made intelligence with the Web of Things (IoT) empowers ongoing information investigation and dynamic in different applications. IoT gadgets, like sensors and actuators, gather information from the actual world and communicate it to human-made intelligence frameworks for examination [27]. In shrewd urban areas, human-made intelligence driven IoT frameworks upgrade traffic stream, oversee energy utilization, and improve public security. In modern robotization, IoT gadgets screen gear execution and natural circumstances, while computer-based intelligence calculations foresee disappointments and streamline processes.

17.7 FUTURE PROSPECTS

The future of AI and ML in engineering is promising, with potential innovations on the horizon. Autonomous systems, such as self-driving cars and drones, are expected to become more prevalent, relying heavily on AI and ML for navigation and decision-making. AI-powered robotics will revolutionize manufacturing, construction, and healthcare, enhancing precision and efficiency. The development of explainable AI, which aims to make AI decision-making processes transparent, will address trust and ethical concerns, facilitating wider adoption in critical engineering applications [28].

17.7.1 Autonomous Vehicles

Autonomous vehicles represent a thrilling development in the application of AI and ML within engineering. These self-driving cars utilize AI to analyze sensor data, make decisions, and manage vehicle dynamics. ML algorithms allow these vehicles to learn from their experiences and enhance their performance over time [29]. The broader implementation of autonomous vehicles could lead to a decrease in traffic accidents, enhanced mobility, and a transformation of transportation systems.

17.7.2 AI-Powered Robotics

AI-powered robotics is set to revolutionize various industries, including manufacturing, construction, and healthcare. In manufacturing, robots equipped with AI can perform complex tasks with high precision and efficiency, enhancing productivity and reducing labor costs. In construction, AI-driven robots can automate tasks such as bricklaying, welding, painting, and improving safety and quality. In healthcare, AI-powered robots assist in surgeries, rehabilitation, and patient care, providing better outcomes and personalized treatments.

17.7.3 Explainable AI

Explainable AI (XAI) aims to make AI decision-making processes transparent and understandable to humans. Traditional AI models, such as DL networks, often operate as "black boxes," making it difficult to interpret their decisions. Explainable AI seeks to address this issue by developing methods to explain how AI models arrive at their conclusions [30]. This transparency is crucial for building trust in AI systems, particularly in critical applications such as healthcare, finance, and autonomous vehicles.

17.8 CHALLENGES AND LIMITATIONS

Despite their potential, AI and ML face several challenges and limitations in engineering. Technical challenges include the need for large amounts of high-quality data, computational resources, and the complexity of developing robust models. Ethical concerns, such as bias in AI algorithms, data privacy, and the impact on employment, pose significant societal challenges [31]. Additionally, the black-box nature of some AI models makes it difficult to understand their decision-making processes, leading to issues with accountability and trust.

17.8.1 Data Quality and Availability

The effectiveness of AI and ML models heavily relies on the quality and accessibility of data. For models to be accurate and dependable, high-quality data is crucial. Nevertheless, acquiring this type of data can be difficult, especially in fields where data may be limited, costly, or sensitive. Data preprocessing techniques such as cleaning, normalization, and augmentation play a vital role in preparing data for model training necessary to prepare raw data for model training. Ensuring data privacy and security is also critical, as the use of personal or sensitive data raises ethical and legal concerns.

17.8.2 Computational Resources

Training and deploying AI and ML models require significant computational resources, including powerful hardware and efficient software frameworks. The availability of GPUs and TPUs has accelerated the development of complex models, but these resources are expensive and may not be accessible to all researchers and practitioners. Cloud computing platforms offer scalable and flexible resources, but the cost and complexity of managing cloud infrastructure can be prohibitive for some users [32].

17.8.3 Model Interpretability

The interpretability of AI and ML models is a critical challenge, particularly for DL models, which often operate as black boxes. Understanding how a model arrives at its decisions is essential for building trust, ensuring accountability, and addressing ethical concerns. Researchers are developing techniques for explainable AI, such as feature importance analysis, visualization methods, and rule-based systems, to make AI models more transparent and interpretable [33]. However, achieving a balance between model complexity and interpretability remains a challenge.

17.8.4 Ethical and Societal Implications

The widespread adoption of AI and ML raises several ethical and societal concerns, including bias, privacy, and the impact on employment. Bias in AI algorithms can result from biased training data or discriminatory practices, leading to unfair and potentially harmful outcomes. Ensuring data privacy and security is critical, as the use of personal or sensitive data can have significant ethical and legal implications. The automation of tasks traditionally performed by humans raises concerns about job displacement and the need for reskilling and education.

17.9 RESEARCH METHODOLOGY

17.9.1 Data Collection

Data collection is the first step in any AI and ML project. Depending on the application, data can be collected from various sources such as sensors, simulations, historical records, or publicly available datasets [34]. Ensuring the quality and relevance of data is crucial, as it directly impacts the performance of AI models. Data preprocessing, including cleaning, normalization, and feature extraction, are essential to prepare the data for analysis.

17.9.2 Sources of Data

Data for AI and ML research can come from various sources, including sensors, simulations, historical records, and publicly available datasets. Sensors, such as accelerometers, cameras, and temperature sensors, collect real-time data from the physical world. Simulations generate data by modelling complex systems, such as climate models or engineering simulations. Historical records, such as maintenance logs or financial transactions, provide valuable information for training predictive models [35]. Publicly available datasets, such as the UCI ML Repository or Kaggle, offer diverse and well-documented datasets for research.

17.9.3 Data Preprocessing

Data preprocessing is a critical step in preparing raw data for analysis. It involves cleaning the data to remove noise and errors, normalizing the data to ensure consistency, and extracting relevant features to reduce dimensionality. Cleaning the data may involve removing duplicates, filling in missing values, and correcting errors. Normalization ensures that the data are on a consistent scale, making it easier to compare and analyze. Feature extraction identifies the most relevant attributes of the data, reducing complexity and improving model performance.

17.9.4 Data Augmentation

Data augmentation techniques, such as oversampling, undersampling, and synthetic data generation, can be used to address issues of data imbalance and scarcity. Oversampling involves duplicating minority class samples to balance the dataset, while undersampling reduces the number of majority class samples. Synthetic data generation creates new data samples by combining or modifying existing data. These techniques help improve model performance by providing more diverse and representative training data.

17.10 MODEL DEVELOPMENT

Creating AI and ML models requires choosing suitable algorithms, training the models on gathered data, and assessing their performance [36]. Frequently used algorithms include linear regression, decision trees, neural networks, and support vector machines. The training process consists of supplying the model with data and fine-tuning its parameters to reduce errors. Performance is evaluated using metrics such as accuracy, precision, recall, and F1-score

17.10.1 Algorithm Selection

Choosing the right algorithm is an essential part of model development. The selection is influenced by the problem type, the nature of the data, and the intended results. Linear regression is ideal for predicting continuous values, whereas logistic regression is designed for binary classification. Decision trees and random forests are versatile, performing well in both regression and classification tasks, while providing interpretability and strength. Neural networks, especially DL models, are particularly effective for high-dimensional data tasks, such as image and speech recognition.

17.10.2 Model Training

Model training involves feeding the algorithm with training data and adjusting its parameters to minimize error. This process requires splitting the data into training and validation sets, with the training set used to fit the model and the validation set used to evaluate its performance. Training algorithms, such as gradient descent, iteratively adjust the model's parameters to reduce the difference between predicted and actual values [37]. Regularization techniques, such as L1 and L2 regularization, prevent overfitting by penalizing complex models.

17.10.3 Model Evaluation

Model evaluation involves measuring a trained model's performance through several metrics, such as accuracy, precision, recall, and F1-score. Accuracy indicates the model's overall correctness, while precision and recall provide insights into its performance concerning specific classes. The F1-score merges precision and recall into one metric, giving a more comprehensive assessment of the model's effectiveness. Furthermore, cross-validation entails splitting the data into multiple subsets and training the model on each one, which improves the robustness and generalizability of the results.

17.11 TOOLS AND FRAMEWORKS

Several tools and frameworks are available for AI and ML research, including TensorFlow, PyTorch, scikit-learn, and Kera's. These tools provide prebuilt functions for data processing, model training, and evaluation, simplifying the development process. Additionally, platforms like Jupiter Notebook offer interactive environments for coding and visualization, enhancing the research workflow.

17.11.1 TensorFlow

TensorFlow, developed by Google, is an open-source framework for ML and DL. It provides a comprehensive ecosystem of tools and libraries for building and deploying AI models. TensorFlow supports a wide range of applications, from simple linear models to complex neural networks. Its flexible architecture allows for deployment on various platforms, including desktops, servers, and mobile devices [38]. TensorFlow also offers TensorFlow Lite for edge computing and TensorFlow Serving for scalable model deployment.

17.11.2 PyTorch

PyTorch, developed by Facebook's AI Research lab, is another popular open-source framework for ML and DL. PyTorch emphasizes flexibility and ease of use, making it a favorite among researchers and practitioners [39]. Its dynamic computation graph allows for real-time model modification, facilitating experimentation and debugging. PyTorch's extensive library of prebuilt models and tools simplifies the development process, enabling rapid prototyping and deployment.

17.11.3 Scikit-learn

is a popular open-source library for ML in Python. It offers straightforward and efficient tools for data mining, data analysis, and ML tasks. The library includes a variety of algorithms for both supervised and unsupervised learning, such as regression, classification, clustering, and dimensionality reduction. With its user-friendly interface and comprehensive documentation, scikit-learn is accessible to both newcomers and seasoned practitioners [40]. Additionally, it integrates seamlessly with other scientific libraries like NumPy and pandas, facilitating data manipulation and analysis.

17.11.4 Keras

Kera's is a Python-based open-source library for neural networks. It offers a high-level interface for constructing and training DL models, streamlining the development process. Kera's is compatible with various backends, such as TensorFlow, Theano, and Microsoft Cognitive Toolkit (CNTK), enabling users to select the platform that best fits their requirements. Its modular and user-friendly design facilitates experimentation with different model architectures and hyperparameters, thereby speeding up the research workflow.

17.12 CASE STUDIES

Case studies provide real-world examples of AI and ML applications in engineering. These examples illustrate the practical challenges and benefits of implementing AI and ML solutions. Case studies may include AI-driven predictive maintenance systems in manufacturing, ML-based traffic management systems in smart cities, or AI-powered robotic surgery systems in healthcare. Each case study highlights the specific techniques and methodologies used, as well as the outcomes and lessons learned.

17.12.1 Predictive Maintenance in Manufacturing

Predictive maintenance systems in manufacturing use AI and ML algorithms to predict equipment failures and recommend maintenance actions. By analyzing historical data and identifying patterns, ML models can estimate the remaining useful life of machinery and schedule maintenance before failures occur. This proactive approach reduces downtime, extends the lifespan of equipment, and improves operational efficiency. Case studies demonstrate the implementation of predictive maintenance systems in various industries, including automotive, aerospace, and energy [41].

17.12.2 Traffic Management in Smart Cities

AI-based traffic management systems in smart cities use ML algorithms to optimize traffic flow and reduce congestion. By analyzing real-time data from sensors, cameras, and GPS devices, these systems can adjust traffic signals, reroute vehicles, and provide drivers with real-time information [42]. Case studies illustrate the deployment of AI-driven traffic management systems in cities around the world, highlighting the benefits of reduced travel times, lower emissions, and improved public transportation.

17.12.3 Robotic Surgery Systems in Healthcare

AI-powered robotic surgery systems in healthcare enhance the precision and control of surgical procedures. ML algorithms analyze preoperative imaging data to plan the surgery and guide robotic instruments during the operation. These systems improve surgical outcomes by reducing invasiveness, minimizing complications, and shortening recovery times. Case studies showcase the use of AI-driven robotic surgery systems in various medical specialties, including cardiology, orthopedics, and neurosurgery, demonstrating their impact on patient care and treatment outcomes.

17.13 RESEARCH GAPS

Despite the significant advancements in AI and ML within engineering, several research gaps remain that warrant further exploration. Addressing these gaps could enhance the effectiveness and applicability of AI and ML technologies in engineering practices.

1. **Integration of AI and ML in Legacy Systems**
   Gap: Many engineering fields still rely on legacy systems that were not designed to accommodate modern AI and ML technologies. Integrating these technologies with existing infrastructure presents technical and operational challenges that are not yet fully addressed.

Potential Research Directions:

- **Development of Integration Frameworks:** Research on frameworks and methodologies for seamless integration of AI and ML into legacy systems.
- **Adaptation Strategies:** Investigate strategies for adapting AI and ML technologies to work effectively with older systems without requiring complete overhauls.

2. **Data Privacy and Security**

   Gap: The use of AI and ML in engineering often involves handling sensitive data, raising concerns about data privacy and security. Current solutions may not fully address the complexities of safeguarding data in AI-driven environments.

Potential Research Directions:

- **Privacy-Preserving Techniques:** Explore new methods for ensuring data privacy while using AI and ML, such as federated learning and differential privacy.
- **Security Protocols:** Develop robust security protocols tailored to the specific needs of AI and ML applications in engineering.

3. **Explain ability and Trustworthiness**

   Gap: AI and ML models, especially DL models, are often criticized for their lack of transparency and interpretability. This can hinder trust and acceptance in engineering applications where understanding model decisions are crucial.

Potential Research Directions:

- **Explainable AI Techniques:** Research methods for improving the explainability of complex models, including visualization tools and model-agnostic approaches.
- **Trust Metrics:** Develop metrics for assessing the trustworthiness of AI and ML models in engineering contexts.

4. **Real-Time Adaptation and Learning**

   Gap: Engineering systems often operate in dynamic environments that require real-time adaptation. Existing AI and ML models may struggle to adapt quickly to new or unexpected conditions.

Potential Research Directions:

- **Online Learning Algorithms:** Investigate online learning algorithms that can continuously adapt to new data and changing conditions in real time.
- **Adaptive Systems:** Explore frameworks for creating adaptive AI systems that can adjust their behavior based on real-time inputs and feedback.

5. **Ethical and Societal Implications**

   Gap: The deployment of AI and ML in engineering raises ethical and societal concerns, including the impact on employment, decision-making biases, and societal inequalities. These implications are not fully understood or addressed.

Potential Research Directions:

- **Ethical Frameworks:** Develop ethical frameworks and guidelines for the responsible use of AI and ML in engineering.
- **Impact Assessment:** Study the societal impacts of AI-driven engineering solutions, focusing on aspects such as job displacement and equitable access.

6. **Cross-Disciplinary Applications**
   Gap: AI and ML research often focuses on specific domains within engineering, with limited exploration of cross-disciplinary applications that integrate insights from multiple fields.

Potential Research Directions:

- **Interdisciplinary Approaches:** Research how AI and ML can be applied across different engineering disciplines to solve complex, multifaceted problems.
- **Collaborative Models:** Develop collaborative models that leverage expertise from various engineering domains to enhance AI and ML applications

17.14 PRACTICAL IMPLICATIONS

AI and ML have profound practical implications for the field of engineering, transforming various aspects of how engineering tasks are performed, optimized, and managed. This section explores these practical implications through case studies, challenges, and emerging trends.

17.14.1 Case studies

1. **Predictive Maintenance in Manufacturing:** Predictive maintenance utilizes ML algorithms to predict equipment failures before they occur. For instance, Siemens uses AI to analyze data from sensors on industrial machinery. By identifying patterns and anomalies, the system predicts potential failures and schedules maintenance proactively. This approach significantly reduces downtime and maintenance costs, enhancing overall operational efficiency.
2. **Automated Design Optimization:** In the aerospace industry, Boeing has implemented AI-driven design optimization tools. These tools use genetic algorithms to explore numerous design configurations and identify the most efficient solutions. The result is a more streamlined design process, reduced time-to-market, and improved performance of aerospace components.
3. **Energy Management Systems:** AI technologies are being used to optimize energy consumption in large industrial plants. For example, Google's DeepMind has successfully applied ML to manage energy usage in data centers. By analyzing historical data and predicting future energy needs, the system adjusts cooling systems in real time, leading to a significant reduction in energy consumption and operational costs [43–45].

17.14.2 Challenges and Solutions

1. **Data Quality and Management:** Effective AI and ML applications require high-quality, relevant data. Engineers often face challenges related to data accuracy, completeness, and consistency. Solutions include implementing robust data collection processes and employing data cleaning techniques to

ensure that the data used for training AI models is reliable and representative [46].

2. **Model Accuracy and Reliability:** Ensuring that AI models perform accurately and reliably in real-world conditions can be challenging. Engineers must validate models through rigorous testing and calibration. Techniques such as cross-validation and real-time monitoring are essential for maintaining model performance and addressing issues as they arise.
3. **Integration with Existing Systems**
   Integrating AI and ML solutions into established engineering systems can be challenging. Issues related to compatibility and the necessity for system adjustments can complicate the implementation process. To overcome these obstacles, engineers can opt for modular and scalable AI solutions that allow for gradual integration with existing systems, thereby reducing disruption and facilitating a smoother transition.

17.14.3 Future Trends

1. **Explainable AI (XAI)**
   As AI systems grow increasingly complex, the demand for transparency in their decision-making processes is also rising. Explainable AI seeks to improve the interpretability of AI models, enabling engineers to comprehend and have confidence in the decisions made by these systems [47]. This movement is anticipated to boost the adoption of AI in essential engineering applications where understanding the reasoning behind decisions is vital.
2. **Edge Computing**
   Edge computing involves processing data locally on devices rather than sending it to centralized servers. In engineering, this means AI models can be deployed on-site for real-time data analysis and decision-making. For example, in autonomous vehicles, edge computing enables immediate processing of sensor data, improving safety and performance [48].
3. **Collaborative Robotics (Cobots)**
   Collaborative robots, or cobots, are designed to work alongside human engineers, enhancing productivity and safety. AI-driven robots can perform repetitive tasks, assist in complex assembly processes, and adapt to changing conditions. This trend is likely to increase as industries seek to improve operational efficiency and leverage human-robot collaboration [49].
4. **Cross-Disciplinary Applications**
   By focusing on these areas, researchers and practitioners can develop more reliable, transparent, and ethically sound AI solutions [50].
5. In conclusion, the journey of integrating AI and ML into engineering is a continuous one, marked by both opportunities and challenges. As these technologies advance, they promise to further transform engineering practices, driving innovation and improving outcomes across various domains. Embracing these advancements while addressing associated challenges will be essential for maximizing the benefits of AI and ML in the engineering field.

17.15 CONCLUSION

AI and ML are transforming the engineering field by implementing advanced methods that improve efficiency, accuracy, and innovation. In this chapter, we have examined the important practical effects of AI and ML, emphasizing their transformative influence across different engineering sectors. AI and ML have demonstrated their potential in optimizing engineering processes through practical applications such as predictive maintenance, automated design optimization, and energy management systems. These technologies not only streamline operations but also drive substantial cost savings and operational efficiencies. Case studies illustrate how organizations are successfully implementing AI and ML to solve complex engineering problems and achieve remarkable outcomes.

However, the integration of AI and ML into engineering practices is not without challenges. Issues related to data quality, model accuracy, and the integration with existing systems need ongoing research and development. Addressing these challenges requires a multifaceted approach, including the development of robust integration frameworks, privacy-preserving techniques, and adaptive learning algorithms.

Future research is crucial for overcoming current limitations and unlocking the full potential of AI and ML in engineering. Key research gaps include enhancing the explainability of AI models, addressing ethical and societal implications, and exploring cross-disciplinary applications. By focusing on these areas, researchers and practitioners can develop more reliable, transparent, and ethically sound AI solutions.

In conclusion, the journey of integrating AI and ML into engineering is a continuous one, marked by both opportunities and challenges. As these technologies advance, they promise to further transform engineering practices, driving innovation and improving outcomes across various domains. Embracing these advancements while addressing associated challenges will be essential for maximizing the benefits of AI and ML in the engineering field.

17.16 SUGGESTION

1. **Summarize Key Insights**: Concisely restate the most impactful findings of the chapter regarding the practical implications of AI and ML in engineering. Highlight how these technologies are revolutionizing various engineering processes and areas.
2. **Address Critical Research Gaps**: Clearly outline the main research gaps identified in the chapter. Emphasize the importance of addressing these gaps for advancing the field and improving AI and ML applications in engineering.
3. **Discuss Future Directions**: Offer a forward-looking perspective on how the integration of AI and ML might evolve. Mention emerging trends and potential breakthroughs that could shape the future of engineering.
4. **Highlight Broader Implications**: Reflect on the broader societal and industrial implications of AI and ML advancements. Discuss how these technologies might influence other sectors or contribute to addressing global challenges.

5. **Encourage Further Exploration**: Invite readers to engage with ongoing research, explore new developments, and consider their own applications of AI and ML. Provide suggestions for resources, journals, or organizations where they can learn more or get involved.

REFERENCES

1. Alpaydin, E. (2020). *Introduction to Machine Learning*. Cambridge, MA: MIT Press.
2. Bishop, C. M. (2006). *Pattern Recognition and Machine Learning*. New York: Springer.
3. Goodfellow, I., Bengio, Y., & Courville, A. (2016). *Deep Learning*. Cambridge, MA: MIT Press.
4. Kelleher, J. D., Mac Carthy, R., & Tierney, B. (2015). *Data Science and Machine Learning: Mathematical and Statistical Methods*. Chichester: Wiley.
5. Mitchell, T. M. (1997). *Machine Learning*. Boston: McGraw-Hill.
6. Russell, S., & Norvig, P. (2016). *Artificial Intelligence: A Modern Approach.* : Upper Saddle River, NJ: Pearson.
7. Zhang, C., & Zhao, H. (2018). *Machine Learning: A Probabilistic Perspective*. Cambridge, MA: MIT Press.
8. Géron, A. (2019). *Hands-On Machine Learning with Scikit-Learn, Keras, and TensorFlow*. Sebastopol, CA: O'Reilly Media.
9. Jang, J.-S. R., Sun, C., & Mizutani, E. (1997). *Neuro-Fuzzy and Soft Computing: A Computational Approach to Learning and Machine Intelligence*. Upper Saddle River, NJ: Pearson.
10. Domingos, P. (2015). *The Master Algorithm: How the Quest for the Ultimate Learning Machine Will Remake Our World*. New York: Basic Books.
11. Bhatt, B., Sharma, H., Arora, K., Joshi, G. P., & Shrestha, B. (2023). Levy flight-based improved grey wolf optimization: A solution for various engineering problems. *MDPI Journal of Mathematics*, 11(7), 1745, https://doi.org/10.3390/math11071745
12. He, K., Zhang, X., Ren, S., & Sun, J. (2016). *Deep Residual Learning for Image Recognition*. Las Vegas: CVPR.
13. Silver, D., et al. (2016). Mastering the game of Go with deep neural networks and tree search. *Nature*, 529, 484–489.
14. LeCun, Y., Bengio, Y., & Hinton, G. (2015). Deep learning. *Nature*, 521, 436–444.
15. Schmidhuber, J. (2015). Deep learning in neural networks: An overview. *Neural Networks*, 61, 85–117.
16. Bzdok, D., & Meyer-Lindenberg, A. (2018). Machine learning for psychiatry: A primer. *NeuroImage*, 180, 348–368.
17. Nguyen, A., Yosinski, J., & Clune, J. (2015). *Deep Neural Networks Are Biased Towards Simple Functions*. San Diego: ICLR.
18. Hastie, T., Tibshirani, R., & Friedman, J. (2009). *The Elements of Statistical Learning: Data Mining, Inference, and Prediction*. New York: Springer.
19. Ioffe, S., & Szegedy, C. (2015). *Batch Normalization: Accelerating Deep Network Training by Reducing Internal Covariate Shift*. Lille, France: ICML.
20. Kingma, D. P., & Ba, J. (2014). *Adam: A Method for Stochastic Optimization*. San Diego: ICLR.
21. Chollet, F. (2018). *Deep learning with Python*. Shelter Island, NY: Manning Publications.
22. Witten, I. H., Frank, E., & Hall, M. A. (2016). *Data Mining: Practical Machine Learning Tools and Techniques*. Burlington, MA: Morgan Kaufmann.

23. Arora, K., Kumar, A., Kamboj, V. K., Prashar, D., Shrestha, B., and Joshi, G. P. (2021), Impact of renewable energy sources into multi area multi-source load frequency control of interrelated power system. *MDPI Journal of Mathematics*, 9(2), 186, https://doi.org/10.3390/math9020186
24. Xu, B., & Wang, J. (2018). A survey of deep learning for big data. *Journal of Computer Science and Technology*, 33, 423–447.
25. Yao, H., & Liu, Y. (2017). A survey of deep learning algorithms and applications in engineering. *Engineering Review*, 37, 1–13.
26. Zhang, Y., & Yang, Q. (2015). A survey on multi-task learning. *IEEE Transactions on Knowledge and Data Engineering*, 28, 1997–2013.
27. Wang, Z., & Liu, X. (2019). Survey of machine learning algorithms for big data. *Journal of Computer Science and Technology*, 34, 634–650.
28. Wu, J., & Wang, W. (2019). Machine learning for engineering: A review. *Journal of Computational Science*, 33, 100–120.
29. L. Zhiqiang et al. (2019). Machine learning for complex engineering systems: A survey. *Advanced Engineering Informatics*, 40, 38–52.
30. Cheng, Y., & Zhang, Y. (2017). A survey on machine learning applications in engineering. *Engineering Applications of Artificial Intelligence*, 64, 101–118.
31. Xie, Y., & Li, L. (2020). Machine learning and its applications in engineering: A comprehensive review. *Journal of Mechanical Engineering Science*, 234, 341–360.
32. Liu, X., & Zhang, X. (2021). Machine learning techniques in engineering: Trends and perspectives. *International Journal of Engineering Research and Applications*, 11, 1–15.
33. Yang, S., & Wu, J. (2020). Deep learning methods for engineering applications: A survey. *Journal of Computational Design and Engineering*, 7, 78–92.
34. Dong, H., & Zhang, Y. (2018). Machine learning in engineering: A review of recent advances and future directions. *Engineering Science and Technology*, 21, 172–185.
35. Li, L., & Liu, S. (2021). Overview of machine learning algorithms in engineering. *Journal of Engineering Science and Technology*, 16, 1–20.
36. Wang, X., & Zhang, J. (2020). Machine learning for predictive maintenance in engineering. *Journal of Intelligent Manufacturing*, 31, 901–915.
37. Chen, X., & Hu, J. (2019). Recent advances in machine learning applications in civil engineering. *Computers and Structures*, 214, 100–115.
38. Zhao, Q., & Zhang, X. (2019). Machine learning for mechanical engineering applications: A survey. *Mechanical Systems and Signal Processing*, 116, 238–255.
39. Yang, Y., & Zhang, L. (2021). AI-driven design optimization in engineering: A review. *Structural and Multidisciplinary Optimization*, 64, 1125–1138.
40. Wu, Z., & Xu, Y. (2021). Machine learning for energy management in engineering systems. *Energy Reports*, 7, 102–116.
41. Zhang, H., & Liu, X. (2020). AI and machine learning in aerospace engineering: A comprehensive review. *Aerospace Science and Technology*, 104, 105–118.
42. Arora, K., Kumar, A., Kamboj, V. K., Prashar, D., Jha, S., Shrestha, B., & Joshi, G. P. (2020). Optimization methodologies and testing on standard benchmark functions of load frequency control for interconnected multi area power system in smart grids. *MDPI Journal of Mathematics*, 8(6), 980. https://doi.org/10.3390/math8060980
43. Li, J., & Wang, X. (2021). Machine learning techniques for optimization in engineering design. *Engineering Optimization*, 53, 1491–1504.
44. Guo, X., & Li, S. (2020). AI and ML applications in chemical engineering: A survey. *Chemical Engineering Research and Design*, 156, 227–244.

45. Zhao, J., & Li, Y. (2021). Machine learning and AI for environmental engineering: A review. *Environmental Science and Technology*, 55, 1452–1468.
46. Zhang, W., & Liu, Y. (2019). AI-driven decision support systems in engineering: Trends and applications. *Decision Support Systems*, 123, 45–59.
47. Lin, X., & Huang, J. (2020). Machine learning applications in industrial engineering: A survey. *Industrial Engineering Journal*, 62, 59–76.
48. Wang, S., & Wu, Z. (2021). Deep learning for engineering systems: Innovations and challenges. *Journal of Engineering Research*, 42, 300–317.
49. Zhang, Q., & Li, J. (2020). AI in engineering design: Methodologies and applications. *Design Science Journal*, 14, 95–113.
50. Zhou, L., & Yang, X. (2021). Machine learning and artificial intelligence in engineering: A systematic review. *Engineering Applications of AI*, 99, 103–120.

18 Impact of Automation and Control Systems on Industry

Praveen Kumar Malik

18.1 INTRODUCTION

Automation and control systems have enabled industry to move towards more sustainable practices. Automation and control systems can be programmed to reduce energy consumption, as they are able to adjust to changing conditions and optimize production processes. Automation and control systems also allow for the collection of data, which can be used to identify areas for improvement, leading to greater sustainability. In conclusion, automation and control systems have revolutionized the modern industrial landscape, providing industries with increased efficiency and precision, cost savings, improved quality, and sustainable practices [1]. Automation and control systems have enabled industry to move from manual processes to automated processes, providing businesses with the opportunity to maximize their efficiency and increase their profitability. Automation and control systems are increasingly being used in the industrial sector to improve production processes, reduce costs, and increase efficiency. These systems allow the automation of processes that were formerly done manually and the control of systems that were previously unmanageable. Automation and control systems are used in a wide variety of industries, such as manufacturing, transportation, energy, and healthcare. The primary benefit of automation and control systems in industry is improved efficiency and productivity. By automating processes and controlling systems, companies can reduce labor costs, improve safety and quality, increase throughput, reduce waste, and improve product quality. Automation and control systems also allow companies to monitor their operations in real time, providing better visibility into processes and helping companies make better decisions.

Automation and control systems also help to reduce the risk of human error and improve accuracy. Automated systems can detect errors or inconsistencies in data and alert staff, allowing for more consistent and accurate data collection. Automation and control systems can also be used to monitor and control hazardous materials, keeping workers and the environment safe. In addition to increased efficiency and improved safety, automation and control systems can also help to reduce costs. Automation and control systems can be used to reduce energy consumption and waste, enabling companies to save money and become more sustainable. Automation and

DOI: 10.1201/9781003563440-18

control systems also enable companies to reduce labor costs by automating processes that were previously done manually. Overall, automation and control systems have a significant impact on industry, increasing efficiency, improving safety, and reducing costs. Automation and control systems can be used in a wide variety of industries, from manufacturing to healthcare, and can help companies to become more efficient and competitive [2].

18.2 ACCURACY AND PRECISION IN PRODUCTION PROCESSES

Accuracy and precision are two important concepts in production processes. Accuracy is the degree to which a measurement is close to the desired value or true value. Precision is the degree to which repeated measurements of a quantity are consistent with each other. Both accuracy and precision are important in production processes to ensure that quality standards are met and products are made within specified tolerances. Accuracy is measured by comparing a measurement or result to the true or desired value. Accuracy is typically expressed as a percentage or ratio. Precision is measured by assessing the consistency of repeated measurements or results. Precision is typically expressed as a standard deviation or coefficient of variation (CV). By monitoring the accuracy and precision of the production process, manufacturers can identify areas that need improvement and ensure that quality standards are met [3]. High accuracy and precision are necessary for production processes that require exact measurements and specifications, such as those used in aerospace and medical industries. Low accuracy and precision are sufficient for production processes that require less exact measurements and specifications, such as those used in the food and beverage industries.

18.3 EFFICIENCY AND PRODUCTIVITY

Efficiency and productivity in mechatronics can be increased through the use of automation, advanced technology, and intelligent systems. Automation can be used to handle repetitive tasks, making them more efficient and allowing more time for creative problem solving. Advanced technology can be used to reduce production times and increase accuracy. Intelligent systems can be used to monitor and adjust processes to optimize performance. This can help to reduce costs, increase reliability, and improve safety. Finally, proper maintenance and training can help to ensure that the equipment is running optimally [4]. Following are some of the entities which can be taken care to increase the efficiency and productivity:

1. Utilize automation and robotic systems to reduce labor costs.
2. Use advanced sensors and machine vision to increase accuracy and precision.
3. Implement predictive maintenance techniques to reduce downtime.
4. Streamline processes to reduce waste and increase throughput.
5. Develop and maintain a comprehensive preventive maintenance program.
6. Invest in quality control systems to ensure consistent product quality.
7. Improve workplace safety and ergonomics with updated equipment.

8. Utilize data-driven decision making for improved efficiency.
9. Regularly review processes and procedures for optimization.
10. Invest in advanced control systems for improved performance.
11. Utilize cloud computing and edge computing for increased scalability.

18.4 FASTER PROCESS CONTROL AND AUTOMATION

Process control and automation can be enhanced through the use of technology. For example, automation systems can be used to control the speed and accuracy of processes, as well as to monitor and adjust parameters such as temperature, pressure, and flow rate [5]. Additionally, the use of robotics can help to streamline processes, reducing the need for manual labor and increasing overall efficiency. Additionally, data analytics can be used to identify potential areas for improvement and better understand customer needs.

1. Use of programmable logic controllers (PLCs): PLCs can be used to control and automate mechatronic systems. They are used to replace electromechanical relays and contactors. PLCs can be programmed to control the speed, position, and force of a mechanical system.
2. Use of motion controllers: Motion controllers are used to control the motion of mechatronic systems. They can be used to control the speed, acceleration, and position of the system.
3. Use of sensors: Sensors can be used to measure the various parameters of the system such as temperature, pressure, flow, etc. This data can be used to control and automate the system.
4. Use of human–machine interface (HMI): HMI is used to provide a user interface to the system. It allows users to interact with the system and control it.
5. Use of artificial intelligence: Artificial intelligence can be used to improve the accuracy and efficiency of mechatronic systems. AI can be used to automate the system and make decisions based on the data collected from the sensors.

18.5 INCREASED FLEXIBILITY AND SCALABILITY

Flexibility and scalability are important considerations in mechatronics. Flexibility refers to the ability to easily change or modify a system to meet new requirements or conditions. Scalability is the ability to increase the complexity or scale of the system to meet the needs of a growing or changing environment. In mechatronics, flexible and scalable systems can be achieved through the use of modular architectures, open system interfaces, and distributed control architectures [6]. Modularity allows components of a mechatronic system to be easily replaced or upgraded without affecting the overall system architecture. Open system interfaces enable the connection of different components from different manufacturers, and distributed control architectures allow for the control of multiple mechatronic systems or components from a single source. These features provide increased flexibility for the system and allow it to be adjusted to meet changing needs. In addition, mechatronics systems can be made more flexible and scalable through the use of advanced control technologies such as artificial intelligence and machine learning [7]. These technologies can be used to enable the system to detect changes in the environment and adjust its behavior accordingly. This allows the system to respond quickly to changing conditions without the need for manual intervention.

18.6 IMPROVED TRACEABILITY AND DOCUMENTATION

Traceability and documentation are important in mechatronics as they provide evidence of the design process and performance of the system. Traceability provides a historical record of the design process, which can be used to identify any changes or modifications made to the system, as well as any issues that may have arisen [8]. Documentation of the design process also allows for future maintenance and modification of the system, as well as providing a record of the system performance. Documentation can include diagrams, drawings, specifications, test reports, and operational manuals, among other documents. Traceability and documentation are essential for ensuring the performance and safety of the mechatronic system. It can be further explained as follows:

1. Streamlining of data collection and analysis: By streamlining the data collection and analysis process, the amount of time needed to document and trace changes in mechatronic systems can be reduced.
2. Automated documentation: Automated documentation systems can be used to store and track data related to mechatronic systems, making it easier to trace changes over time.
3. Standardized data formats: By standardizing data formats, it becomes easier to identify and trace errors, as well as providing a consistent way to document changes.
4. Automated testing: Automated testing helps to ensure that changes to mechatronic systems are documented and traceable. This can also help to quickly identify potential problems before they become costly.
5. Real-time monitoring: By monitoring mechatronic systems in real-time, it becomes easier to identify potential problems and trace changes over time.
6. Electronic documentation: By using electronic documentation, any changes that are made to mechatronic systems can be documented and tracked quickly and easily.
7. Integration of data sources: By integrating data sources, it becomes easier to trace changes over time and identify potential problems.

18.7 POTENTIAL FOR DATA-DRIVEN DECISION MAKING

Data-driven decision making in mechatronics is becoming increasingly important as the complexity of mechatronic systems increases. Data-driven decision-making allows for better understanding of the system, more accurate predictions on system performance, and more effective maintenance. By leveraging data collected from sensors, machine learning algorithms, and other sources, mechatronic engineers can gain insight into the system's behavior and make informed decisions about how to optimize performance and increase efficiency [9]. Data-driven decision-making can also be used to detect and diagnose faults, as well as to predict potential failures and maintenance needs. Additionally, data-driven decision-making can be used to identify opportunities for cost savings and cost reduction that can result in improved system performance. Further it can be defined with the help of following points:

1. Improved accuracy of predictive maintenance models
2. Increased efficiency of automated production lines
3. Enhanced ability to identify and address system weaknesses
4. Improved overall performance of mechatronic systems
5. Reduced costs associated with maintenance and repair
6. Improved safety and reliability of mechatronic systems
7. Optimized energy consumption in mechatronic systems
8. Reduced environmental impact of mechatronic systems
9. Increased automation of production processes
10. Enhanced product quality and consistency
11. Improved customer satisfaction and experience

18.8 CONCLUSION

The impact of automation and control systems on industry is undeniable. Automation and control systems have allowed companies to streamline their production processes and increase their efficiency. This has resulted in improved product quality and cost savings for companies. Moreover, automation and control systems provide a platform to collect and analyze data that can be used to further improve production processes and operations. Automation and control systems are essential for companies to remain competitive in the current economic landscape. Additionally, automation and control systems can improve safety in the workplace by reducing the number of mistakes made and the frequency of accidents. Overall, automation and control systems have had a significant impact on industry, allowing companies to improve their operations and increase their competitiveness.

REFERENCES

1. Y. Jiang, J. Fan, T. Chai, F. L. Lewis and J. Li, "Tracking Control for Linear Discrete-Time Networked Control Systems With Unknown Dynamics and Dropout," *IEEE Transactions on Neural Networks and Learning Systems*, vol. 29, no. 10, pp. 4607–4620, Oct. 2018, doi:10.1109/TNNLS.2017.2771459
2. Y. Jiang, D. Shi, J. Fan, T. Chai and T. Chen, "Set-Valued Feedback Control and Its Application to Event-Triggered Sampled-Data Systems," *IEEE Transactions on Automatic Control*, vol. 65, no. 11, pp. 4965–4972, Nov. 2020, doi:10.1109/TAC.2020.3004348
3. P. Vrba et al., "Rockwell Automation's Holonic and Multiagent Control Systems Compendium," *IEEE Transactions on Systems, Man, and Cybernetics, Part C (Applications and Reviews)*, vol. 41, no. 1, pp. 14–30, Jan. 2011, doi:10.1109/TSMCC.2010.2055852

4. S. Xiao and J. Dong, "Distributed Adaptive Fuzzy Fault-Tolerant Containment Control for Heterogeneous Nonlinear Multiagent Systems," *IEEE Transactions on Systems, Man, and Cybernetics: Systems*, vol. 52, no. 2, pp. 954–965, Feb. 2022, doi:10.1109/TSMC.2020.3002944
5. Q. Wei, L. Wang, J. Lu and F. -Y. Wang, "Discrete-Time Self-Learning Parallel Control," *IEEE Transactions on Systems, Man, and Cybernetics: Systems*, vol. 52, no. 1, pp. 192–204, Jan. 2022, doi:10.1109/TSMC.2020.2995646
6. K. Arora, A. Kumar, V. K. Kamboj, D. Prashar, S. Jha, B. Shrestha and G. P. Joshi, "Optimization Methodologies And Testing on Standard Benchmark Functions of Load Frequency Control for Interconnected Multi Area Power System in Smart Grids," *MDPI Journal of Mathematics*, vol. 8, no. 6, p. 980, 2020, doi:10.3390/math8060980
7. S. Xiao and J. Dong, "Cooperative Fault-Tolerant Containment Control for Nonlinear Multiagent Systems With Switching Directed Topologies Based on Hierarchical Mechanism," *IEEE Transactions on Systems, Man, and Cybernetics: Systems*, vol. 52, no. 9, pp. 5424–5433, Sept. 2022, doi:10.1109/TSMC.2021.3125772
8. J. Lu, Q. Wei, Y. Liu, T. Zhou and F. -Y. Wang, "Event-Triggered Optimal Parallel Tracking Control for Discrete-Time Nonlinear Systems," *IEEE Transactions on Systems, Man, and Cybernetics: Systems*, vol. 52, no. 6, pp. 3772–3784, Jun. 2022, doi:10.1109/TSMC.2021.3073429
9. T. Samad and G. Stewart, "Perspectives on Innovation in Control Systems Technology: Compatibility With Industry Practices," *IEEE Transactions on Control Systems Technology*, vol. 21, no. 2, pp. 284–288, Mar. 2013, doi:10.1109/TCST.2013.2240531

Index